PEAK BAGGING MUNROS

VOLUME 1

THE DEFINITIVE GUIDE

PEAK BAGGING MUNROS

VOLUME 1

HIGHLANDS & SOUTHERN CAIRNGORMS:
64 ROUTES DESIGNED TO COMPLETE THE 141 SOUTHERLY MUNROS IN THE MOST EFFICIENT WAY

NICOLA HARDY

Vertebrate Publishing, Sheffield
www.adventurebooks.com

PEAK BAGGING MUNROS

VOLUME 1

HIGHLANDS & SOUTHERN CAIRNGORMS: 64 ROUTES DESIGNED TO COMPLETE THE 141 SOUTHERLY MUNROS IN THE MOST EFFICIENT WAY

First published in 2025 by Vertebrate Publishing.

VERTEBRATE PUBLISHING
Omega Court, 352 Cemetery Road, Sheffield S11 8FT, United Kingdom.
www.adventurebooks.com

A CIP catalogue record for this book is available from the British Library.

ISBN 978-1-83981-258-3 (Paperback)
ISBN 978-1-83981-259-0 (Ebook)

Front cover: Sgùrr a' Mhaim from Ben Nevis (route 39). © Craig Aitchison, *www.landandlight.co.uk*
Back cover (L–R): view from the summit of Stùc a' Chroin (route 05); Allt Coire Dhomhain with Sgairneach Mhòr beyond (route 53); view across Loch Lomond to Loch Arklet and Loch Katrine from the mountain path on Ben Vane (route 03); Càrn Mòr Dearg Arête and Ben Nevis from Càrn Mòr Dearg (route 39); Buachaille Etive Beag and Buachaille Etive Mòr from Beinn Fhionnlaidh (route 31); the cairn marking the saddle between Beinn Dòrain and Beinn an Dothaidh with the south-western crags of the latter shrouded in fog (route 18); view from the summit of Lochnagar - Cac Carn Beag to Loch nan Eun, Carn a' Choire Bhoidheach and Càrn an t-Sagairt Mòr (route 63); Ben Alder, Bealach Dubh, Geal-Chàrn (Loch Ossian) and Aonach Beag from Geal-charn (Drumochter Pass) (route 53).

Photography by Nicola Hardy unless otherwise credited.

Mapping contains data from OS © Crown copyright and database right (2025) and © OpenStreetMap contributors, *Openstreetmap.org/copyright*
Relief shading produced from data derived from U.S. Geological Survey, National Geospatial Program.
Cartography by Richard Ross, Active Maps Ltd.
www.activemaps.co.uk

Edited by Helen Parry, cover design, layout and production by Jane Beagley.
www.adventurebooks.com

Printed and bound in Slovenia by Latitude Press.

Vertebrate Publishing is committed to printing on paper from sustainable sources.

FSC www.fsc.org MIX Paper | Supporting responsible forestry FSC® C105997

Contents

For overview map see inside front cover.

Download the GPX files from
www.adventurebooks.com/munros-vol-1-GPX

6▲ = Number of Munros on route

Introduction

I can't quite believe this is real. Tears sting my eyes, a mix of happiness and relief. Just 50 metres left to complete the adventure of a lifetime. Is this really the end of my Munro-bagging journey across Scotland? Family and friends have formed a tunnel of walking poles, a guard of honour leading to summit number 282. My partner, James, and I hike triumphantly through the arch and touch our final trig point amid whoops, cheers and the spray of champagne. Our parents beam back at us; our last Munro is their first. It's the perfect ending.

Forty-eight hours earlier we're two exhausted, weather-battered peak-baggers struggling through the Rough Bounds of Knoydart. We've not seen another human for days. Our feet are sodden, our skin is midge-ravaged, and our clothes are dank and stinky. Over the previous six months, we've walked, climbed and sometimes bum-shuffled our way up and down 282 Scottish mountains: a journey of over 2,000 kilometres with more than 100,000 metres of ascent, pushing ourselves to the very edge of our comfort zones and occasionally beyond.

It started with a mini mid-life crisis in my 30s. I was living and working in my home county of South Yorkshire. I'd somehow found myself in IT project management, a job I'd never dreamed about, and I couldn't quite work out how I'd got there. I craved more adventure, but how could I make it happen? After lots of soul-searching, I took a leap of faith. I quit my job, sold my house and, with a healthy dose of trepidation, set off with my partner up Ben More on the Isle of Mull. My first ever Munro and the start of an epic journey. Our goal? To summit Sir Hugh Munro's iconic list of 282 Scottish mountains in just six months.

At the time of writing, nearly 8,000 people have officially recorded having achieved the same goal, each in their own style. Currently the fastest is Jamie Aarons: she finished a full self-propelled Munro round in a remarkable 31 days, 10 hours and 27 minutes, cycling and kayaking the vast distances between the mountains. By contrast, 'Golden Munroists' are those whose first and last Munros are over 50 years apart – quite literally the project of a lifetime. Martin Moran and Anna Wells both finished speedy winter Munro rounds in 83 days, 40 years apart. Steven Fallon and Hazel Strachan have taken it to the next level by recording a whopping 16 and 12 rounds respectively.

But why should **you** take on this grand challenge? You'll be pleased to know there is no right or wrong method to bag the Munros – it's a blissfully rule-free zone. As long as you touch all 282 summits, you're all good. It's up to you if you take the chairlift down Meall a' Bhùiridh or hitch-hike down a long access road. This is your journey.

The experience is yours too. The physical and mental health benefits of hillwalking are well documented, but – rather than the views, the escapism or the fresh air – for many Munroists it's the random, chucklesome and serendipitous moments along the way that make the journey so life-affirming. I have a few golden nugget memories that always make me smile: a stand-off with a badger intrigued by my bag of snacks, being dunked head first into a river in slow motion by a rickety wire bridge, and enjoying the warm hospitality of new Scottish friends. What memories will you forge along your journey?

It won't all be plain sailing and happy memories, however. This is Scotland, after all, and you're sure to experience some atrociously unpredictable weather en route. During my six-month Munro challenge I endured unforecasted thunderstorms, brutal winds, horizontal rain and snow blizzards in June. Yet I also enjoyed heatwaves, dreamy temperature inversions above marshmallow-like clouds, countless rainbows and two brocken spectres. I have no regrets – the good times were far sweeter thanks to the tough times.

Historically, Munro-bagging has been a male-dominated pursuit: it's estimated that only 20 per cent of Munroists are women. But times are changing, more women are getting involved and, personally, I think it's a fantastic thing. When I agreed to write this book, inspiring other women was one of my big motivations. If I can do it, an IT project manager from Sheffield with no outdoorsy pedigree, then so can you!

Another motivation, of course, was the chance to revisit my beloved Scotland. Since becoming a 'compleatist', I'd ventured north to walk routes like the Sutherland Trail and the Ring of Fire in the Galloway Hills, but I'd perhaps neglected the Munros. This book was my opportunity to indulge yet again in their delights.

In researching this book, I've fallen in love with Scotland's mountains all over again and I'm excited to share a collection of routes which are varied, efficient and charming. True to my project management background, I've also provided bucket loads of practical notes, logistical tips and handy advice to help you plan your own Munro walks.

Not everyone has the niche compulsion to both climb mountains and meticulously tick them off a list – and that's absolutely fine. I've designed this book so that you can use it to plan an adventurous weekend away or a fabulous hillwalking holiday, even if you know you'll never bag the full set of Munros.

But, having said that, I'd encourage anyone to climb them all. Why not? It might just be the journey that changes your life for the better, like it has for me. You can do this. It's an achievable goal – and there are 282 majestic reasons to give it a go.

Nicola Hardy
February 2025

Feedback and updates

Access and services change over time so if you find any inaccuracies or have any feedback, I'd love to hear from you: ***peakbaggingmunros@gmail.com***

Thank you and happy Munro-bagging!

Trio of cairns on the Ben Nevis plateau (route 39).

Acknowledgements

I'm indebted to all who have joined me on Munro walks over the years, especially those who walked with me while I researched this book – Chris H., Liz, Jess, Ben, Anna, Jordan, Graeme, Bryony, Adrian, Mark, Sally, Matt and Chris G. I'm grateful for your varied insights into the trail difficulty gradings, for allowing me to photograph your walks and for your patience during my notetaking on our stop/start ascents. I'm not at all bitter about all the times I sat in the pub editing images and tidying up my route notes while you all enjoyed a relaxing post-walk pint!

To the strangers I met on the Munros who sat with me and shared a summit view or walked with me a while, thank you for brightening my day. Sharing smiles or chats with like-minded hillwalkers is one of the reasons I love adventuring in Scotland. To those who were walking the West Highland Way through Tyndrum one minute and then found themselves helping to jump-start my car the next, you are the best!

I'd also like to acknowledge those who made my first Munro round so special. Thank you to old friends and new who joined James and me on multiple summits – Eddie, Matt, Adrian, Lachlan, Andrew, Neil, Matthew, Becky, Kate, Sean, Saul, Humphrey and everyone who joined us on our 282nd Munro, sadly too many to mention by name, but you made it a day James and I will never forget.

I have a great deal of gratitude for everyone at Vertebrate Publishing for supporting this project. Commissioning editor Kirsty first approached me via email in the summer of 2023. I'd just finished my solo hut-to-hut trek in the Pyrenees on the Coronallacs trail in Andorra. I read the email over an Aperol spritz and tapas in sunny Andorra la Vella – a world away from the wilds of Scotland – but somehow, I knew I was going to say yes and change the course of the next few years of my life. Thank you to Helen who has done an incredible job editing the book and also to Jon, John, Lorna, Sophie and Vicky for their hard work and support. The creative work of designer Jane, cartographer Richard and photographer Craig has really brought the book to life – thank you.

A big thank you to my partner, James, who I frequently leaned on to turn a blethering mess of words into a concise sentence, and to Corinne and the Braziers, my parents, Liam, Jess, James's family and all of our friends – thank you for being you. I couldn't have done this without you.

The biggest thank you of all goes to you, the reader. May the wind be always at your back.

View down Loch Ericht from the approach to Ben Alder Lodge en route to Culra (route 51).

About the book

What is a Munro?

To qualify as one of the 282 Munros of Scotland, a mountain must measure over 3,000 feet or 914.4 metres with sufficient prominence and separation from neighbouring peaks. Sir Hugh Munro (1856–1919), mountaineer and founder member of the Scottish Mountaineering Club (SMC), recorded the original list in 1891. The SMC maintains the current list of Munros; mountains have been added to the list when remeasured with modern equipment. Similarly, Munros have been declassified after remeasurement. The current list records 282 Munros with the last significant change being in 2012 when Beinn a' Chlaidheimh was demoted.

Mountains over 3,000 feet with insufficient prominence and separation are known as Munro Tops, while mountains between 2,500 and 3,000 feet (762 and 914.4 metres), with a drop of at least 500 feet (152 metres) to the next hill, are known as Corbetts.

Munro-bagging is a hobby which suits a niche subsection of the population, myself included, who enjoy adventuring in the outdoors alongside a nerdy compulsion to tick summits off a list. Ben Nevis is the highest Munro at 1,344.5 metres, and the smallest is Beinn Teallach at 914.6 metres. Ben Lomond is the most southerly Munro; Ben Hope the most northerly; Sgùrr na Banachdich on the Isle of Skye is the most westerly; and Mount Keen the most easterly.

Those who have bagged all 282 Munros are called *Munroists* or 'compleators'. At the end of your Munro-bagging journey, you can register your round on the SMC website, ***www.smc.org.uk/hills/compleators*** and join the Munro Society, ***www.themunrosociety.com***

About the routes

This book contains 64 efficient routes which visit the southernmost 141 Munros. Most routes are designed as day walks, which should take approximately nine hours or less to hike – although trekkers, fastpackers or trail runners will complete them more quickly, or might choose to tackle more than one route in a day. Some routes are designed to be completed over multiple days or benefit greatly from the use of a mountain bike. These include **48: Munros North-East of Loch Ossian** (page 241), **51: Ben Alder Range** (page 259) and **63: Lochnagar Horseshoe** (page 325). When completing walks over multiple days, a bothy stay or wild camp is required.

The majority of routes visit more than one Munro, and ten take in four or more; **61: East Glen Shee Munros** (page 315) visits six Munros. The routes can all be tackled by a competent hillwalker, fastpacker or trail runner and require no climbing. However, technical terrain, short scrambles and exposed edges are sometimes encountered, and some of these routes are best tackled only in good weather.

How to use this book

The 64 routes in this book are divided into 10 Highland hubs which are described at the beginning of each chapter – more details on page x. In addition to detailed directions and inspiring photography, the following information is provided for each route.

- A short summary.
- An overview map highlighting the Munros en route.
- Details of the start location, route length and ascent.
- Essential information on public transport, parking and refreshments.
- An elevation profile.
- A detailed 1:40,000-scale route map, showing the route as well as parking, public transport, refreshments, hostels and bothies.
- Timings for each route, for walkers, trekkers, fastpackers and trail runners.
- Gradings which summarise the route as a whole, the terrain, the distance and the navigational difficulty of each route.

- An indication as to whether a mountain bike can be used for part of the route.
- Pronunciation guide and name meanings for the Munros.
- Suggested escape routes.

Within each Highland hub, the routes are usually presented in order of their position from west to east and south to north.

Introducing the Highland hubs

The 64 Munro walks recommended in this book have been divided into 10 Highland hubs. The aim of this is for you to use each hub as a springboard to embark on a selection of Munro walks, returning to the hub after each route.

The collection of Highland towns and villages that make up the 10 hubs each has its own charm and appeal:

- **Arrochar** – gateway to the Argyll Forest Park and close to the Arrochar Alps.
- **Comrie** – 'Shaky Toun' close to the Highland Boundary Fault.
- **Oban** – the Seafood Capital of Scotland is a buzzing harbour town.
- **Tyndrum** – a village at a vital crossroads with road and rail connections to Oban and Fort William.
- **Killin** – a small village beside the mesmerising Falls of Dochart.
- **Glencoe** – a village at the mouth of the famous glen formed by a glacier during the last ice age.
- **Fort William** – the largest town in Lochaber, known as the Outdoor Capital of the UK.
- **Corrour** – of *Trainspotting* fame, this quirky hub is only accessible by train.
- **Pitlochry** – the 'place of the Sentinel Stone' is a thriving town in the heart of Scotland.
- **Braemar** – an Aberdeenshire village in the magnificent Cairngorms National Park.

Each of the hubs has a selection of nearby accommodation to suit all budgets – where possible, this includes a campsite and a hostel, plus a bed and breakfast or hotel option.

On average, car journeys from each of the hubs to the Munro walks are between 20 and 30 minutes; in some cases, they are longer due to the remoteness of the mountains.

For each walk the closest public transport options, including bus, train and ferry, are listed, alongside a selection of rest day activities, enabling you to dive into the culture and hospitality of each area on stormy days, or when you fancy a break from hillwalking.

An extended stay in each hub provides an opportunity to support local businesses who provide essential services to tourists in the Scottish Highlands.

Route gradings

Each route has been assigned a series of grades for: the route as a whole; the terrain; distance; and the navigational hazards you are likely to encounter (see Table 1, opposite). While no Munro walk is 'easy', these grades can be used as a decision-making aid when considering the order in which to undertake the walks. Note that the grade 5 terrain score is split between rock hazards and water hazards. There are four routes in this book which involve river crossings and thus carry the highest terrain score – **11: Queen of the Southern Highlands** (page 57), **17: Beinn Mhanach** (page 83), **19: Achallader Munros** (page 91) and **51: Ben Alder Range** (page 259). The highest terrain score due to rock is not needed for the southern Munros.

GRADE	TERRAIN	DISTANCE	NAVIGATION
1	Wholly well-maintained footpaths Steady gradient on ascent and descent Grass, gravel, stone-pitched paths or woodland trails	**5–10km**	**Straightforward:** 100% clear path network; even in thick mist, the path is visible High footfall, popular route with walkers
2	Mostly good footpaths Short sections of steep gradient Short sections of rocky or boggy terrain	**10–15km**	**Moderate:** up to 20% of the route has indistinct paths
3	Some good footpaths Multiple sections of steep gradient Realistic possibility of sustained rocky or boggy terrain	**15–20km**	**Tough:** 20–40% of the route has indistinct paths or no path
4	Sustained sections of steep gradient Challenging varied terrain which could include: loose terrain (scree slopes), boulder fields, rocky pinnacles in exposed positions with good hand and footholds or boggy terrain A slip may result in a fall	**20–25km**	**Challenging:** 40–60% of the route has indistinct paths or no paths Alternatively, there is a critical point in the route where a navigational error could result in venturing on to dangerous terrain
5	**Rock** Great care is needed as sustained graded scrambling is involved – hand and footholds may be difficult to find A slip may result in a fall **Water** River crossings	**Over 25km**	**Exceptionally challenging:** over 60% of the route has indistinct paths or no paths Alternatively, there is a critical point in the route where a navigational error could result in venturing on to extremely dangerous terrain

Table 1

An alphabetical list of Munros is available on the inside back cover. A series of helpful tables available at ***www.adventurebooks.com/peak-bagging-munros-tables*** organise the Munros by height and the Munro walks by length, by ease of navigation and by complexity of terrain.

Naturally, your competency and confidence will increase as you progress along your journey to becoming a Munroist. Consider employing the services of a mountain guide for the most challenging Munro routes.

Walk, trek, fastpack, run

To give you an indication of how long each route may take, we have identified four user groups: **Walkers**, **Trekkers**, **Fastpackers** and **Trail runners**. The **Jones–Ross formula** has been developed that modernises and develops Naismith's rule to provide realistic timings for each user group.

All user groups move at different speeds and have a higher flat speed than ascent and descent speeds, but they also move at different rates in ascent and descent.

Walkers move at a flat speed of around five kilometres per hour and move more quickly while descending than ascending.

Trekkers move at a flat speed of around six kilometres per hour and move more quickly while descending than they do while ascending.

Fastpackers, who typically run on some sections of routes and walk on others, move at a flat speed of around eight kilometres per hour and move more quickly while descending than they do while ascending.

Trail runners move at a flat speed of around 10 kilometres per hour and move far more quickly while descending than they do while ascending.

The Jones–Ross formula assumes average speeds – heavy kit, adverse weather conditions, your particular strengths in ascent or descent and tired legs may all impact on your actual pace. As you follow the routes, you will quickly learn to adjust the timings provided by the Jones–Ross formula to fit your qualities of movement out on the hills.

The Jones–Ross formula in detail

Expressed in words, the Jones–Ross formula is:

$$\text{time} = \frac{\text{distance}}{\text{flat speed}} + \text{adjustment for ascent} + \text{adjustment for descent}$$

More precisely, it can be expressed as:

$$\underset{\text{(minutes)}}{\text{time}} = \frac{\text{distance (km)}}{\text{flat speed (km per hour)/60}} + \frac{\text{vertical ascent (metres)}}{\text{vertical ascent speed (metres per hour)/60}} + \frac{\text{vertical descent (metres)}}{\text{vertical descent speed (metres per hour)/60}}$$

The flat speeds and vertical ascent and descent speeds for the four user groups are shown in Table 2.

	Flat speed (km per hour)	Vertical ascent speed (metres ascended per hour)	Vertical descent speed (metres descended per hour)
Walker	5	425	600
Trekker	6	450	750
Fastpacker	8	600	1000
Trail runner	10	1000	2000

Table 2

Navigation

The route descriptions and maps provided in this book give detailed information on the route, but you may want to download the GPX files for each route to your navigation device. Do not rely on a single form of navigation. Mobile signals may be intermittent and electronic devices may be broken or lose charge; paper maps may be damaged or lost. It is advisable to always carry a map (along with a compass), which will also provide you with a fuller picture of the surrounding area, and can be useful if you need to find an alternative route. Paper maps should be compact, but not short on detail. Examples include Ordnance Survey's 1:25,000 Explorer series and Harvey's 1:40,000 British Mountain Map and Ultramap series.

The Devil's Ridge leading to Sgùrr a' Mhàim's summit on the Ring of Steall (route 41). © Jon Barton

Map legend

Start	Ferry terminal	Route
Route direction	Pub	Optional route
Timing point	Cafe	Path
DRIESH ▲ Munro	Hostel	Track
Parking	Campsite	Roads
Bus stop	Bothy	Wood/forest
Train stop		Marsh
		Rock outcrops
		300 260 Contours at 20m

0 1 Kilometre

Safety

Before beginning any of these routes, it is important to ensure you are fully prepared for changeable conditions both in terms of weather and underfoot. In addition to appropriate footwear and waterproof and windproof outerwear, take sufficient clothing, a first aid kit, extra food and drink, and a bivvy bag or emergency shelter. Trekking poles can provide stability and security on steep ground or slippery footpaths. Leave details of your route with a person who knows what to do if you have not returned by a certain time.

Beware of ticks and carry a removal tool. Always seek medical advice if you think you have been bitten and develop a rash: ***www.lymediseaseaction.org.uk***

From May to October midges will seek to ruin your enjoyment of the Munros. Consider investing in a midge net, wear repellent and, most importantly, keep moving. The adder is the only venomous snake in the UK and can be spotted in Scotland. They will only bite as a last means of defence so if you spot one on a walk, simply give it space.

Mountain rescue

Always carry a **mobile phone** in the mountains in order to alert emergency services in the event of an accident. **999** is the British emergency number; you should ask for **POLICE – MOUNTAIN RESCUE**. Phone signal can be intermittent over much of the Highlands. Although primarily aimed at deaf and speech impaired people, **EmergencySMS** can be used by anyone, if your service provider supports it, to contact the emergency services – useful if you have low battery or intermittent signal. It requires registration: text 'register' to 999 and then follow the instructions in the reply. **EmergencySMS should only be used when voice call contact with the emergency services is not possible.**

Mountain and other countryside rescue services in Great Britain and Ireland are provided as part of national emergency services and by voluntary organisations. Organisations such as Mountain Rescue are charitable bodies, financed by public donation and reliant entirely on volunteers. In the event of needing to call for rescue, be prepared to provide details about what has happened and your location, preferably as a six-figure grid reference; make sure you know how to locate your UTM coordinates using your mobile phone or smartwatch.

Consider supporting our mountain rescue volunteers by making a donation: ***www.scottishmountainrescue.org/donate-today***

The remote nature of many Munro walks may prompt you to consider investing in a device on the global Iridium® network which uses satellites to track your position. Devices like the Garmin inReach® Mini have an in-built SOS button which alerts mountain rescue services in the event of activation. Subscriptions also incorporate handy features like messaging and tracking services which provide peace of mind to loved ones during your adventures.

Distress signal

The international (European) distress signal is **six blasts** of a **whistle** evenly spaced over **one minute**, followed by a break of **one minute**. Then **repeat**. The response that confirms that your signal has been received is three blasts of a whistle over one minute followed by a break of one minute. At night, flashes of a torch in the same sequence can be used instead. **Always carry a torch and a whistle.**

River crossings

Crossing rivers or even some streams in Scotland can be a serious undertaking which requires dynamic risk assessment and great care. The volume of recent rainfall will dictate whether or not you should attempt a crossing.

Always look both upstream and downstream for a suitable crossing point. A good crossing point will be where the water is shallow, flat along the riverbed, with no obstacles downstream and slowly moving water. This is often the widest point of the river. A poor crossing point would involve deep water, large rocks or fallen trees downstream, an uneven riverbed, steep banks or fast-flowing water.

Before you cross, check the map. Can you re-route to avoid the crossing? Can you re-route to the nearest bridge? If you must cross, secure your personal possessions in dry bags, keep your boots or shoes on, and walk slowly and diagonally while facing upstream. Use your walking poles in front of you to add stability.

Winter advice

This is intended as a summer hillwalking guidebook. That said, snow can hit the Munros unexpectedly even during the height of summer. If you are planning to walk the Munros in wintery conditions, consider the following tips.

- Attend a winter skills course where you will develop a deep understanding, practical knowledge and experience of winter navigation, avalanche avoidance, how to use an ice axe and crampons, and how to self-arrest in the result of a fall under the tutorship of trained professionals.
- Build up slowly. Start with lower hills and Munros with a shallow elevation profile.
- Learn how to interpret Scottish Avalanche Information Service data: ***www.sais.gov.uk***

The Scottish Outdoor Access Code

See ***www.outdooraccess-scotland.scot/practical-guide-all*** for more details. Best practice in line with the Scottish Outdoor Access Code is as follows:

Pair of walkers starting the short descent on to the narrowest part of the ridge to Stob Ghabhar (route 20).

RESPECT EVERYONE

- Be considerate to those living in, working in and enjoying the countryside.
- Leave gates and property as you find them.
- Do not block access to gateways or driveways when parking.
- Be nice, say hello, share the space.

PROTECT THE ENVIRONMENT

- Take your litter home – leave no trace of your visit (including fruit peel).
- Avoid lighting fires/making firepits and only use BBQs that are raised off the ground to avoid scorching the earth.
- Always keep your dogs under control and in sight.
- Dog poo – bag it and bin it in any public waste bin or take it home.
- Care for nature – do not cause damage or disturbance (do not move stones, damage ruins or plants and trees, or disturb wildlife).

ENJOY THE OUTDOORS

- Check your route (make sure you have the relevant maps) and local conditions.
- Plan your adventure – know what to expect and what you can do.
- Wild camp responsibly, away from roads/public areas (see code for full details).
- Enjoy your visit, have fun, make a memory.

Bothies and wild camping

Bothies are shelters which are usually in remote outdoor locations. They are unlocked and can be used for shelter.

A number of routes in this book pass bothies maintained by the Mountain Bothies Association charity; these include Lairig Leacach bothy (**43: The Grey Corries**, page 213), Allt Scheicheachan bothy (**55: Beinn Dearg**, page 281) and Glas-allt-Shiel bothy (**63: Lochnagar Horseshoe**, page 325).

If you build a bothy stay into your adventure, you must follow the Bothy Code which requires you to:

- Respect other users – leave the bothy clean and tidy, carry out any waste, make other visitors welcome and be considerate.
- Respect the surroundings – use the spade provided to bury human waste out of sight and away from the bothy and water sources.
- Respect the agreement with the estate – observe any restrictions upon usage including during lambing or stalking season and stay for short periods only.
- Respect the restriction on numbers – large groups (of six or more) should not use a bothy.

Never assume you'll be able to sleep in a bothy. As you cannot book a sleeping platform in a bothy you can never be sure how many other people will have the same idea as you, so always carry a tent as a backup.

Consider making a donation to the Mountain Bothies Association or become a member: ***www.mountainbothies.org.uk***

If you wish to wild camp, please follow the Scottish Outdoor Access Code and leave no trace. You have the right of responsible access and to camp on most unenclosed land. Note that in Loch Lomond and The Trossachs National Park there are by-laws meaning that camping in some areas at certain times of year is only permitted within campsites or with a permit: ***www.lochlomond-trossachs.org/things-to-do/camping/go-wild***

Glossary

allt – a stream

bealach – a saddle or low point on a mountain ridge, often where a pass crosses or reaches the ridge

bivvy bag – a lightweight, waterproof one-person shelter

blaeberry – a bilberry

bothy – an unlocked shelter in a remote outdoor location

burn – a stream

compleat – in relation to a Munro round, to have visited the top of all of the Munros of Scotland

compleator – in relation to a Munro round, someone who has visited the top of all of the Munros of Scotland

Corbett – a mountain between 2,500 and 3,000 feet (762 and 914.4 metres), with a drop of at least 500 feet (152 metres) to the next hill

corrie or **coire** – a large hollow on the side of a mountain (often a U-shaped amphitheatre created by glaciation)

emergency shelter – a pole-free shelter which provides protection from the wind and rain; occupants sit on the edges of the shelter which locks in heat

glen – a valley (which usually has a watercourse running through it)

loch – a freshwater lake or sea inlet

lochan – a small loch

Munro – a mountain in Scotland measuring over 3,000 feet (914.4 metres) with sufficient prominence and separation from neighbouring peaks

Munroist – someone who has visited the top of all of the Munros of Scotland

Munro Top – a mountain in Scotland measuring over 3,000 feet (914.4 metres) with insufficient prominence and separation from neighbouring peaks

shoulder – a high ridge projecting laterally from the mountain

A lone tree on the ascent of Bidean nam Bian with Am Bodach in the distance (route 33).

HIGHLAND HUB 1
ARROCHAR

The first of the Highland hubs is Arrochar, the gateway to the Argyll Forest Park and the perfect place to base yourself when walking the Arrochar Alps and surrounding Munros.

Arrochar and Tarbet railway station is just over a kilometre east of the village. From Glasgow you can reach the most southerly hub in just over an hour. Similarly, you can arrive using Scottish Citylink buses in near enough the same timeframe.

Braeside Stores has all the basics if you've forgotten something, plus Tarbet Village Shop near the station operates out of a small container but manages to stock the essentials. With a bistro, a couple of pubs, a chip shop and places to buy lunch, you can easily stay fuelled for a long weekend of Munro-bagging.

Arrochar's position at the head of Loch Long makes it ideally placed as a base from which to walk in the Arrochar Alps. Ben Lomond is a little more complex to reach but building in a jaunt on the waterbus across Loch Lomond adds a little novelty to an otherwise short and straightforward day on the hill.

The Mary Queen of Scots Way long-distance trail starts from Arrochar with a walk into Glen Loin to the north of the hub. If you want to reduce your travel time, it is possible to follow this track out of Arrochar to the foot of Ben Vane and Ben Vorlich, cutting out the need to drive to Inveruglas, but adding around six kilometres of pleasant walking to route 03.

WHERE TO STAY

£	**££**	**£££**
Glenloin Caravan Park & Campsite	Arrochar Hotel	Village Inn

TRAVEL TIMES

01 Beinn Bhuidhe	**20 minutes**
02 South-West Arrochar Alps	**1km walk**
03 North-East Arrochar Alps	**10 minutes**
04 Ben Lomond	**1 hour**

Rest day ideas

- Wander around the northern tip of Loch Long. This sea loch connects Arrochar to the Firth of Clyde; it was used as a torpedo testing area during World War II.
- Cycle or drive up to the Rest and Be Thankful viewpoint, which divides Glen Kinglas and Glen Croe.
- Kayak or paddleboard on Loch Lomond, Scotland's largest freshwater loch: ***www.lochlomond-trossachs.org/things-to-do/water-activities*** Or if you're feeling more adventurous, try water skiing from Loch Lomond Water Ski Club in Balloch.

Opposite Ben Lomond in the distance and Cruach nam Miseag in the foreground from the ascent of Beinn Narnain (route 02).
Overleaf View down Loch Lomond from the descent of Ben Vorlich. Ben Lomond is on the left (route 03).

Loch Awe
BEINN BHUIDHE
01
BEN VORLICH
03
BEN VANE
Inveraray
BEINN IME
BEINN NARNAIN
02
BEN LOMOND
Arrochar
04
Loch Lomond

01 Beinn Bhuidhe

HUB **Arrochar – 20-minute drive**

OVERALL 1 2 (3) 4 5 TERRAIN 1 2 3 (4) 5
DISTANCE 1 2 3 (4) 5 NAVIGATION 1 (2) 3 4 5

DISTANCE **20.4km** ASCENT **1,060m**
TIMES **8:20hrs** (Walker) **7:10hrs** (Trekker)
5:25hrs (Fastpacker) **3:40hrs** (Trail runner)

START LOCATION **Walkers' car park, Glen Fyne**
START LATITUDE/LONGITUDE **56.2730, -4.9189**
START GRID REFERENCE **NN193127**
START W3W **///wrenching.pitchers.televise**

	PRONUNCIATION	NAME MEANING
Beinn Bhuidhe	**'beyn vooyer'**	**yellow hill**

Introduction
A 20-kilometre hike for just one Munro may seem woefully inefficient to determined baggers, but the long walk-in from Loch Fyne's north-eastern tip to visit this far-flung summit is unavoidable. The good news is, once you've smashed out those hard-won miles, Beinn Bhuidhe's rocky upper reaches are delightful. You'll be rewarded for your efforts with expansive views to the Arrochar Alps (pages 9 and 13), Ben Cruachan (page 45) and, on clear days, even as far as the Glen Coe Munros, along with Jura and Mull.

Public Transport
Scottish Citylink bus services link Glasgow with the Loch Fyne Oyster Bar, which is a seven-minute walk from the start point.

Parking
The Ardkinglas Estate offers free parking for around 20 cars in the walkers' car park, accessed off the A83 on the bend in the road at the north-eastern end of Loch Fyne. The car park is nestled in woodland behind the main road. The nearest public toilet is 13.5 kilometres away in Inveraray but there are sometimes Portaloos in the car parks around Arrochar.

MTB-ability
The best way to bike 'n' hike this route is to ignore the descent instructions and return via the route of ascent as you can leave your bike at a gate 5.7 kilometres into Glen Fyne. Alternatively, experienced mountain bikers can cycle up using the route of descent to leave a bike after 8.1 kilometres at NN215184, at almost 600 metres elevation.

Escape Route
None.

Approaching Inverchorachan, just before the ascent of Beinn Bhuidhe begins.

Refreshments

Offering food to suit all budgets, the Tree Shop Cafe and Loch Fyne Oyster Bar are both close to the car park. Alternatively, you could pop into the Fyne Ales Farm Brewery Taproom and take advantage of their beer garden on a sunny day.

The Route

S Leave the car park via the vehicle entrance and turn right to cross a bridge over the River Fyne. Turn immediately left after the bridge and follow the road into Glen Fyne, passing the Fyne Ales Farm Brewery Taproom on the way. After 2.5km, cross the bridge over the River Fyne and continue walking north-east up the glen. After another 2.5km, cyclists are requested to leave bicycles at the gate at NN229164.
2 Pass through the gate and walk uphill on a rough track. Pass the cottage at Inverchorachan and shortly afterwards pass through a gate across the track. Immediately after this, pass through another gate on the left. This marks the abrupt start of a steep mountain path surrounded by bracken which can reach past head height in summer. The path climbs over tree roots and rocks as you ascend a gully with the Allt na Faing down to your right. In a couple of spots, a short, simple scramble is required and there are a few moments where a slip could lead to a fall down the drop to the right – some may struggle with this exposure.
3 As you exit the gully, the path swings to the left up the hillside, away from the stream and its pretty waterfalls, before swinging back around to meet it further up. At this point, Munros Ben Oss and Beinn Dubhchraig (page 61) are visible to the north-east. The gradient soon eases, and a wetter path leads to a track.
4 Turn right here and walk on the track briefly before taking a path leading west up the hillside from NN215184. After 500m, the path reaches a boggy depression which is crossed before you ascend steeply over loose scree and rock to top out on the ridge between Beinn Bhuidhe and its north-east 901m top. Once you've gained the ridge, the piercing towers of the Ben Cruachan group (page 45) appear.
5 Head south-west towards the summit of Beinn Bhuidhe on the path which runs along the left-hand side of the ridge, enjoying the view of Loch Fyne beyond a wind farm, unfurling towards Kintyre. The path meanders north and then west towards the summit, where the remnants of a trig pillar can be found. There are substantial views south-east to the Arrochar Alps and also to the easily identifiable pyramidal shapes of Ben More and Stob Binnein (page 79) to the east-north-east. Ben Lui to the north-east (page 57) has a great profile from this angle too. On a clear day the views north to the Glen Coe Munros showcase mountain layering at its best.
6 Retrace your steps from the summit to the track junction at NN215184 (**4**). Cyclists will retrace their steps to their bikes from here. For a bit of variety, walkers can follow the track south which makes for a more gentle descent on the knees and maintains a higher vista of the Arrochar Alps for longer. Walk for 2.3km down the track to pass between two reservoirs.
7 After the reservoirs fork left twice in close succession to continue descending the right-hand side of the south-western shoulder of Newton Hill to the bottom of the glen. Pass through a gate at the bottom of the track and turn right on to the road that services the north-west side of the glen. Walk beside the working quarry and, when you reach Clachan Power Station, follow the road to the left and return to the start.

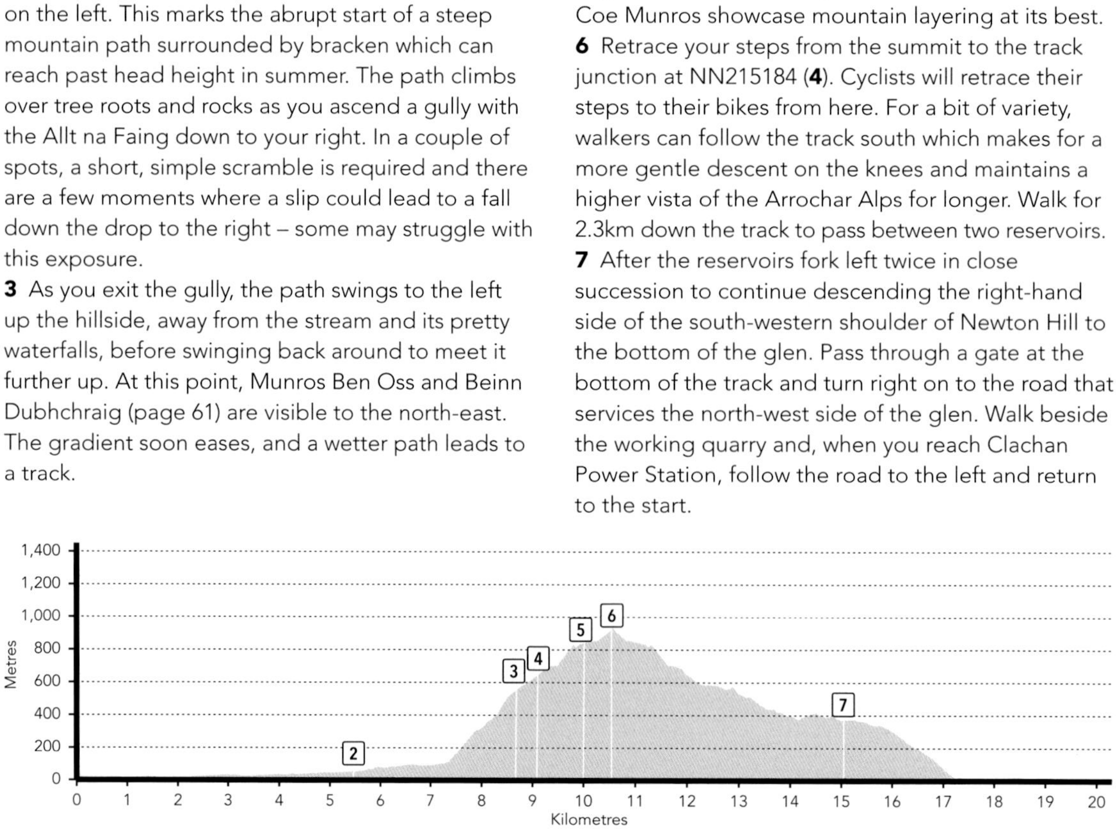

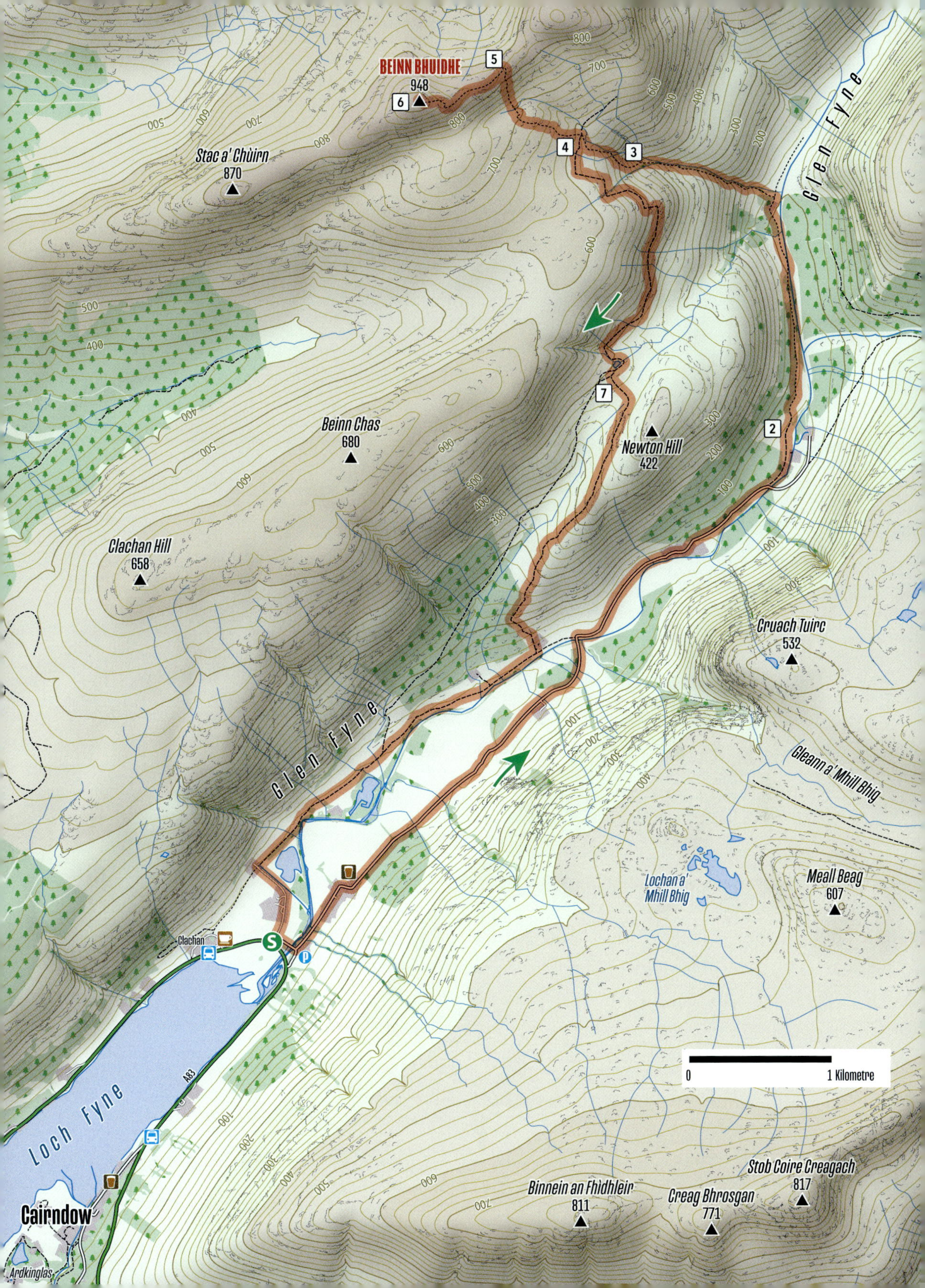

BEINN BHUIDHE
948
Stac a' Chùirn
870
Beinn Chas
680
Clachan Hill
658
Newton Hill
422
Cruach Tuirc
532
Glen Fyne
Gleann a' Mhill Bhig
Lochan a' Mhill Bhig
Meall Beag
607
Clachan
Loch Fyne
A83
Cairndow
Ardkinglas
Binnein an Fhìdhleir
811
Creag Bhrosgan
771
Stob Coire Creagach
817
0
1 Kilometre
1
2
3
4
5
6
7
S
P

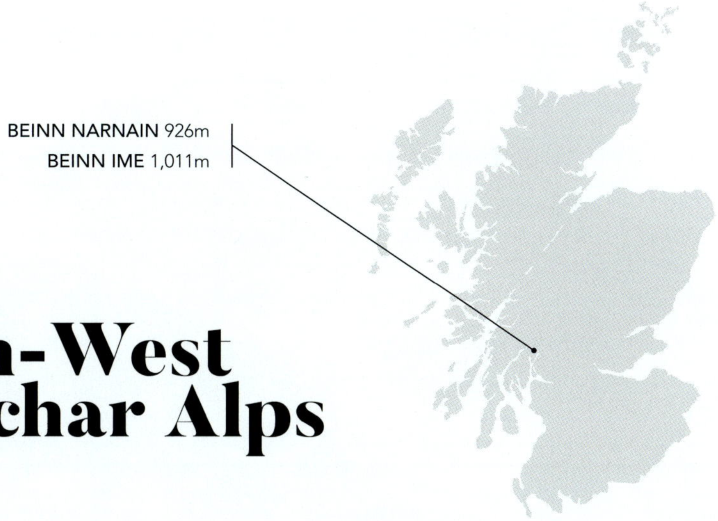

02 South-West Arrochar Alps

HUB **Arrochar – 1km walk**

OVERALL 1 2 **(3)** 4 5 TERRAIN 1 2 3 **(4)** 5
DISTANCE 1 **(2)** 3 4 5 NAVIGATION 1 **(2)** 3 4 5

DISTANCE **14.1km** ASCENT **1,350m**
TIMES **8:15hrs** (Walker) **7:10hrs** (Trekker)
5:20hrs (Fastpacker) **3:25hrs** (Trail runner)

START LOCATION **Glenloin No 2 car park, Succoth**
START LATITUDE/LONGITUDE **56.2061, -4.7507**
START GRID REFERENCE **NN294049**
START W3W **///notebook.overnight.seducing**

	PRONUNCIATION	NAME MEANING
Beinn Narnain	**'beyn narnyn'**	**notched mountain**
Beinn Ime	**'beyn eemer'**	**butter mountain**

Introduction

They might not have the towering height, permanent snows and majestic glaciers of their European namesakes, but the Arrochar Alps are worthy of the name. Complex, steep and dramatic, with distinctive profiles, this compact group of peaks in Loch Lomond and The Trossachs National Park is a superlative hillwalking destination. This route ticks off two of the Alps' four Munros – Beinn Narnain and Beinn Ime – and delivers first-rate views of The Cobbler, one of Scotland's most iconic Corbetts.

Walker descending Beinn Ime with Beinn Narnain in the distance.

Public Transport

Scottish Citylink bus services link Glasgow with the start point of this walk in Succoth, stopping right outside the car park. In addition, Arrochar and Tarbet railway station on the West Highland Line is a 40-minute walk from the start point.

Parking

Glenloin No 2 car park on the A83 at Succoth offers chargeable parking for upwards of 60 cars beside Loch Long. The nearest public toilets are four kilometres away at Tarbet on Loch Lomond but there are sometimes Portaloos in the car parks around Arrochar.

Escape Route

Descend from Beinn Narnain following the route instructions and leave the out-and-back of Beinn Ime for another day.

Refreshments

The Olive Branch Arrochar offers a simple menu that hits the spot.

The Cobbler, from the approach to Beinn Narnain.

The Route

S Leave the car park via the south-west corner and cross the busy A83. Pick up a footpath on the other side of the road which leads into the woodland. After only 50m turn right on to a muddy path with protruding tree routes. Follow the path as it ascends through the woodland. The path exits out on to a track; cross this and continue on the path on the other side, starting with a step up a steep, often wet, wooded banking before weaving up through woodland and out on to the open hillside on a firm gravel path heading north-west.

2 The path is then stone-pitched as it leads towards craggier ground. When the manicured path ends, an eroded line leads you to the left of a knoll and over to an epic view of The Cobbler and Loch Long.

3 Reach point 660m on the OS map and the bottom of the first real scramble of the day on Cruach nam Miseag. It amounts to no more than a couple of simple scramble moves with adequate handholds and footholds before the path continues up the ridge. As you approach the summit of Cruach nam Miseag, Ben Vane and Ben Vorlich (page 13) come into view to your right.

4 Descend north-west from the top of Cruach nam Miseag before continuing your ascent from the bealach. There are a couple more simple scrambling moves required here but the majority of the climb is a 'hands off' approach. While the start of this climb from the bealach is on the left, the path swings right and up a short gully next to The Spearhead crag. The path then leads around to the trig pillar atop Beinn Narnain. To the north are the craggy brothers and sisters of the Arrochar Alps; to the south, Loch Long twists and turns towards Glasgow.

5 Head to the large cairn and follow the path west off the summit plateau. The path leads over a small boulder field and continues downhill. At 800m elevation the path becomes stone-pitched and splits at NN266068. Take the right-hand fork down to the bottom path.

6 At Bealach a' Mhaim, turn right and walk north passing through a gate to begin your ascent of Beinn Ime which is mostly stone-pitched. This makes navigation really straightforward as the path stretches northwards. Towards the top of the Munro, follow an eroded line to the summit which has a shelter on the top.

7 Retrace your steps from the summit of Beinn Ime to Bealach a' Mhaim (**6**) and continue along the main path, ignoring the turn-offs left to Beinn Narnain and right to The Cobbler, as you descend between the two mountains. The path is well maintained and leads down into Coire a' Bhalachain.

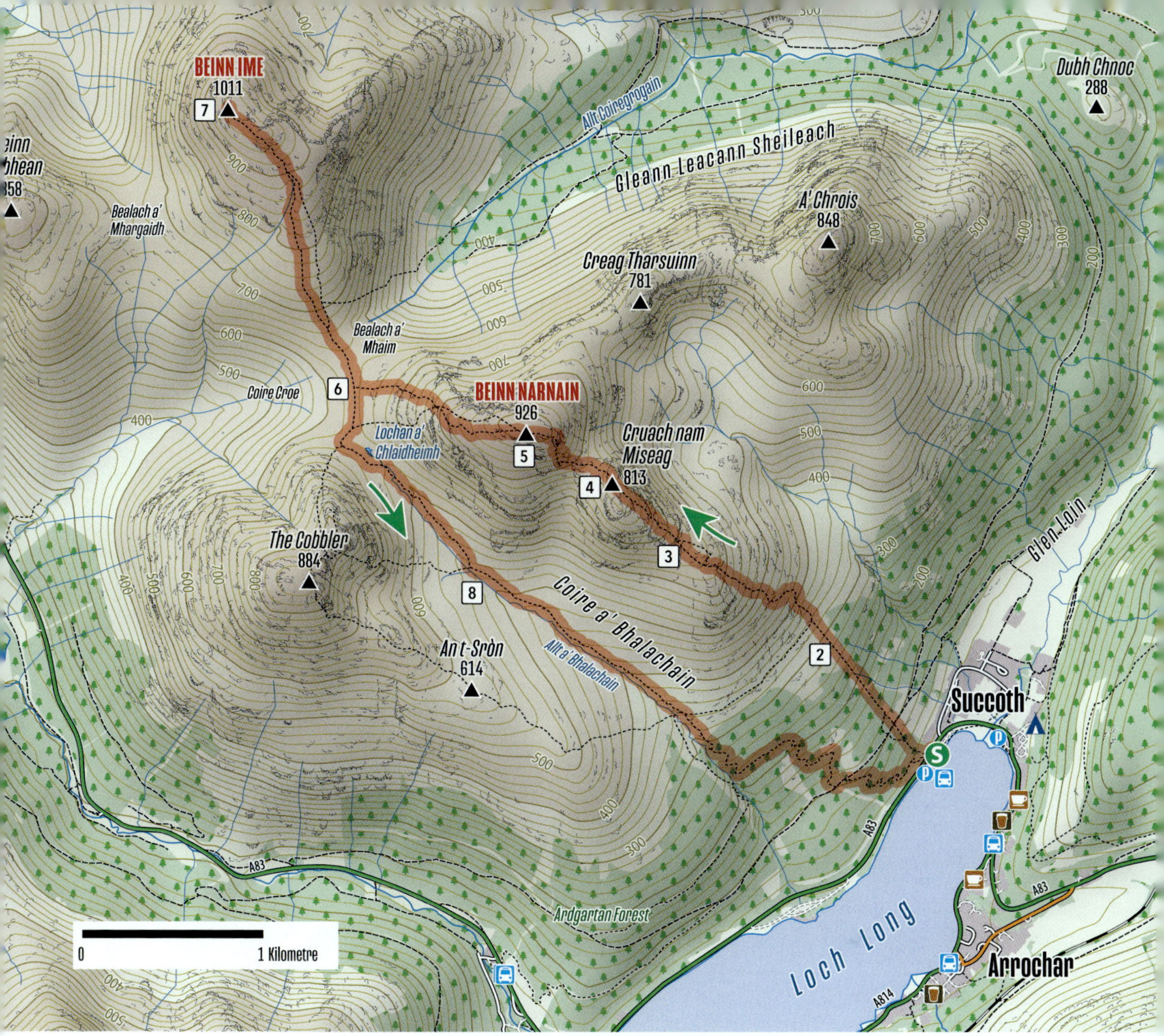

8 Continue south-east beside the Allt a' Bhalachain with the distinctive profile of Ben Lomond (page 17) now directly ahead in the distance. Stick to the main path which leads into woodland. Switchbacks enable a more pleasant descent for the knees. Exit left on to a track and walk for 60m before taking the next right turn to finish your descent back to the start.

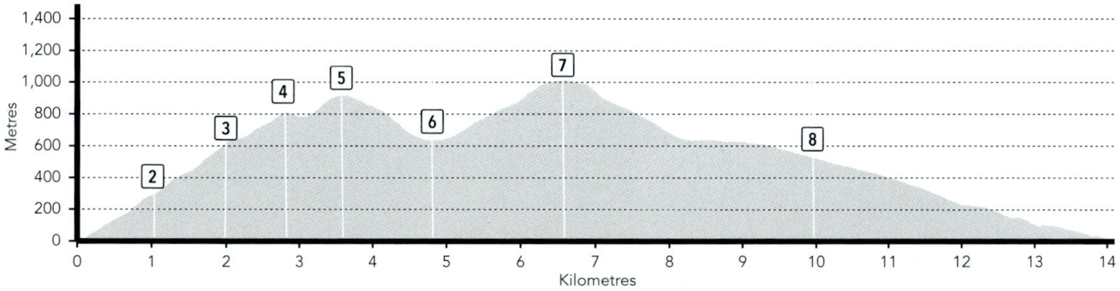

BEN VANE 916m
BEN VORLICH 943m

03 North-East Arrochar Alps

HUB **Arrochar – 10-minute drive**

OVERALL 1 2 **(3)** 4 5 TERRAIN 1 2 **(3)** 4 5
DISTANCE 1 2 **(3)** 4 5 NAVIGATION 1 **(2)** 3 4 5

DISTANCE **19.5km** ASCENT **1,740m**
TIMES **10:55hrs** (Walker) **9:25hrs** (Trekker)
7:05hrs (Fastpacker) **4:35hrs** (Trail runner)

START LOCATION **Inveruglas Visitor Centre, A82**
START LATITUDE/LONGITUDE **56.2519, -4.7091**
START GRID REFERENCE **NN322099**
START W3W **///abolish.urge.infringe**

	PRONUNCIATION	NAME MEANING
Ben Vane	**'ben vain'**	**mountain of the bay**
Ben Vorlich	**'ben vorlick'**	**middle mountain**

Introduction
Ben Vane and Ben Vorlich are forever tied in a fraternal bond – two brothers standing proudly side by side, their union eternal and unbreakable. That's why it seems only appropriate to climb them both on the same day. Sadly, this involves two big ascents. Loch Sloy splits the pair and there are no connecting ridgelines. But, if you're energetic enough, doubling up makes for a memorable Arrochar Alps adventure. If not, Vane and Vorlich are more than worthy of separate day hikes.

Public Transport
Scottish Citylink bus services connect Glasgow to Fort William and Oban. These buses stop outside Sloy Power Station which is opposite the start point for this walk.

Parking
Inveruglas Visitor Centre off the A82 offers chargeable parking for around 60 cars beside Loch Lomond. Toilet facilities are provided.

MTB-ability
The first 3.5 kilometres, the middle 1.3 kilometres and the last four kilometres can be mountain biked.

Escape Route
Retrace your steps after the first Munro and return another day for Ben Vorlich.

Refreshments
After your walk visit Ben Arthur's Bothy bar in Arrochar for pub grub and a drink overlooking Loch Long.

View across Loch Lomond to Loch Arklet and Loch Katrine from the mountain path on Ben Vane.

Ben Vorlich from Ben Vane.

The Route

S Leave the car park and cross the A82. Walk south on the footpath beside the A82 passing the front of Sloy Power Station. This path is part of both the Three Lochs Way and the Loch Lomond and Cowal Way, both of which are waymarked. The path leads to a gated access track which sits on the north side of Inveruglas Water. Pass through the gate and ascend on tarmac under the railway line.

2 Pass a large grey building, the Sloy Switching Station, and continue west on the track. Here you get the sensation that you're heading into the heart of the Arrochar Alps, with Ben Vane in front of you and the slopes of Ben Vorlich up to your right. Ignore a major turn-off to the left which is signposted *Glen Loin Loop Arrochar* and continue north-west up the track.

3 After 500m, fork left at a major track junction, noting that you'll return to this point after summiting Ben Vane. Cross the bridge over Inveruglas Water and ignore a left-hand fork in the track. Gently ascend for 500m before crossing a stream and taking a stone-pitched path on the right.

4 The path leads up to a flatter boggy section with a great view of the dam before the stone-pitched path continues again as the route steepens. The path becomes rougher as it weaves westwards up the mountain, but it stops short of becoming technical. A trio of lochs becomes visible over in the east – Loch Lomond, Loch Arklet and Loch Katrine.

5 Crags on the higher reaches of Ben Vane loom but technical scrambling can be avoided by choosing the leftmost path. Ben Vane is the 'are we there yet?' equivalent of a car journey with children because there are multiple false tops. The summit always feels close, but never quite is.

6 Ben Vane has a distinctive plateau top and the rocky outcrop beyond a small pond is the highest point. The distinctive profiles of Beinn Narnain (page 9) and the knobbly profile of The Cobbler can be seen to the south. Retrace your steps to the track and the junction beyond the bridge over Inveruglas Water (**3**). From here it's an additional 7.5km with 780m of ascent to include Ben Vorlich. If Ben Vane was more than enough for you for one day, retrace your steps to the start from here. If continuing, turn left to walk on tarmac to NN295102 where a stone-pitched path leads north-east up the second Munro.

7 Pass over a boggy section and ascend steeply with a stream on your left before crossing over the stream at NN299105. Navigate over a wet section of trail over grass before the stone-pitched path is picked up again. The route weaves north then north-north-west over eroded ground to gain the lumpy summit ridge. The gradient is then gentler as you approach the summit.

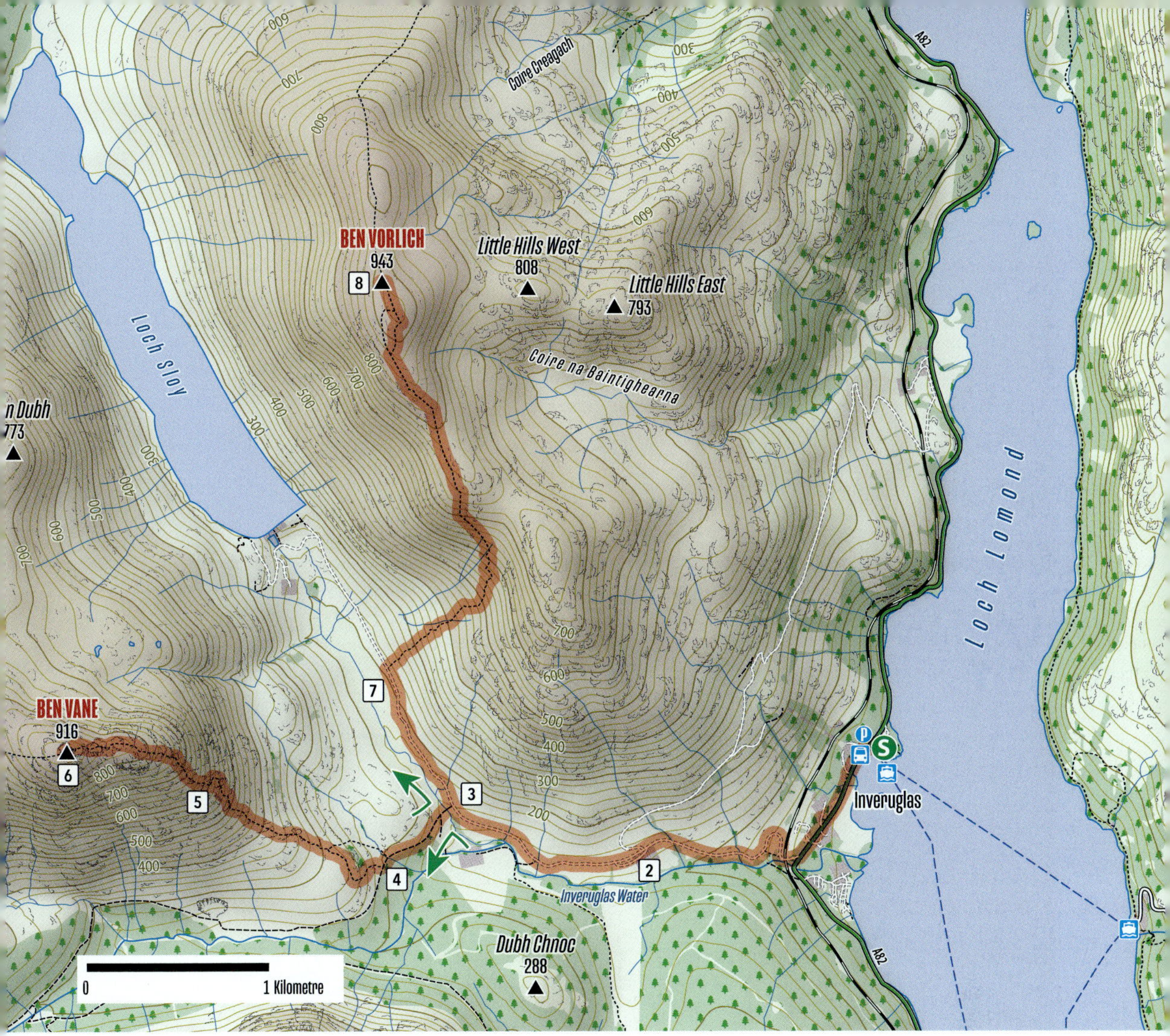

Pass the trig pillar which sits atop a knoll on your way to the highest point of the mountain, which is just beyond that – it is marked by a cairn on a rocky outcrop.

8 There are expansive views from the summit of Ben Vorlich to giants like Ben Lui (page 57) and Ben More (page 79). Retrace your steps while the stunning view to the south of Loch Lomond stretches out in front of you. When you reach the track walk the final 4km back to the start.

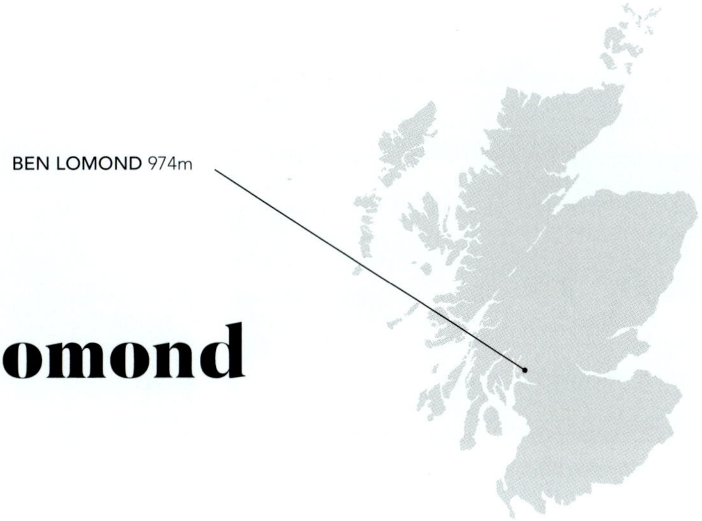

04 Ben Lomond

HUB **Arrochar – 1-hour drive to Rowardennan or travel to Tarbet and take a 45-minute waterbus crossing to Rowardennan**

OVERALL **1** 2 3 4 5 TERRAIN **1** 2 3 4 5
DISTANCE 1 **2** 3 4 5 NAVIGATION **1** 2 3 4 5

DISTANCE **12.6km** ASCENT **970m**
TIMES **6:25hrs** (Walker) **5:30hrs** (Trekker)
4:10hrs (Fastpacker) **2:45hrs** (Trail runner)

START LOCATION **Rowardennan car park, Rowardennan**
START LATITUDE/LONGITUDE **56.1519, -4.6421**
START GRID REFERENCE **NS359986**
START W3W **///frizz.purified.stuffing**

	PRONUNCIATION	NAME MEANING
Ben Lomond	**'ben lowmond'**	**beacon hill**

Introduction

Ben Lomond is, arguably, the most straightforward of all 282 Munros. It's also the most southerly. These two attributes make it an excellent choice as a peak-bagger's first – or indeed last – summit. For beginners it offers a confidence-building first foray into the world of Munro-bagging, with an excellent path, simple navigation and superb views. For 'compleatists' it's a mountain of perfect compromise: easy enough for friends and family to join in the celebrations, but dramatic enough to feel like a fitting finale to a grand adventure.

Ben Lomond from the approach to Beinn Narnain (see route 02).

Public Transport

There are no bus services to Rowardennan, however there is a Cruise Loch Lomond waterbus from Tarbet across Loch Lomond to the jetty half a kilometre north of the car park from May to September on Saturdays, Sundays and Mondays.

Parking

Chargeable parking in the large Rowardennan car park, also known as the Ben Lomond car park, on the eastern shore of Loch Lomond. Toilets available. If this car park is full, there is an overflow car park, 800 metres south of the main car park at NS361979, which provides access to the hill via an alternative path which merges with the described route after one kilometre.

Escape Route

None.

Refreshments

The dog-friendly Clansman Bar in the Rowardennan Hotel is a five-minute walk from the car park. The beer garden overlooks Loch Lomond.

Pair of walkers ascending Ben Lomond.

On the Ben Lomond summit ridge.

Descending Ben Lomond with the islands on Loch Lomond in the distance.

The Route

S Leave the car park on the path behind the toilet block – this wide footpath leads through woodland heading initially south-east. A fingerpost reads *Ben Lomond*. Pass through a gate in the deer fence and continue uphill. At a major path junction, continue straight on.

(If starting the walk from the overflow car park, head north-east out of the car park and walk for 1km before reaching this same junction – it will be a right turn for you.)

2 Cross a bridge and continue uphill as the path winds north-east. The footpath is mainly gravel or stone-pitched and is very well maintained. There is the occasional short segment of natural rock which presents no technical or navigational challenge and adds a bit of spice to what is otherwise a very manicured walk compared to most other Munro routes.

3 Pass through a gate and continue uphill, exiting out of the woodland. Pass through a final gate and continue ascending on the clear path. Pause to admire the southern end of Loch Lomond with all its little islands. Across the loch you'll notice Glen Douglas which stretches west from Inverbeg on the Loch Lomond shore between Doune Hill and Beinn Bhreac. The trail becomes rockier towards the very end, providing the beginner Munro-bagger with a very gentle insight into the terrain that will become commonplace as you continue your journey towards becoming a Munroist.

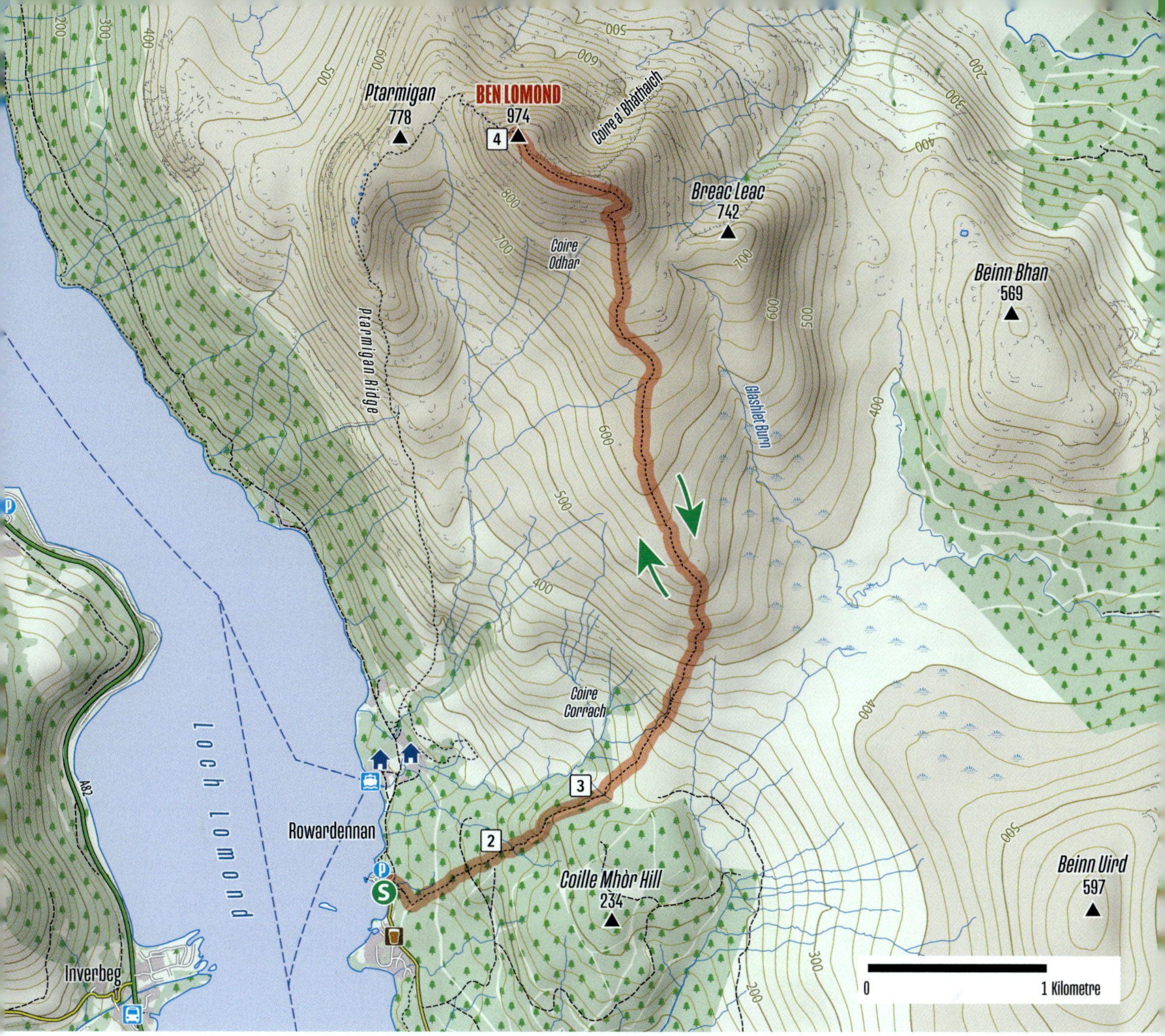

4 There is a trig pillar on the summit along with incredible views to the surrounding Arrochar Alps, Ben Vorlich and Ben Vane (page 13) and Beinn Narnain and Beinn Ime (page 9). Marvel at the expanse of Loch Lomond with its waters speckled with tiny islands. The Ptarmigan Ridge can also be appreciated from here. As this is a beginner's walking route, we retrace our steps back to the car. Confident scramblers may wish to continue along the Ptarmigan Ridge and use that as the route of descent.

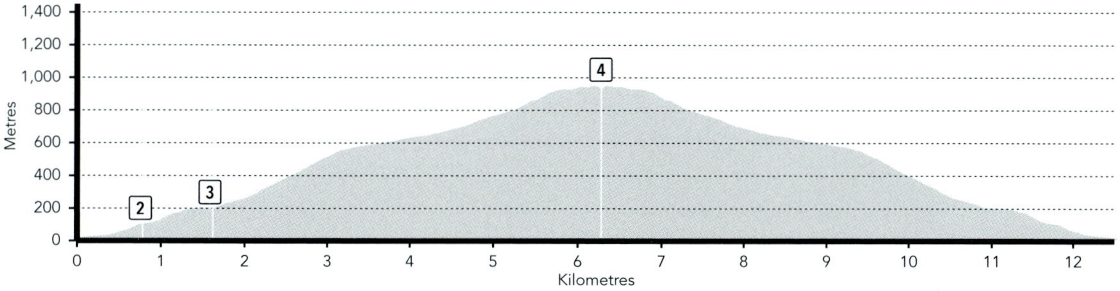

HIGHLAND HUB 2

COMRIE

The second of the Highland hubs is Comrie. This 'Shaky Toun' is close to the Highland Boundary Fault and as such the residents have felt more tremors here than anywhere else in the UK. It's the smallest of the hubs as it is used as a base for only two Munro walks in the Southern Highlands, perfect for a Munro-bagging weekend break.

The small village has a compact supermarket and a couple of drool-inducing bakeries and cafes. There's also a chip shop which is great for grabbing some fast food to eat on a bench beside the River Earn, plus a couple of restaurants and wine bars if that's more your scene.

Comrie used to have a railway station, but it closed in the 1960s so the only way to reach the hub by public transport is by local bus services. The Munros in the hub sit in between the two main arterial roads which extend northwards into Glen Coe (A82) and the Cairngorms (A9). From the south, one of the easiest ways to reach Comrie is via Perth on the bus. The operator may choose to reduce the frequency of bus services during the winter months.

The Three Saints Way, a Scottish pilgrim route, passes through Comrie on the way from Killin to St Andrews.

WHERE TO STAY

£	££	£££
Comrie Croft Campsite	Eco-Lodge at Comrie Croft Campsite	Comrie Hotel

TRAVEL TIMES

05 Loch Earn Munros	**20 minutes**
06 Ben Chonzie	**10 minutes**

Rest day ideas

- Visit the site of the Earthquake House – a small research station in a listed building situated on a rocky outcrop in a field on the western side of the village (NN765217, ///positions.boosted.wired).
- Wander north from Comrie to see the Deil's Caldron waterfall which runs through a tight gorge in the River Lednock.
- Explore Scotland's only remaining prisoner of war camp which used to house thousands of prisoners during World War II. Not only can visitors explore the history of Cultybraggan Camp, but thanks to the Comrie Development Trust, multiple community groups use the site and hundreds of volunteers have contributed towards positive local initiatives which include planting native woodland, maintaining an orchard and cultivating allotments.

Opposite Ascending Ben Vorlich with Loch Earn in the distance and the Ben Lawers range beyond (route 05).
Overleaf View from the summit of Stùc a' Chroin with the route of ascent clear in the distance (route 05).

BEN CHONZIE
06
Loch Earn
05
BEN VORLICH
STÙC A' CHROIN
Comrie
Crieff
Callander

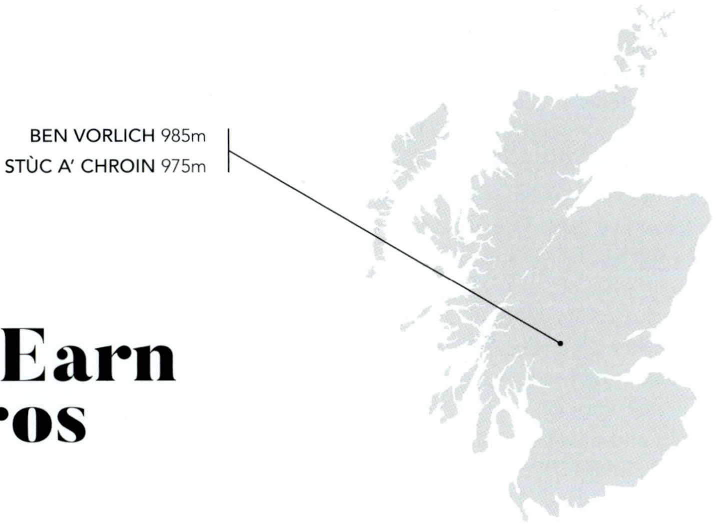

05 Loch Earn Munros

HUB **Comrie – 20-minute drive**

OVERALL 1 2 **3** 4 5 TERRAIN 1 2 3 **4** 5
DISTANCE 1 **2** 3 4 5 NAVIGATION 1 2 3 **4** 5

DISTANCE **14.7km** ASCENT **1,175m**
TIMES **7:40hrs** (Walker) **6:35hrs** (Trekker)
5:00hrs (Fastpacker) **3:15hrs** (Trail runner)

START LOCATION **Roadside parking on the western side of the bridge at Ardvorlich, southern shore of Loch Earn**
START LATITUDE/LONGITUDE **56.3814, -4.2164**
START GRID REFERENCE **NN632232**
START W3W **///flicks.lawns.dentures**

	PRONUNCIATION	NAME MEANING
Ben Vorlich	**'ben vorlick'**	**mountain of the bay**
Stùc a' Chroin	**'stook a kroin'**	**peak of danger**

Introduction

Not to be confused with its namesake in the Arrochar Alps, this Ben Vorlich rises to the south of Loch Earn in the north-eastern Trossachs. By Munro standards, it's a simple hike from Ardvorlich to the 985-metre summit – a popular route suitable for energetic kids, sprightly grandparents and everyone else in between. But beyond Ben Vorlich the ascent to Stùc a' Chroin gets far more serious, as the name meaning implies. You can avoid the scramble, but confident route-finders should give it a go. It's an exciting way to the top.

Public Transport

There are bus stops along the A85 on the northern side of the loch between Lochearnhead and St Fillans but there are none on the southern side. The closest bus stop is in Lochearnhead – a six-kilometre walk, mostly along a singletrack road, from the start of the route. Scottish Citylink services from Edinburgh stop here, alongside other local services.

Parking

Free parking for around 12 cars on the roadside on the western side of the bridge over the Ardvorlich Burn on the road which runs along the southern edge of Loch Earn. The nearest toilet facilities are in Lochearnhead.

Escape Route

Retrace your steps after the first Munro and return another day for the second. Alternatively, from the bealach between the two Munros, it is possible to drop north-west on to the lower path and ascend Stùc a' Chroin via the descent route from here. This avoids all of the steep scrambling.

Refreshments

Back at the hub in Comrie, Cafe Comrie will serve a slap-up Scottish breakfast to power you up those hills.

The summit trig pillar on Ben Vorlich with Stùc a' Chroin in the distance.

The Route

S Walk east along the road over the bridge. Turn right on to the long driveway for Ardvorlich House and walk south with the burn on your right. At the junction 250m later, fork right and cross the bridge before taking your next left. Follow this track south, passing through a gate. The burn is now on your left. Follow the track uphill with woodland including alder and rowan on your left and fields on your right. Pass over a stile and continue uphill on the main track. Pass over another two stiles after which the track fords a stream. Continue on the track and cross a wooden bridge after which the track splits.

2 Fork right and continue uphill crossing two more bridges and a further ford on the track. The track leads to a larger bridge. Cross this and the track now ends and becomes a clear path which leads southwards up Ben Vorlich.

3 At NN629200 note the path which joins from the right for later as this is your descent route. For now, continue south. The path becomes more eroded as it winds uphill to reach the Sgiath nam Tarmachan ridge.

4 Behind you in the distance is the Lawers range (pages 123 and 127). The trail becomes steeper and more eroded as it zigzags towards the summit of Ben Vorlich. To the west is Ben More (page 79) and the Crianlarich Munros (pages 69 and 75).

5 Leave the summit to the west, descending the grassy ridge. A third of the way down, the route becomes rocky and steep but is non-technical. Next, follow an eroded path which is loose underfoot down to a boggy section on the upper part of the bealach. From here, iron fence posts lead the way.

6 From the bottom of the bealach, beginner Munro-baggers, or those nervous about tough route-finding or scrambling, should drop north-west on to the lower path and ascend Stùc a' Chroin via the descent route from here. This avoids all of the steep scrambling. Experienced Munro-baggers and good navigators can continue on the stony path, ignoring a grassy fork to the right.

7 At the foot of a boulder field, the path swings to the left. Follow the path to NN619179 and cross the boulder field. Ascend steeply, keeping to the right of a large crag beside a slabby rock before traversing over to the northern side of the buttress. From here, a path leads to the foot of some straightforward scrambling. The scrambling itself is minimal both in terms of the number of moves and the difficulty, but the key is correctly identifying the safest way on to this part of the buttress. Top out at a cairn and walk south following iron fence posts to the summit of Stùc a' Chroin, keeping to the left when the path splits just before the summit mound. Ben Vorlich and Coire Garbh, its broad south-facing aspect, can be admired from the summit cairn.

8 Retrace your steps from the summit initially. After 200m at NN617176, the path splits. Take the left fork which eases downhill, following an eroded line over grassy terrain. At NN615180, the path leaves the ridge and heads steeply downhill over loose, eroded ground into the bowl.

9 At the bottom of the steep path, cross two streams and begin traversing below the north-western edge of Bealach an Dubh Choirein. The ground is wet but route-finding is straightforward. The path mainly hugs the 700m contour line over soft, wet ground before it becomes firmer underfoot and traverses around the north-west shoulder of Ben Vorlich. Here, the path begins to drop gently and becomes boggier once more, until you meet the ascent path at NN629200 (**3**). From here, follow the main path back to the start.

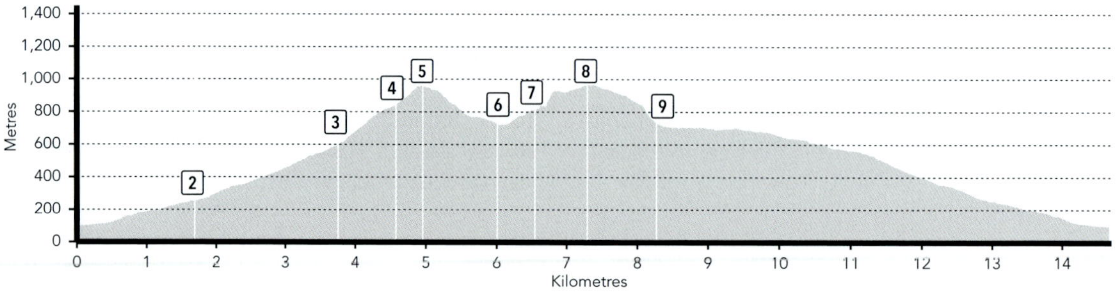

Glen Beich
A85
Lochearnhead
Loch Earn
A84
Glen Ample
Glen Vorlich
Ben Our
740
Creagan nan Gabhar
725
Coire Mheobhith
Coire Buidhe
Creagan an Lochain
685
Lochan na Mna
Beinn Domhnuill
739
Sgiath nam Tarmachan
BEN VORLICH
985
Coire Fhuadaraich
Coire Garbh
Bealach an Dubh Choirein
Creag Chasain
STÙC A' CHROIN
975
Gleann an Dubh Choirein
Lochan a' Chroin
0
1 Kilometre

06 Ben Chonzie

HUB **Comrie – 10-minute drive**

OVERALL 1 2 3 4 5 TERRAIN 1 2 3 4 5
DISTANCE 1 2 3 4 5 NAVIGATION 1 2 3 4 5

DISTANCE **12.6km** ASCENT **720m**
TIMES **5:25hrs** (Walker) **4:40hrs** (Trekker)
3:30hrs (Fastpacker) **2:20hrs** (Trail runner)

START LOCATION **Parking area at Coishavachan**
START LATITUDE/LONGITUDE **56.4209, -4.0393**
START GRID REFERENCE **NN742272**
START W3W **///frail.from.nursery**

	PRONUNCIATION	NAME MEANING
Ben Chonzie	**'ben honzie'**	**mossy hill**

Introduction

Ben Chonzie is an isolated Munro with an uncomplicated ascent from Glen Lednock along a wide track for the bulk of the route. For this reason, it is popular with beginner Munro-baggers. Strong walkers will make incredible progress and find themselves down in time for lunch. On a quiet day, look out for mountain hares on the plateau – they're easier to spot in early spring when they're still white from winter.

Descending Ben Chonzie.

Public Transport

Glen Lednock is not serviced by public transport. The nearest bus stop is a seven-kilometre walk from the start of the route in Comrie, where local bus services stop in the village.

Parking

Free parking for over 30 cars at Coishavachan, Glen Lednock. The nearest public toilet is in Comrie.

MTB-ability

An experienced mountain biker could cycle up to 710 metres in elevation on the first and last 4.5 kilometres of the route.

Escape Route

None.

Refreshments

Hansen's Kitchen is a deli in Comrie which sells a variety of locally sourced and home-baked treats.

Descending Ben Chonzie.

View from the summit of Ben Chonzie to Loch Turret Reservoir.

The Route

S Leave the parking area following a track to the north. After 250m turn right in front of a row of cottages. The track continues north with a mild incline. Cross the bridge over the Invergeldie Burn and pass through a gate, continuing uphill along the wide track. Ignore two left turns and pass through another gate at NN745285.

2 Continue walking north above the Invergeldie Burn which is down to your right. Ignore a further left turn and continue on the main track, crossing the Invergeldie Burn again just below a dam. Continue walking up the track on the other side of the burn as it leads south for a short time, ignoring a right-hand fork. The track then weaves north-east and climbs more steeply up the hillside. Ideally, walk this route when the heather is in bloom as it brightens up both sides of the track considerably. Cross a stream, after which the track steepens and leads east-north-east.

3 At NN767293, fork left on to a peaty footpath and walk north crossing a small burn. The route is eroded and stony in places with one main patch of boggy ground to overcome. The trail becomes grassier as it flattens, and the path becomes indistinct. Fortunately, this coincides with the appearance of a line of old iron fence posts. Follow these north-north-west to NN767303.

4 Here, the line of fence posts turns to point

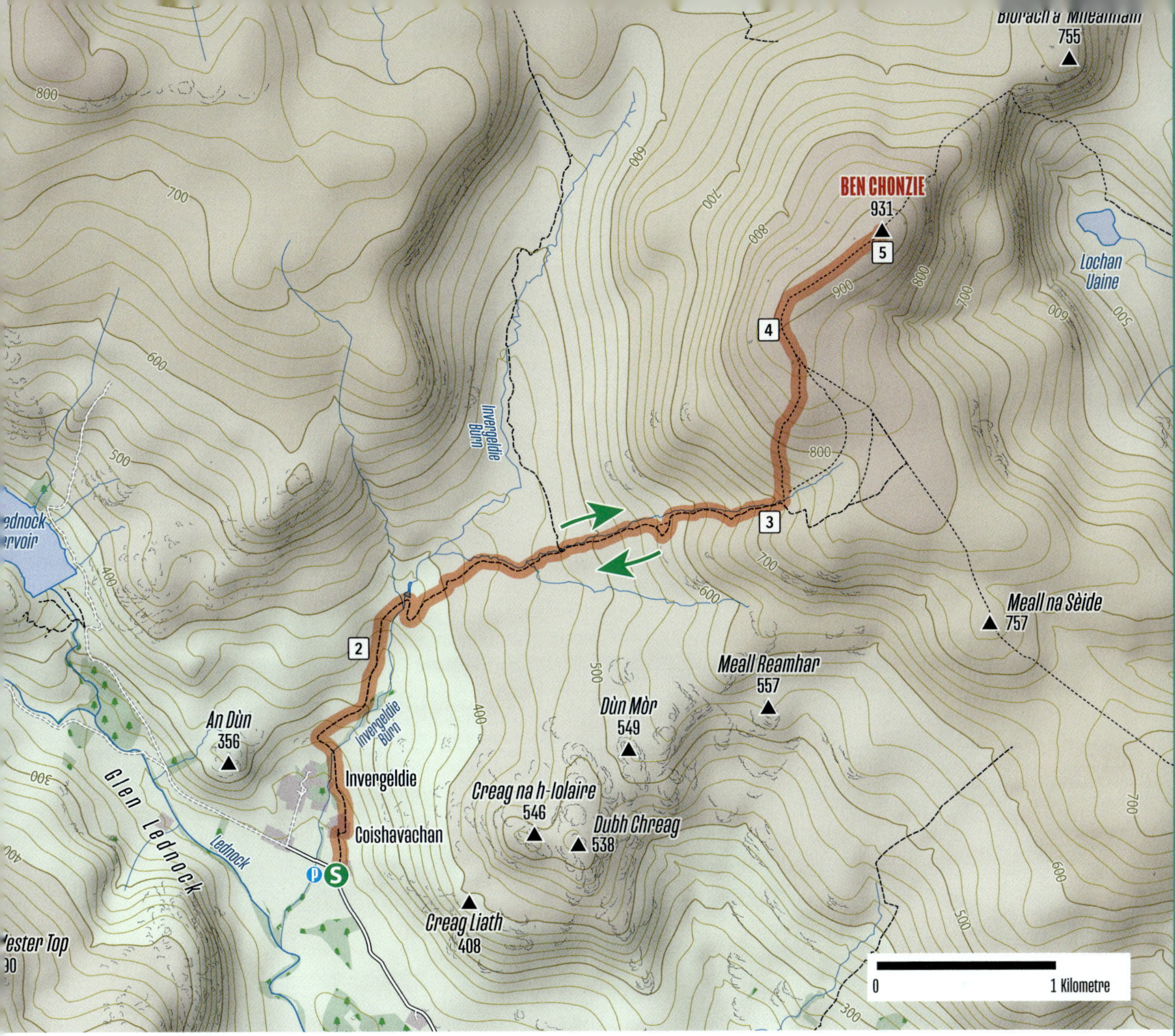

north-east. In poor visibility the fence posts can be handrailed to the summit shelter. In good visibility you can make your way across the broad summit plateau while enjoying views to the Ben Lawers range (pages 123 and 127) to the north-west.

5 Delight in a well-earned snack in the summit shelter while taking in the view south-east down Loch Turret Reservoir. Retrace your steps to the start, while appreciating the view to the distant Munros of Stùc a' Chroin and Ben Vorlich (page 25).

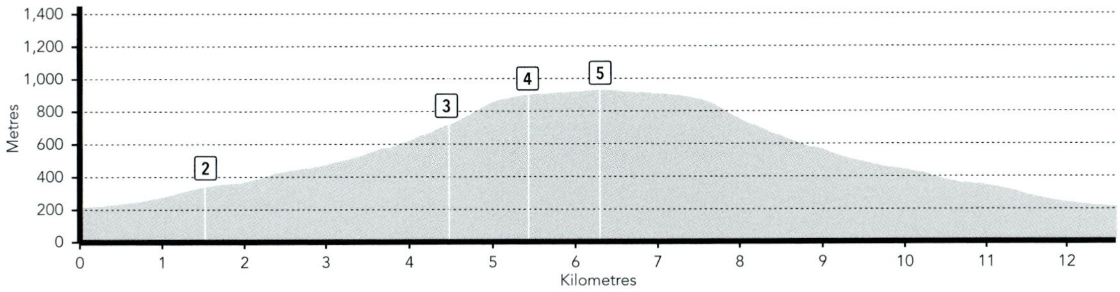

HIGHLAND HUB 3

OBAN

The third of the Highland hubs is Oban – this buzzing harbour town is known as the Seafood Capital of Scotland. During the summer season the population of Oban swells to accommodate thousands of visitors. It is the last big town on the west coast before Fort William which makes it a popular place to stop on a road trip north.

Oban has a railway station which connects it to the Central Belt. A train from Glasgow takes just over three hours to reach the town, with buses taking a similar amount of time.

There are large supermarkets in Oban, so you'll never be short on places to resupply, along with multiple eateries and attractions. I was once fortunate enough to be invited on to *Zuza*, an Oban-based VentureSail yacht for a skippered sailing weekend around Mull and Lismore. I'll never forget eating sweet and delicate hand-dived scallops while admiring the view across the bay.

While it is possible to find pockets of quiet in the area, Oban is also a busy port with tourists and locals alike relying on ferries from Oban to carry them to the Inner and Outer Hebridean islands of Mull, Lismore, Colonsay, Coll, Barra and South Uist. If you're not prepared to kayak across the Firth of Lorn to reach Ben More, the only island Munro in this volume, you'll need to use a Caledonian MacBrayne (CalMac) ferry: ***www.calmac.co.uk*** Luckily, they're frequent and very efficient.

WHERE TO STAY

£	££	£££
Oban Holiday Park	Oban Youth Hostel	Whisky Vaults Hotel

TRAVEL TIMES

07 Ben More (Mull)	**1 hour and 35 minutes**
08 Beinn Sgulaird	**30 minutes**
09 Cruachan Horseshoe	**30 minutes**
10 Dalmally Munros	**35 minutes**

Rest day ideas

- Take a tour of a whisky distillery. Oban Distillery was first established in 1794 and is one of the smallest distilleries in Scotland. It's right in the middle of town.
- Walk for three kilometres along Gallanach Road to a small ferry terminal which links the area to the small island of Kerrera. There are a couple of short walks which can be undertaken on this beautiful island and if you can't stop yourself aiming for another summit, the highest point is Carn Breugach (189 metres).
- Take in a glorious sunset from the headland around Dunollie Point, just north of town.

Opposite View down to Cruachan Reservoir and Loch Awe from the ascent of Ben Cruachan (route 09).
Overleaf View towards Drochaid Ghlas and Stob Daimh from Ben Cruachan's summit (route 09).

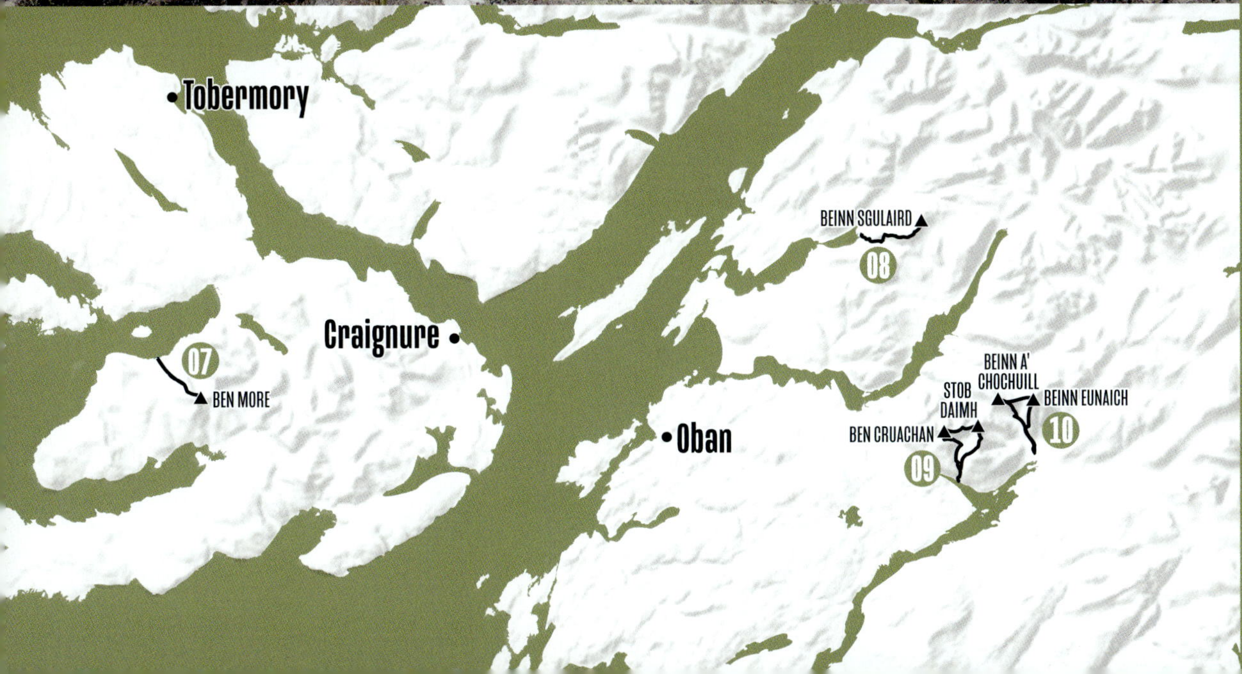

Tobermory
Craignure
Oban
BEINN SGULAIRD
08
07
BEN MORE
BEINN A'
CHOCHUILL
BEINN EUNAICH
STOB
DAIMH
BEN CRUACHAN
09
10

07 Ben More

HUB **Oban – 1-hour ferry followed by a 35-minute drive**

OVERALL 1 ② 3 4 5 TERRAIN 1 ② 3 4 5
DISTANCE ❶ 2 3 4 5 NAVIGATION 1 ② 3 4 5

DISTANCE **9.6km** ASCENT **965m**
TIMES **5:45hrs** (Walker) **5:00hrs** (Trekker)
3:45hrs (Fastpacker) **2:25hrs** (Trail runner)

START LOCATION **Grassy roadside parking on the B8035 opposite the bottom of the track to Dhiseig**
START LATITUDE/LONGITUDE **56.4491, -6.0674**
START GRID REFERENCE **NM494359**
START W3W **///ahead.impulsive.unhappily**

	PRONUNCIATION	NAME MEANING
Ben More	**'ben more'**	**big hill**

Introduction

Ben More is the only island Munro in this book and requires a ferry from the mainland on to the Isle of Mull. The ferry in itself makes for an enjoyable escapade which helps to make Ben More a popular and memorable 'final Munro'. It also makes for a special first ever Munro, as I can personally attest. For this reason, I recommend the simplest route, from which you can absorb the unparalleled coastal panorama.

Public Transport

There are no public transport services close to this walk.

Parking

There is free parking for around 12 cars just off the road on the B8035 opposite the bottom of the track to Dhiseig on a firm grassy area between the road and Loch na Keal. The nearest public toilets are in Salen, halfway between the ferry port at Craignure and the start point of the walk.

Escape Route

None.

Refreshments

For yummy cakes which are baked on the premises, stop at The Coffee Pot in Salen on your way to or from the ferry port.

View back down the ascent trail of Ben More towards the island of Eorsa in Loch na Keal.

View to Ben More from Ben Cruachan (see route 09).

The Route

S Leave the parking area and head south up the gravel track to Dhiseig, where a couple of properties managed by the Benmore Estate sit with handsome views over Loch na Keal. When the track swings left towards the cottages, walk straight ahead on the wide grassy track and pass through a gate. Turn right and walk along the fence towards the stream and then follow the stream south-east on the footpath.

2 At NM499353, beside a small waterfall, cross the fence using the stepping stone and when the path forks just after this, continue to follow the line of the stream over slabby ground with a mild incline initially. Handily positioned stones then assist passage over boggy ground before the path becomes grassy once more.

3 At NM502348, cross a small stream and continue along the northern bank of the Abhainn Dhiseig towards Ben More, avoiding the temptation to veer left towards the higher ground of An Gearna.

4 The path leads to a crossing point of the Abhainn Dhiseig in a shallow clough at NM505345. The path loosely follows the line of the stream on the other side and passes some pretty waterfalls. Beyond this, the path begins to peel away from the stream following an eroded line up on to stonier ground. The stream disappears into a steep gorge down to your left as you walk south-east up the hillside.

5 The path curves southwards to reach the first in a series of cairns which mark the approach to the west-north-west branch of the north-west ridge of Ben More via a stony path which becomes more pronounced as you gain higher ground. On a clear day, the highest Munros on the mainland are in view to the north-east, including Ben Nevis (page 195). Ben More was an active volcano several million years ago and the terrain has a distinctly volcanic feel to it as you ascend the ridge on dark, dusty scree. Keep your eyes peeled for resident white-tailed eagles,

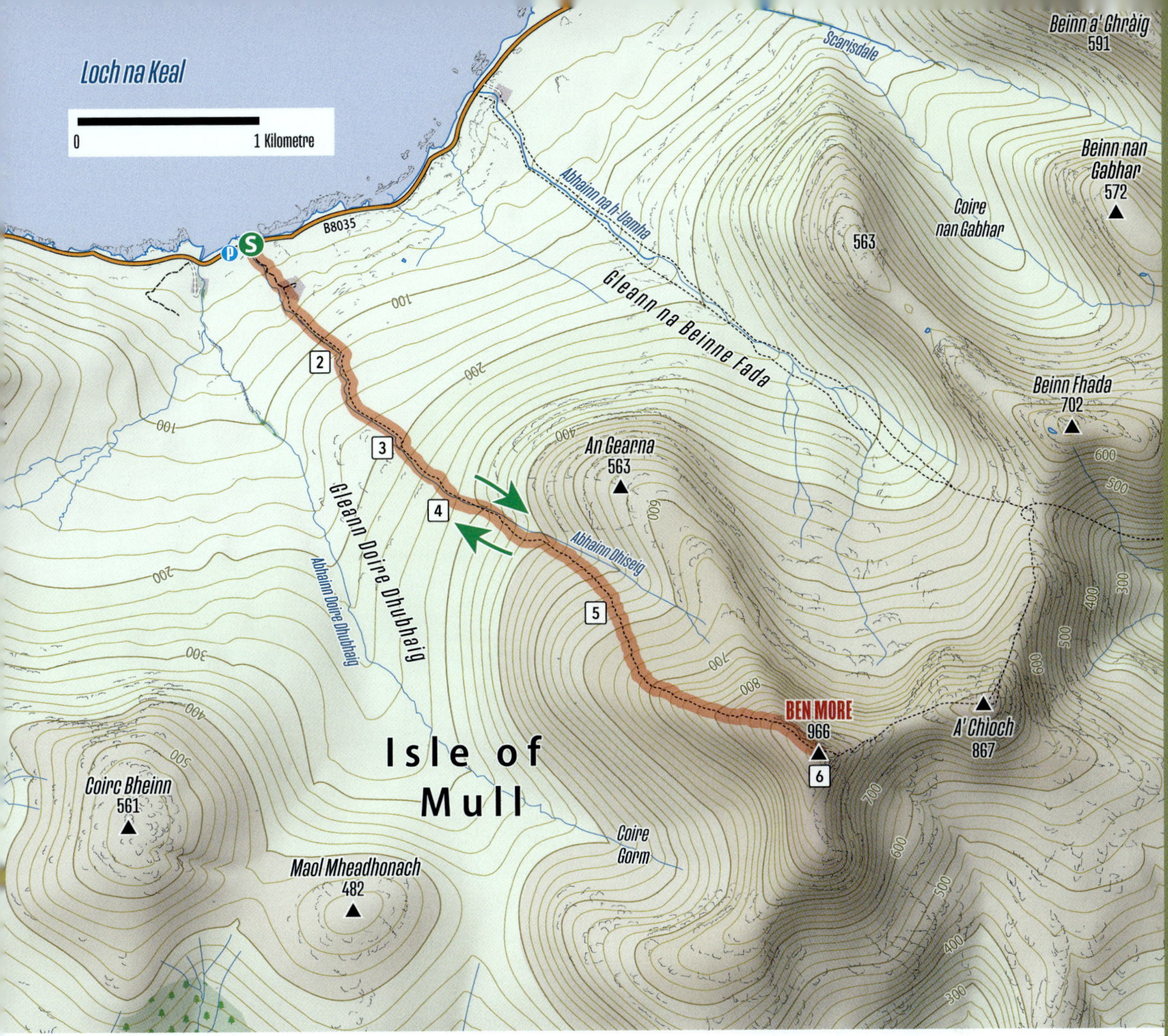

the UK's largest bird of prey, which are more likely to be spotted near coastal saltwater areas. The gradient levels out towards the top of the Munro and a large stone shelter can be found at the summit.

6 Retrace your steps to the start, enjoying the stunning views towards the Ardnamurchan peninsula, Loch na Keal and the nearby islands of Eorsa and Ulva.

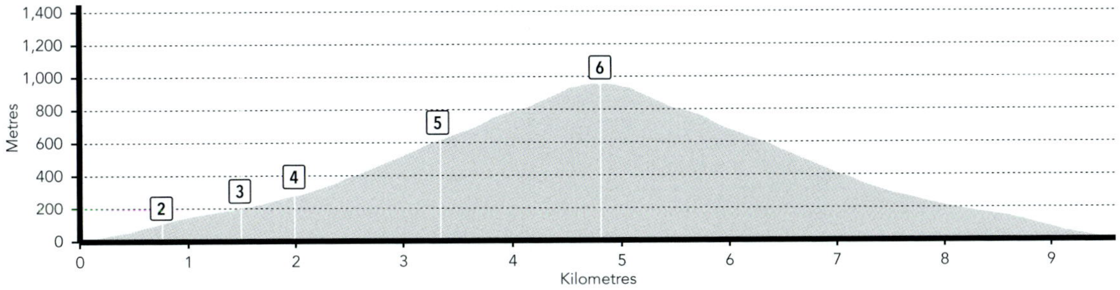

08 Beinn Sgulaird

HUB **Oban – 30-minute drive**

OVERALL 1 2 **(3)** 4 5   TERRAIN 1 2 3 **(4)** 5
DISTANCE 1 **(2)** 3 4 5   NAVIGATION 1 2 **(3)** 4 5

DISTANCE **13.6km**   ASCENT **1,245m**
TIMES **7:45hrs** (Walker)   **6:40hrs** (Trekker)
5:00hrs (Fastpacker)   **3:15hrs** (Trail runner)

START LOCATION **Lay-by north of Druimavuic, Glen Creran**
START LATITUDE/LONGITUDE **56.5556, -5.2411**
START GRID REFERENCE **NN009451**
START W3W **///blatantly.inserting.germinate**

	PRONUNCIATION	NAME MEANING
Beinn Sgulaird	**'beyn skoolershd'**	**hat-shaped hill**

Introduction
At 937 metres in height, you'd be forgiven for assuming that the quickest and easiest ascent of this solitary Munro would be achievable before lunchtime. On the contrary, this Munro makes for a rewarding full day out and requires a surprising 1,245 metres in ascent due to the start point being close to sea level and the requirement to walk over multiple tops along the long and knobbly ridge which boasts exceptional views to several Hebridean islands.

Public Transport
There are bus stops at the northern end of Creagan Bridge, which is 4.8 kilometres from the start of the route.

Parking
Free parking for around six cars just off the minor road north of Druimavuic in Glen Creran. The nearest public toilets are in Oban.

MTB-ability
The first and last two kilometres of the route on the access track are cyclable but steep in places.

Escape Route
None.

Refreshments
A 10-minute drive south-west on the edge of Loch Creran is the Racer Cafe which offers hot breakfasts and lunches in a creatively decorated venue.

Beinn Sgulaird from the descent of Beinn Fhionnlaidh (see route 31).

View down Loch Creran from the ascent of Beinn Sgulaird.

The Route

S Walk south-west down the road to the first bend. Here there are two turn-offs to the left. Take the first one, pass through a gate and walk south on the gravel track. Pass through a second gate and remain on the track for a further 1.5km, ignoring minor turn-offs to the right and left until you reach point NN022444 which is marked by a small cairn.

2 Turn off the track to the left, following a steep grassy trail which leads to a bealach on the ridge above. There are some tall crags on your left and a fence on your right. There's a large flat rock at the base of the wire fence at NN025450; cross here and follow the trail east-south-east uphill.

3 The trail is grassy at first before becoming stony for a short time and then grassy once more. Progress is speedy along the flatter, grassier sections and slows on the brief steeper, rockier sections, although at this stage of the walk no scrambling is required. After a period of wet weather, it's great to watch the veins of white water streaming down Coire Dubh between Meall Garbh and Meall nan Caorach on the other side of the valley – on your right as you ascend.

4 A cairn at NN036447 marks a fork in the trail. Fork left here and head east uphill. Walk to the side of a series of small knolls, one of which is marked as point 856m on the OS map. Pass to the right of a small lochan and up and over point 863m.

5 From here undulate north-east to a cairn before descending carefully down a section of red-tinged boulder scree, followed by a grassier undulating path to the bealach. From the bealach, ascend Meall Garbh weaving up the eroded trail, in a mostly northerly direction, around large slabby rocks. One simple scrambling move is required, but with the exception of this, it's a straightforward ascent.

6 From the summit of Meall Garbh, undulate over stony ground to the north-east before descending to the bealach over another two short sections of boulder scree. Put aside the feeling that you should have at least hiked over the summit of a Munro by now and begin the ascent of Beinn Sgulaird up the ridge, which is slightly narrower now, but without exposure. The trail weaves up and around large slabs, with a couple more straightforward scrambling moves required towards the top to gain the summit

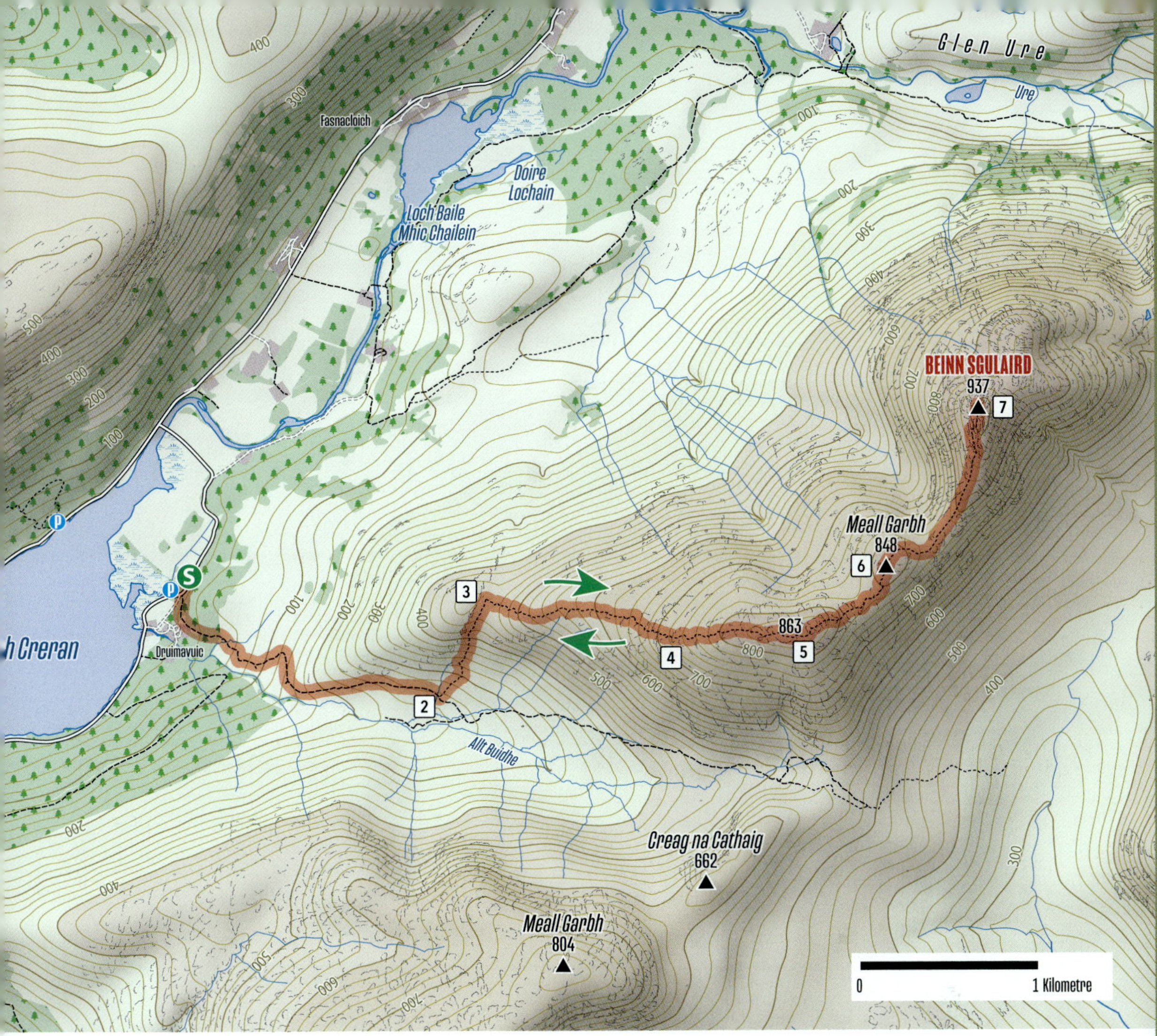

plateau. The summit itself is marked by a large cairn. The treat at the summit is the expansive view down Loch Creran, to the Lynn of Lorn and the long, thin island of Lismore, then beyond to Mull and its lone Munro, Ben More (page 37).

7 From the summit of Beinn Sgulaird, the safest and most enjoyable way down is to retrace your steps to the start.

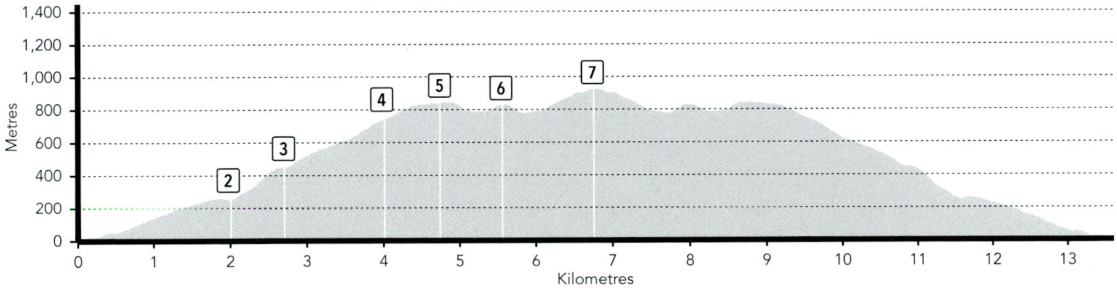

09 Cruachan Horseshoe

HUB **Oban – 30-minute drive**

OVERALL 1 2 (3) 4 5 TERRAIN 1 2 3 (4) 5
DISTANCE 1 (2) 3 4 5 NAVIGATION 1 2 (3) 4 5

DISTANCE **14.2km** ASCENT **1,450m**
TIMES **8:40hrs** (Walker) **7:30hrs** (Trekker)
5:35hrs (Fastpacker) **3:35hrs** (Trail runner)

START LOCATION **Falls of Cruachan railway station**
START LATITUDE/LONGITUDE **56.3936, -5.1116**
START GRID REFERENCE **NN080267**
START W3W **///voltage.fluffed.prickly**

	PRONUNCIATION	NAME MEANING
Ben Cruachan	**'ben krooerkan'**	**heaped or conical hill**
Stob Daimh	**'stob dyv'**	**peak of the stag**

Introduction

The ridge-walking between Ben Cruachan and Stob Daimh presents the run-of-the-mill hillwalker with the biggest buzz. Progress along the ridge is surprisingly non-technical which contrasts with the sensation of being high among jagged, lofty peaks which rise sharply from the loch-scattered landscape far below. This Cruachan Horseshoe can be extended to include Beinn a' Chochuill and Beinn Eunaich (page 49); I walked these four Munros during a two-day adventure many years ago and a photo from that trip remains as my laptop screensaver to this day.

On the ridge between Ben Cruachan and Stob Daimh.

Public Transport

During the summer months, this is one of the most well-served Munro walks in the book when it comes to public transport. The walk starts from Falls of Cruachan railway station where rail (April to October only) or bus connections can be made to Oban, Glasgow and Edinburgh.

Parking

There is free parking for around eight cars on the roadside on the A85 near Falls of Cruachan railway station. The nearest toilet facilities are in Taynuilt, 10 kilometres from the start.

Escape Route

Retrace your steps after Ben Cruachan; once you're committed to the ridge to Stob Daimh there are no escape routes.

Refreshments

On your way back to the hub, detour to Sìtheag, the Gaelic word for 'fairy', where you'll find a tranquil community coffee shop and art space.

The Route

S Take the path which leads up a staircase and through a gate under the railway line. Continue up the steps on the other side for 50m before turning left off the staircase on to a rough trail which leads into the woodland. The trail weaves to the right as it steepens; the sound of the Falls of Cruachan can be heard down to the left as you ascend over rocks and exposed tree roots. Follow the eroded path, a deep channel in parts, as it rises up through silver birch to reach a ladder stile. Pass over the stile on to grassier ground. A path leads between pylons and over a patch of boggier terrain before meeting a track below the dam.

2 Turn left on to the tarmac track which crosses a bridge over the Allt Cruachan and walk for 500m to the foot of the dam. From here, take the path to the left which leads first to a series of grassy steps and then to the foot of a metal staircase. Ascend this staircase to top out on the dam wall. Turn left, pass through a gate at the end of the dam wall and immediately leave the track to walk north on a rocky path. When the path rejoins the track, turn right and walk along the western edge of Cruachan Reservoir.

3 At the end of the track, turn left at a small turbine on to a footpath which follows the right-hand side of a stream which runs down from Coire Dearg. Glance to your left on the ascent to see Loch Awe come into view over the dam. At NN080296, fork right to stick to the higher path for just over 500m until NN075298. Here you should cross a stream and rejoin the lower path as the route steepens. The path increases in roughness as it rises to the bealach between Meall Cuanail and Ben Cruachan where you'll be greeted with a view west towards Mull.

4 Turn right from the bealach to ascend loose stony ground and latterly follow a path up to the foot of a boulder field which must be crossed to gain the summit of Ben Cruachan. There is an incredible view from the summit down Loch Etive, with Ben Starav (page 145) recognisable at the north-eastern tip of the loch and, on a clear day, Ben Nevis (page 195) can be seen to the north-north-east. You may also be lucky enough to see the Rùm Cuillin skyline to the north-west from the summit of Ben Cruachan, despite Askival on Rùm being nearly 100km away as the crow flies.

Ben Cruachan from the descent of Stob Daimh.

5 Leave the summit to the north-east down the rocky ridge. Look out for the path which keeps to the right towards the start of the ridge and avoids much of the scrambling. Descend to a low point on the ridge before traversing slabby rock to a short gully. Ascend here to regain the crest. There are steep drops to the left, but the ridge is just broad enough to not feel overly exposed. Continue along the ridge over non-technical, undulating rocky ground. This ridge is a real pleasure to walk along and is one of my personal favourites. The path meets a junction just below the summit of Drochaid Ghlas. Here the route heads right to walk south-east downhill but this Munro Top can be added in as a short out-and-back if you wish.

6 The terrain on the short descent to the bealach is loose so take care; however, the path up Stob Daimh is clear and uncomplicated. As you near the top, be sure to look behind you as Ben Cruachan's shark-fin-like summit is something to be marvelled at from this angle. At the summit of Stob Daimh, I particularly like the view down to Dalmally with Ben Lui (page 57) beyond.

7 Leave the summit to the south following a path down to the bealach from which you can ascend Stob Garbh. Although Stob Garbh is not a Munro it still offers a great vantage point across most of the route you've walked so far. Continue on to Stob Garbh's grassy south summit at 947m.

8 From here descend Stob Garbh's south-west shoulder. From just above the bealach at NN094292 head west over a very brief 100m section of grassy pathless ground to pick up a lower path. Follow this grassy path as it leads south-west in the direction of the reservoir, past a small obelisk on terrain which is mostly grassy, interspersed with a few wet and rocky sections. The path crosses the burn at the bottom of a high-sided gorge on the 450m contour and continues along the eastern side of Cruachan Reservoir.

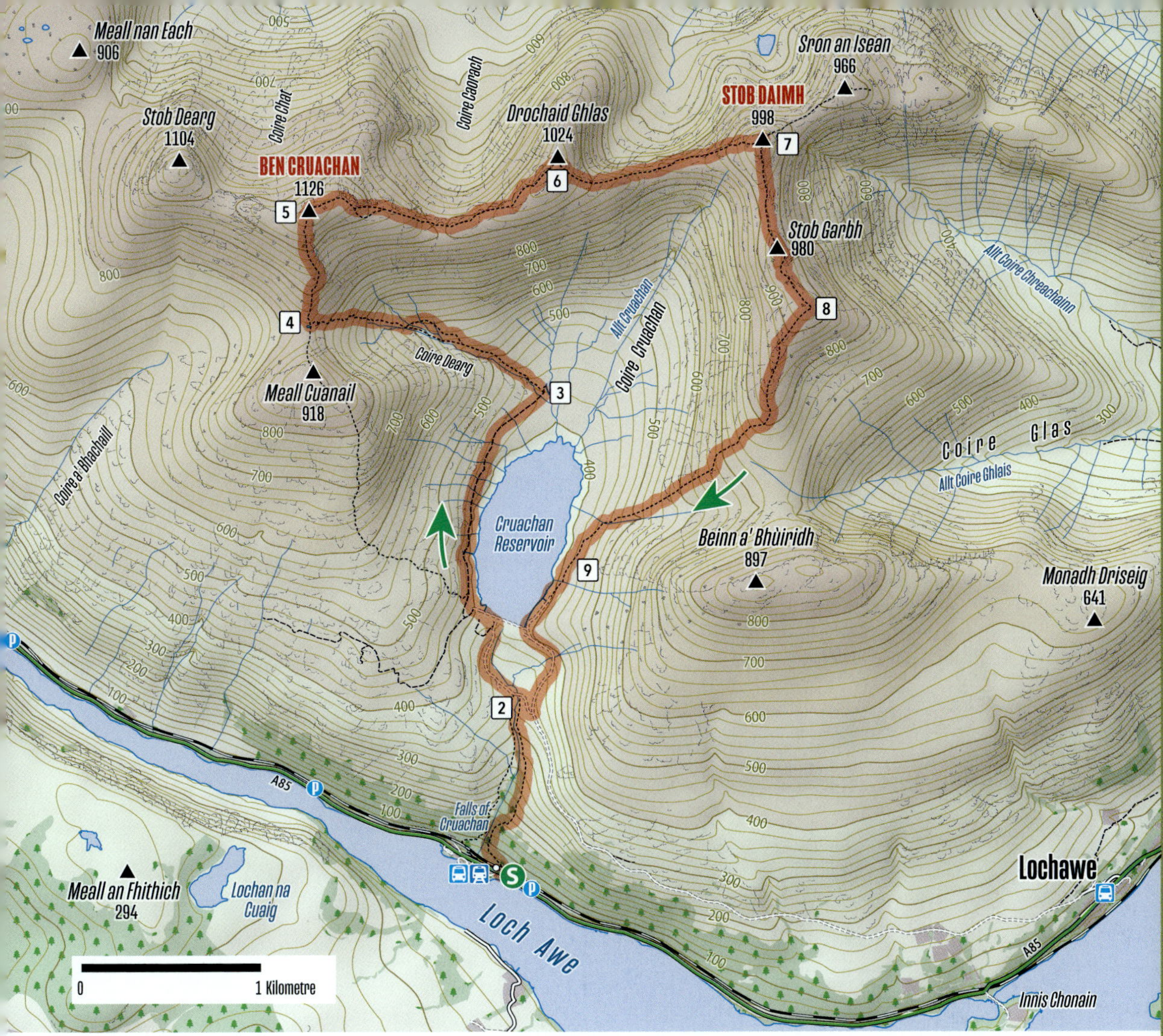

9 On the terrace path above the reservoir, a short wet section over bog and slabby rock is the only obstacle before the path merges with a substantial track. Follow the track as it ascends slightly before descending on tarmac. Pass through two gates and walk south-east down the access track for 500m before meeting a major track junction. Turn right here and descend to the bridge over the Allt Cruachan that you'll remember from the ascent. Retrace your steps to the start from here.

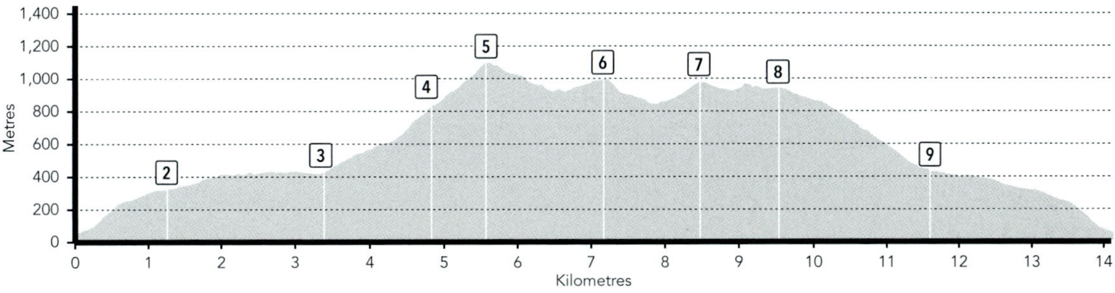

10 Dalmally Munros

HUB **Oban – 35-minute drive**

OVERALL 1 **(2)** 3 4 5 TERRAIN 1 2 **(3)** 4 5
DISTANCE 1 **(2)** 3 4 5 NAVIGATION 1 **(2)** 3 4 5

DISTANCE **13.6km** ASCENT **1,220m**
TIMES **7:35hrs** (Walker) **6:35hrs** (Trekker)
4:55hrs (Fastpacker) **3:10hrs** (Trail runner)

START LOCATION **Roadside parking on the B8077, north-east of Loch Awe**
START LATITUDE/LONGITUDE **56.4142, -5.0239**
START GRID REFERENCE **NN135287**
START W3W **///polishing.mainland.diplomat**

	PRONUNCIATION	NAME MEANING
Beinn a' Chochuill	**'beyn a cho-chil'**	**hill of the shell or hood**
Beinn Eunaich	**'beyn eeyaneech'**	**fowling hill**

Introduction

If you're a lover of Highland cattle, this is the walk for you. These long-horned, shaggy-coated hardy cows are commonly observed in the lower fields near the Allt Mhoille, but they're often spotted further up the valley on the ascent track too. Once up high, the view from Beinn a' Chochuill across the incredible skyline of Stob Daimh and Ben Cruachan is well worth the climb.

Public Transport

There are bus stops just over half a kilometre from the start of the walk on the A85. Local services between Oban and Dalmally stop here, as do national Scottish Citylink services which connect Glasgow and Edinburgh to Oban via the A85.

Parking

There is space for around three cars on the roadside of the B8077, on the western side of the Allt Mhoille bridge. The nearest toilet facilities are in Taynuilt, 17 kilometres from the start.

MTB-ability

The first and last 2.3 kilometres of the route on the access track are cyclable but steep in places.

Escape Route

None – retrace your steps after the first Munro and return another day for the second.

Refreshments

It's worth making an early start on the walk so that you can enjoy a meal at the Ben Cruachan Inn afterwards. The staff give an exceptionally warm welcome to non-residents and residents alike.

Beinn a' Chochuill and Beinn Eunaich from Stob Daimh (see route 09).

The Route

S From the parking area, walk north-east along the road across the Allt Mhoille. After 85m turn left over a cattle grid and on to a track which immediately splits. Fork left here and walk north. At the next split in the track at NN135292, continue straight on. At the next split, before Castles Farm, turn left, pass through a gate and begin to ascend the track which leads up the south-western lower slopes of Stob Maol. A cairn at NN129306 marks your descent path, so take note of this as you continue north-west.

2 The track crosses the Allt Lairig Ianachain using Mhoille Bridge. Shortly after this the track splits; stay right here and remain on the main track. Walk uphill towards a junction with another track at NN121315. Turn right here, and after 200m turn left at a small cairn on to a path up the south-east ridge of Beinn a' Chochuill.

3 The terrain is mostly grassy with eroded steps that have been kicked into the hillside over years gone by. If you lose this line in poor weather head north-west to gain the ridge at the bealach south-west of point 896m. There are a small number of rocky sections along the ascent, but no scrambling is required.

4 Top out on the ridge at NN118325 and turn left to start the out-and-back to Beinn a' Chochuill. The ridge gently undulates; a path leads primarily to the left-hand side of some minor tops until you reach the summit which is marked by a cairn on a rock slab. Pause to enjoy the outstanding view to Ben Cruachan (page 45).

5 Retrace your steps to NN118325 (**4**) and continue along the flat grassy ridge until the trail becomes rockier and drops unproblematically to a bealach at point 728m. Ascend the grassy slopes of Beinn Eunaich. As you approach the top, the path weaves expertly around the boulder scree that forms the upper reaches of the mountain. The summit is marked by a cairn with a small stone shelter to the side.

Sheep with the east shoulder of Sron an Isean in the background from the track up Beinn a' Chochuill.

6 Descend to the south over grassy terrain initially. There are a couple of rockier sections, but the gradient is gradual, so they pose no difficulty. Beinn Eunaich has many spurs that fall away southwards from the mountain; there is a path down the rightmost spur. The descent to 550m is gentle and then the path begins to steepen to the right of some peat hags.

7 The grassy trail drops towards the track and is stony and loose in places due to erosion. At the track, turn left and retrace your steps back to the start.

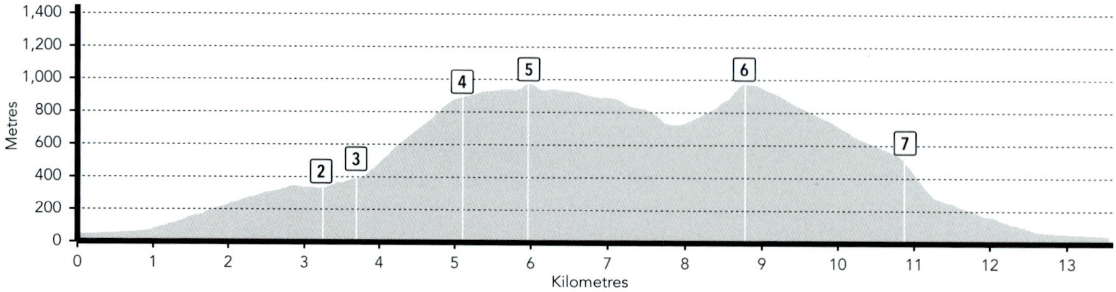

0
1 Kilometre
Meall Voirlich
561
Meall Copagach
810
Allt Easan
BEINN A' CHOCHUILL
980
BEINN EUNAICH
989
896
728
Allt Lairig Ianachain
Allt Mhoille
Sron an Isean
966
STOB DAIMH
Stob Garbh
980
Stob Maol
Allt Coire Chreachainn
Coire Glas
Allt Coire Ghlais
Castles Farm
Barran an Tuirc
Monadh Driseig
641
Stronmilchan
B8077
A85
A819
Lochawe
Loch Awe
Dalmally
Strae

HIGHLAND HUB 4

TYNDRUM

The fourth of the Highland hubs is Tyndrum, a village at a vital crossroads where major A-roads split to loosely follow two railway lines towards Oban to the west and Glen Coe and Fort William to the north. Pronounced 'tyne-drum', in Scottish Gaelic Tyndrum translates to *Taigh an Droma* which means 'the house on the ridge'. It's the hub with the highest concentration of Munro walks within a 10-minute drive (or short train ride), so once you've made the effort to reach the hub you won't have to travel much further before your mountain adventures can begin.

Trains from Glasgow take in the region of two hours to reach the hub and Tyndrum boasts two railway stations. One on the Glasgow to Mallaig line, known as Upper Tyndrum, and one on the Glasgow to Oban line, known as Tyndrum Lower. Both stations are less than a ten-minute walk to the A82 which slices the village in two.

The Green Welly Stop sells fuel and has a small supermarket alongside a cafe, restaurant, outdoor store and gift shop. They also have a shower facility.

Tyndrum also has a pub and a couple more restaurants which offer takeaway options. West Highland Way walkers will pass through during the day, many choosing to stay overnight at By The Way Hostel and Campsite. This gives the place a feeling of transience with each day bringing a different set of patrons and stories, but Munro-baggers are also welcomed – I spent ten happy nights on the campsite myself one summer.

WHERE TO STAY

£	££	£££
By the Way Hostel and Campsite (camping)	By the Way Hostel and Campsite (hostel)	Tyndrum Lodges

TRAVEL TIMES

11 Queen of the Southern Highlands	**10 minutes**
12 Dalrigh Munros	**5 minutes**
13 Beinn Challuim	**5 minutes**
14 South-West Crianlarich Munros	**10 minutes**
15 Middle Crianlarich Munros	**10 minutes**
16 South-East Crianlarich Munros	**10 minutes**
17 Beinn Mhanach	**5 minutes**
18 Bridge of Orchy Munros	**10 minutes**
19 Achallader Munros	**15 minutes**
20 South Black Mount Munros	**20 minutes**

Rest day ideas

- Wander south on the West Highland Way following a well-maintained path beside the Crom Allt past the Lochan of the Lost Sword. Legend speaks of Robert the Bruce and his army throwing their swords into the lochan as they fled from the Battle of Dalrigh.
- Take a day trip on the train. The West Highland Line is quite possibly the most scenic in the country and en route to Mallaig trains cross the Glenfinnan Viaduct of *Harry Potter* fame.

Opposite Beinn a' Chreachain summit plateau (route 19).
Overleaf Beinn Challuim summit view towards the Bridge of Orchy with Beinn Dòrain on the right (route 13).

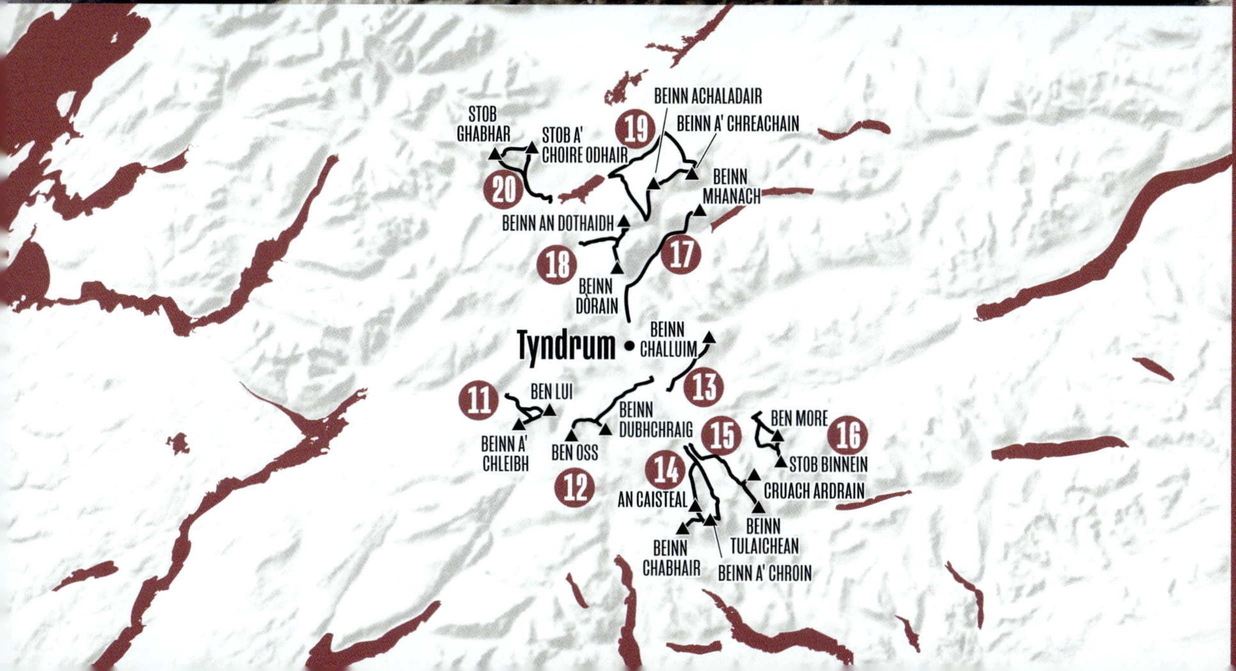

STOB GHABHAR
STOB A' CHOIRE ODHAIR
20
19
BEINN ACHALADAIR
BEINN A' CHREACHAIN
BEINN MHANACH
BEINN AN DOTHAIDH
18
17
BEINN DÒRAIN
Tyndrum
BEINN CHALLUIM
13
11
BEN LUI
BEINN A' CHLEIBH
BEINN DUBHCHRAIG
BEN OSS
12
15
BEN MORE
16
STOB BINNEIN
14
AN CAISTEAL
CRUACH ARDRAIN
BEINN TULAICHEAN
BEINN CHABHAIR
BEINN A' CHROIN

11 Queen of the Southern Highlands

HUB **Tyndrum – 10-minute drive**

OVERALL 1 2 **3** 4 5   TERRAIN 1 2 3 4 **5** (Water)
DISTANCE **1** 2 3 4 5   NAVIGATION 1 2 **3** 4 5

DISTANCE **9.6km** ASCENT **1,095m**
TIMES **6:20hrs** (Walker) **5:30hrs** (Trekker)
4:05hrs (Fastpacker) **2:35hrs** (Trail runner)

START LOCATION **Glen Lochy Forestry and Land Scotland car park, A85**
START LATITUDE/LONGITUDE **56.4103, -4.8560**
START GRID REFERENCE **NN238278**
START W3W **///sobs.geology.regret**

	PRONUNCIATION	NAME MEANING
Ben Lui	**'ben looee'**	**hill of the calf**
Beinn a' Chleibh	**'beyn a clave'**	**hill of the chest**

Introduction

Blessed with a regal elegance and grandeur unmatched elsewhere in the southern half of Scotland, Ben Lui is widely celebrated as the 'Queen of the Southern Highlands' – a majestic mountain with a beautiful profile. This is all a bit ironic, because you can't enjoy Lui's royal opulence without suffering like a lowly peasant through the indignity of a nightmarish bog. Commonly hiked from Glen Lochy, the ascent of Ben Lui not only involves a river crossing (don't attempt when in spate), but there's also a swampy quagmire of thick Scottish goop to negotiate – the kind of ground that swallows up unsuspecting Munro-baggers for breakfast. You'll be left soiled, soaked and shell-shocked, but the suffer-fest is a type-2 kind of fun and something of a rite of passage. It won't take too long before you're reminiscing fondly about your muddy adventures to visit the Queen of the Southern Highlands.

Public Transport

Railway stations and bus stops are a significant distance away in Tyndrum and Dalmally and, while buses have been known to stop at the start point, this would be at the discretion of the driver and so cannot be relied upon. Instead, look into walking these two Munros from Tyndrum up Glen Cononish. This approach is significantly longer than my recommended route, but it does have public transport links.

Parking

Chargeable parking for 30 cars in the Glen Lochy Forestry and Land Scotland car park off the A85 between Tyndrum and Dalmally. The nearest public toilets are in Tyndrum.

Crossing the River Lochy.

Escape Route

Pick up the descent instructions early from the bealach between the two Munros and return for Beinn a' Chleibh another day.

Refreshments

A trip to the Highlands is never complete without a visit to The Green Welly Stop in Tyndrum. It has a restaurant, a snack bar, an outdoor store and a gift shop.

The Route

S Leave the car park using the gravel path which leads to the riverside. Follow the River Lochy downstream for 150m until you reach the ford crossing where the river is at its widest and most shallow. You will usually get your feet wet crossing here and when the river is in spate after heavy rain it may be impassable.

2 Once on the opposite bank, use the crude metal grid bridge to pass under the railway – you'll have to duck so as not to bump your head. Follow the path from here in a south-easterly direction uphill through a deforested area. The path rises up alongside the Eas Daimh, which is on your right, and a deer fence on your left. The path soon begins to run along the right-hand side of a plantation. Ben Lui is directly in front of you and Beinn a' Chleibh is to your right. With shorter walks it's nice to have your targets in sight from the beginning.

3 At a confluence of two large streams, cross the leftmost stream picking up the trail on the other side. Ignore a right-hand turn and continue following the stream in a south-easterly direction to a track junction beside a large bridge. Cross the track and head up the steep stony path with thick woodland on your left and a cascading stream on your right.

4 The stony path soon gives way to what can only be described as a swampy bog-fest. It's not uncommon to lose a foot (or even a whole leg) on this section so pick your way carefully. This sloppy obstacle course is something all Munro-baggers should embrace at least once in their lifetime – it's certainly character-building! Keep the stream on your right as you continue southwards until you reach a metal gate in the deer fence.

5 Turn left on the other side of the gate to walk alongside the deer fence heading east-north-east. You are now 'out of the woods' and on to the open hillside. At the fence corner, continue straight ahead over soft, grassy terrain up Ben Lui's broad western flanks. On a previous visit I chose to sit on a rock here and eat my lunch looking back towards Ben Cruachan (page 45).

6 At around 700m in elevation the path peters out as you walk east-north-east towards Ben Lui's north-west ridge. The terrain is steep, strenuous and mostly pathless. Aim for NN262267, a spot on the far side of the ridge. Aiming for this point reduces the risk of walking on steep scree or craggy ground and a path can be picked up here which leads south-east towards the summit.

7 The path is now dry and stony, but it gets steep and requires a short scramble, up to grade 1, at a couple of points. There are plenty of hand and foot-holds and the sense of exposure is limited. There are two high points on Ben Lui, both very close together. The spot marked 1,130m on the OS map is the one you come to first. Continuing along the ridge you dip and then rise to the south-east summit which is higher by a whisker. Your second Munro, Beinn a' Chleibh, is dwarfed by Ben Lui, which lives up to the nickname, the 'Queen of the Southern Highlands'.

8 Descend back to the depression between Ben Lui's two summits before turning left to head down loose but dry terrain to the south-west. On the way to the bealach, pass a very large boulder. Two distant Munros command your attention here: Beinn Bhuidhe (page 5) is visible to the left of Beinn a' Chleibh and Ben Cruachan (page 45) can be appreciated to the right.

9 From the bealach, continue straight ahead for just over 500m to reach the summit of Beinn a' Chleibh over dry and firm terrain which doesn't feel half as steep as the slopes of Ben Lui did. On a clear day, enjoy the fantastic view north to Bidean nam Bian (page 163) and the Nevis Range (page 195). Ben Lomond (page 17) and the Arrochar Alps (pages 9 and 13) can be seen to the south.

10 Retrace your steps to the bealach (**9**) to complete the short out-and-back. From here, a pair of small cairns mark the point at the col where you fork off to

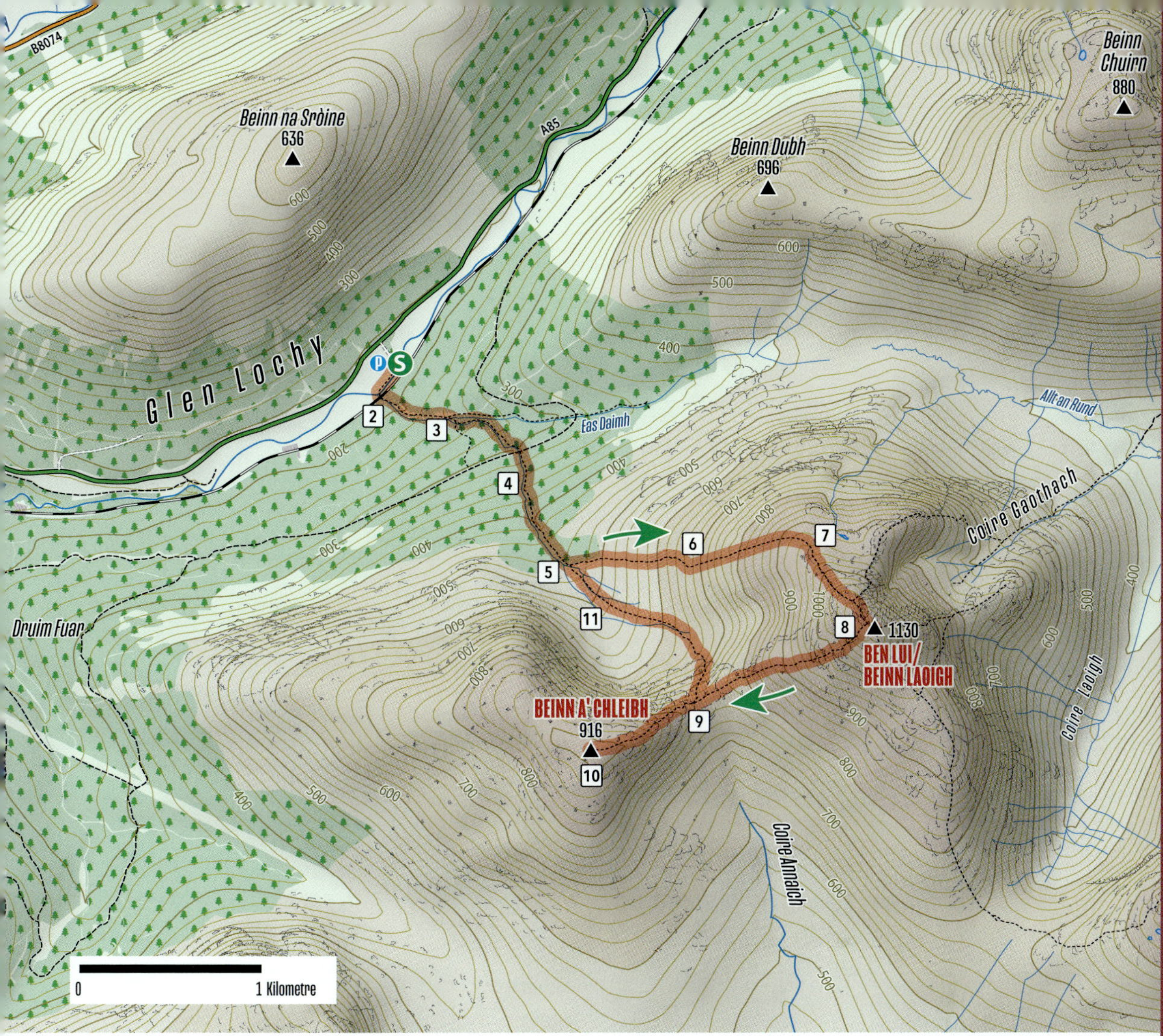

the left to drop northwards into the corrie on a slight diagonal. At first, the path is rather eroded, loose and steep, so descend cautiously. The worst is soon behind you though as the path becomes grassy and wet once again after around 100m of vertical height is lost.

11 The path becomes indistinct as the ground flattens and becomes more saturated, but if you lose the trail here head north-west towards the forest. Exit through the same gate that you entered through earlier and retrace your steps to the start from here.

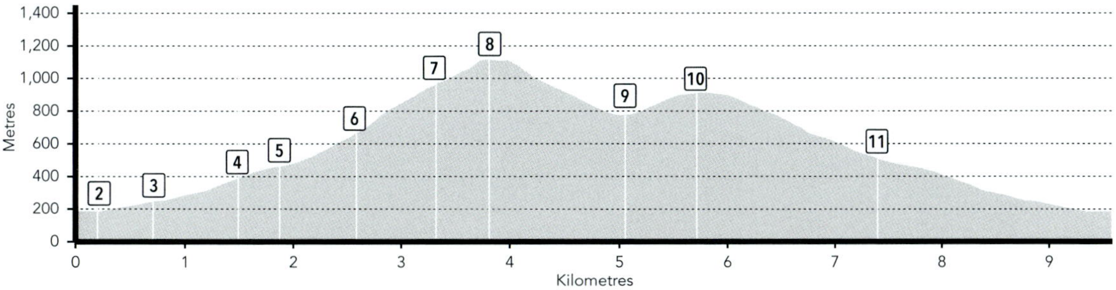

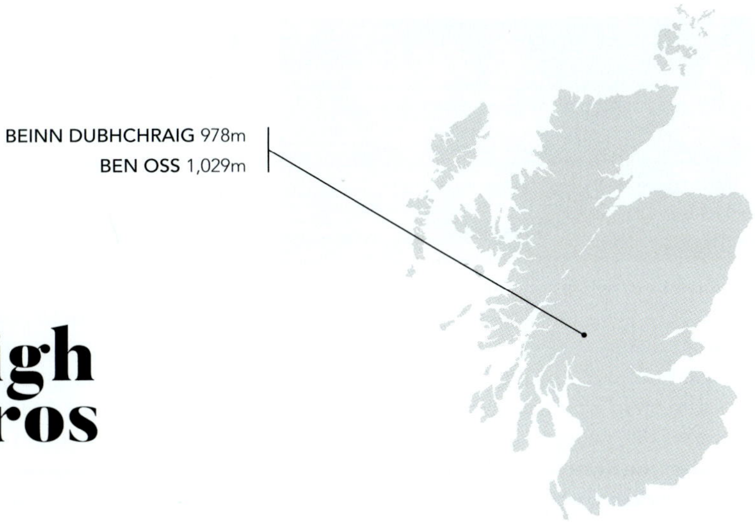

12 Dalrigh Munros

HUB **Tyndrum – 5-minute drive**

OVERALL 1 2 **3** 4 5 TERRAIN 1 2 **3** 4 5
DISTANCE 1 2 **3** 4 5 NAVIGATION 1 **2** 3 4 5

DISTANCE **17.5km** ASCENT **1,285m**
TIMES **8:40hrs** (Walker) **7:30hrs** (Trekker)
5:35hrs (Fastpacker) **3:40hrs** (Trail runner)

START LOCATION **Dalrigh car park, off the A82, south-east of Tyndrum**
START LATITUDE/LONGITUDE **56.4251, -4.6871**
START GRID REFERENCE **NN343291**
START W3W **///wide.obey.graph**

	PRONUNCIATION	NAME MEANING
Beinn Dubhchraig	**'beyn dookrig'**	**black rock hill**
Ben Oss	**'ben oss'**	**elk mountain**

Introduction

Everyone who walks this route is invariably left with the same enduring memory – the eye-catching profile of Ben Lui as seen from Ben Oss. From this perfect viewing platform, Ben Lui looks simply resplendent and noble, and every bit worthy of her title as 'Queen of the Southern Highlands'. You also get pleasant views across Loch Lomond snaking towards the horizon. The route passes through the ancient Caledonian pinewood of the Coille Coire Chuilc.

Ben Lui from Ben Oss.

This Site of Special Scientific Interest is a small remaining window into the temperate forest of Scots pine, birch, rowan, aspen and juniper which once covered vast areas of Scotland.

Public Transport

The nearest bus and rail services are both in nearby Tyndrum. Follow the West Highland Way from Tyndrum to the bridge over the River Fillan and pick up the route from there – this adds a total of around six kilometres to the route.

Parking

There is space for around 30 cars in Dalrigh car park. The operator asks for a donation via an honesty box. The nearest public toilets are in Tyndrum.

Escape Route

No escape routes; retrace your steps after the first Munro and return another day for the second.

Refreshments

You'll find the Artisan Cafe & Deli occupying the site of an old church around one kilometre from the start and finish point of the walk at Tom na Croiche.

Ptarmigan on a rock overlooking Loch Oss.

The Route

S Exit the car park to the south-east and walk through a gate on to a tarmac track. At the cross-roads at NN345288, walk straight ahead and cross the bridge over the River Fillan. Take the right-hand turn after the bridge to walk along a wide track which ascends gently uphill to the west-south-west alongside the railway line. The track leads to a bridge over the railway line. Cross the bridge and at the path junction on the other side of the bridge, turn right on to a gravel footpath which leads west for a period of exceptionally pleasant walking which makes you forget you are in fact en route to climb two Munros.

2 The path leads to a bridge over the Allt Gleann Auchreoch. After crossing the bridge, follow the path southwards. From here, the trail is boggy in places, but otherwise you find yourself walking over a blanket of pine needles or tree roots to the calming sound of the stream and tweeting birds which is marvellous. The path eases south-west.

3 Cross the stream at NN322276 and pick up the path on the other side. The water in the Allt Coire Dubhchraig on your left cascades dramatically as the path leads up to a clearing in the woodland. Once in the clearing, turn to appreciate the view back to Beinn Challuim (page 65).

4 Cross another small burn and follow a path that runs close to the Allt Coire Dubhchraig. Continue uphill to the south-west, keeping the stream on your left. On a hot day you may be tempted to take a dip in the many picturesque pools that intersect the stream at various points on the ascent. As the gradient steepens, the softer terrain is left behind and the path is stony with no technical difficulty.

5 At NN304258 a fork in the path is reached. Take the left-hand turn to climb south-east up the final short portion of vertical ascent that remains to reach the summit of Beinn Dubhchraig. The ascent is straightforward and soon offers gorgeous views down Loch Lomond to the south. The highest point of the Munro is marked by a cairn. Your next target, Ben Oss, can be appreciated to the west, with Ben Lui (page 57) looming ominously behind it.

6 Retrace your steps back to NN304258 (**5**) before turning left to walk towards Ben Oss. The path passes to the left of a small lochan and the view to Ben Oss and Ben Lui remains incredible as the path doesn't drop to the bealach straight away. Loch Oss appears in the coire to your left as you make your way along the eroded trail towards Bealach Buidhe. The path is steep and rocky in a couple of places but presents no technical difficulty.

7 Fork right at NN296258 (this is marked by a small cairn), to approach Ben Oss from the north. The terrain is loose and eroded in places but continues to present no technical difficulty if care is taken. The path leads down and up one last dip before the final 100m of vertical ascent to the summit of Ben Oss. Ben Lui now towers over the Munro from the right in dramatic style. The summit of Ben Oss is marked by a cairn. The view to the crags on the south-east face of Ben Lui is truly a sight for sore eyes.

8 Retrace your steps to the start.

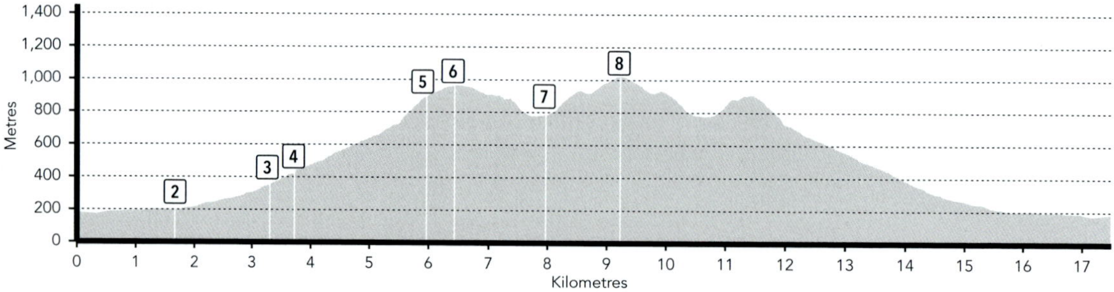

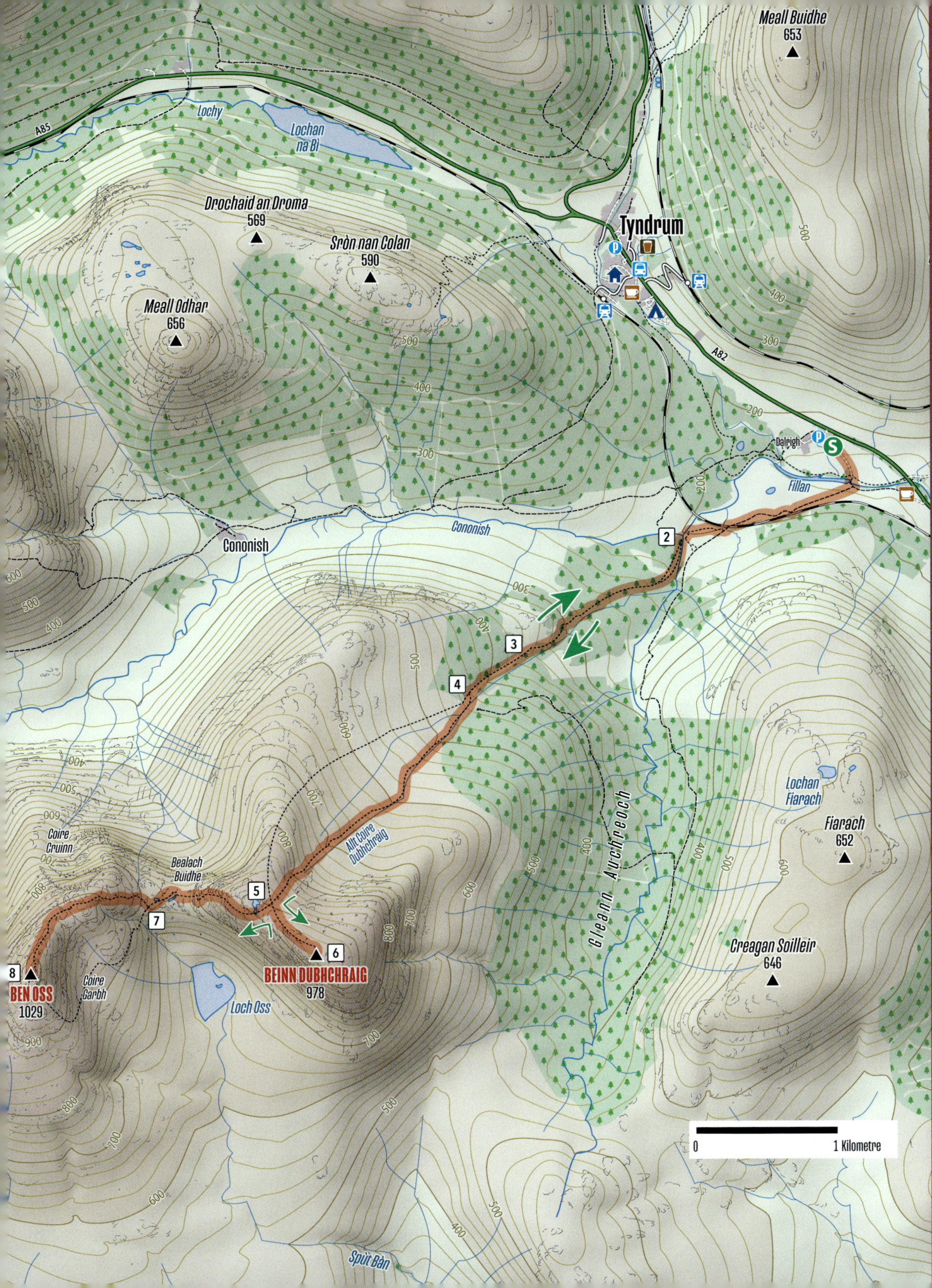
Meall Buidhe
653
A85
Lochy
Lochan
na Bì
Tyndrum
Drochaid an Droma
569
Sròn nan Colan
590
Meall Odhar
656
A82
Dalrigh
Fillan
Cononish
Cononish
2
3
4
5
6
7
8
Allt Coire
Dubhchraig
Gleann Auchreoch
Lochan
Fiarach
Fiarach
652
Creagan Soilleir
646
Coire
Gruinn
Bealach
Buidhe
Coire
Garbh
BEN OSS
1029
BEINN DUBHCHRAIG
978
Loch Oss
Spùt Bàn
0
1 Kilometre

13 Beinn Challuim

HUB **Tyndrum – 5-minute drive**

OVERALL 1 2 **(3)** 4 5 TERRAIN 1 2 **(3)** 4 5
DISTANCE 1 **(2)** 3 4 5 NAVIGATION 1 2 **(3)** 4 5

DISTANCE **11.7km** ASCENT **985m**
TIMES **6:15hrs** (Walker) **5:25hrs** (Trekker)
4:05hrs (Fastpacker) **2:40hrs** (Trail runner)

START LOCATION **Lay-by on the A82 between Tyndrum and Crianlarich**
START LATITUDE/LONGITUDE **56.4166, -4.6664**
START GRID REFERENCE **NN356281**
START W3W **///undertone.removable.meaty**

	PRONUNCIATION	NAME MEANING
Beinn Challuim	**'ben callum'**	**Callum's hill**

Introduction
Beinn Challuim (or Ben Challum) is one of those Munros bestowed with the honour of almost always being the sole goal of a hillwalker. Located to the east of Tyndrum, it's often climbed alone from the A82 – an out-and-back delivering impressive bang for your buck. It only takes a few hours, but you're rewarded with vast views across the Bridge of Orchy Munros, the Ben Lui group, Ben Cruachan in the distance and the Crianlarich Munros. Arguably, however, a more dramatic profile of Beinn Challuim is served up from the Glen Lochay side to the north-east – a wonderful alternative if you fancy diverging from the norm.

Public Transport
The closest bus and rail services are both in nearby Tyndrum. Follow the West Highland Way from Tyndrum to St Fillian's Church and pick up the route from there. Depending on whether you choose the bus or the train, this will add on around 10 kilometres of flat walking to the route.

Parking
Along the A82, halfway between Tyndrum and Crianlarich, you'll find a lay-by with room for around eight cars. Public toilets can be found in both Tyndrum and Crianlarich.

Escape Route
None.

Refreshments
You'll find the Artisan Cafe & Deli occupying the site of an old church around one kilometre from the start and finish point of the walk at Tom na Croiche.

View from Beinn Challuim's summit towards the south summit.

The Route

S Cross the A82 and walk south-east for 50m before you reach a turn-off marked *SRUC Hill & Mountain Research Centre, Kirkton.* Take this turning and walk along the tarmac road which leads to a bridge over the River Fillan. After crossing the bridge, follow the road north-east with farmers' fields on either side. This small section of the walk is shared with the West Highland Way. Follow the footpath to the left in front of Kirkton Farm and pass over a small bridge. At the crossroads walk straight on to follow a track uphill, splitting away from the West Highland Way.

2 The track soon passes Tyndrum (Kirkton) Cemetery with its towering pair of cypress trees in a walled enclosure. This would be a fine final resting place indeed. Follow the track to the railway crossing and continue on the other side for 80m before peeling off the track and walking north-north-east to reach a path which runs parallel to a deer fence.

3 Keep the deer fence on your right as you ascend. The ground is soft and wet in multiple places on the lower slopes of Beinn Challuim, especially after rain. Pass through a perpendicular deer fence using the gate. When you reach the first large rocky outcrop, be sure to turn around and admire the view back to the double-topped peak of Ben Lui (page 57).

4 Continue walking beside the deer fence until you reach your next deer fence obstacle. The stile here consists of a wooden structure on either side of the fence with foot and handholds jutting out to the sides. You'll need a head for heights to use it and many will find this to be the most challenging element of this walk. Follow the fence north-eastwards until you reach the last of three deer fence crossings, this time using a gate.

Ben More and Stob Binnein from the ascent of Beinn Challuim.

5 The gradient eases off for the next section, but this opens the door to peat hags and bog. Cross over the fence on your right and, with the fence now on your left, ascend to join a more defined and notably drier path up the hillside.

6 At NN377305 there is one mild undulation where you descend slightly but you soon regain the height. The fence is no longer there but old fence posts on your left still guide the way and are useful in poor visibility.

7 The terrain becomes stonier in the higher reaches as you seek to gain the ridge, but the ground is not technical. Pass a large cairn and take the decision to visit the 998m south summit – the highest ground on your right. This is worth it for the impressive view southwards towards the unmistakable Munro pairing of Ben More and Stob Binnein (page 79).

8 The path leads north along the crest of the ridge, which is narrow at first but is not exposed and soon broadens. From the saddle along the ridge, all that is left to do is to hike 300m uphill to the main summit. Enjoy the vast views up here to the nearby Bridge of Orchy Munros (page 87), the Ben Lui group (page 57) and the distant Ben Cruachan (page 45).

9 Retrace your steps back to the start.

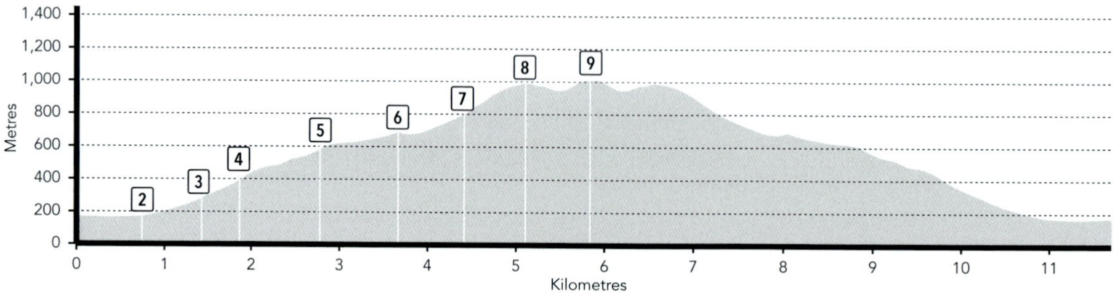

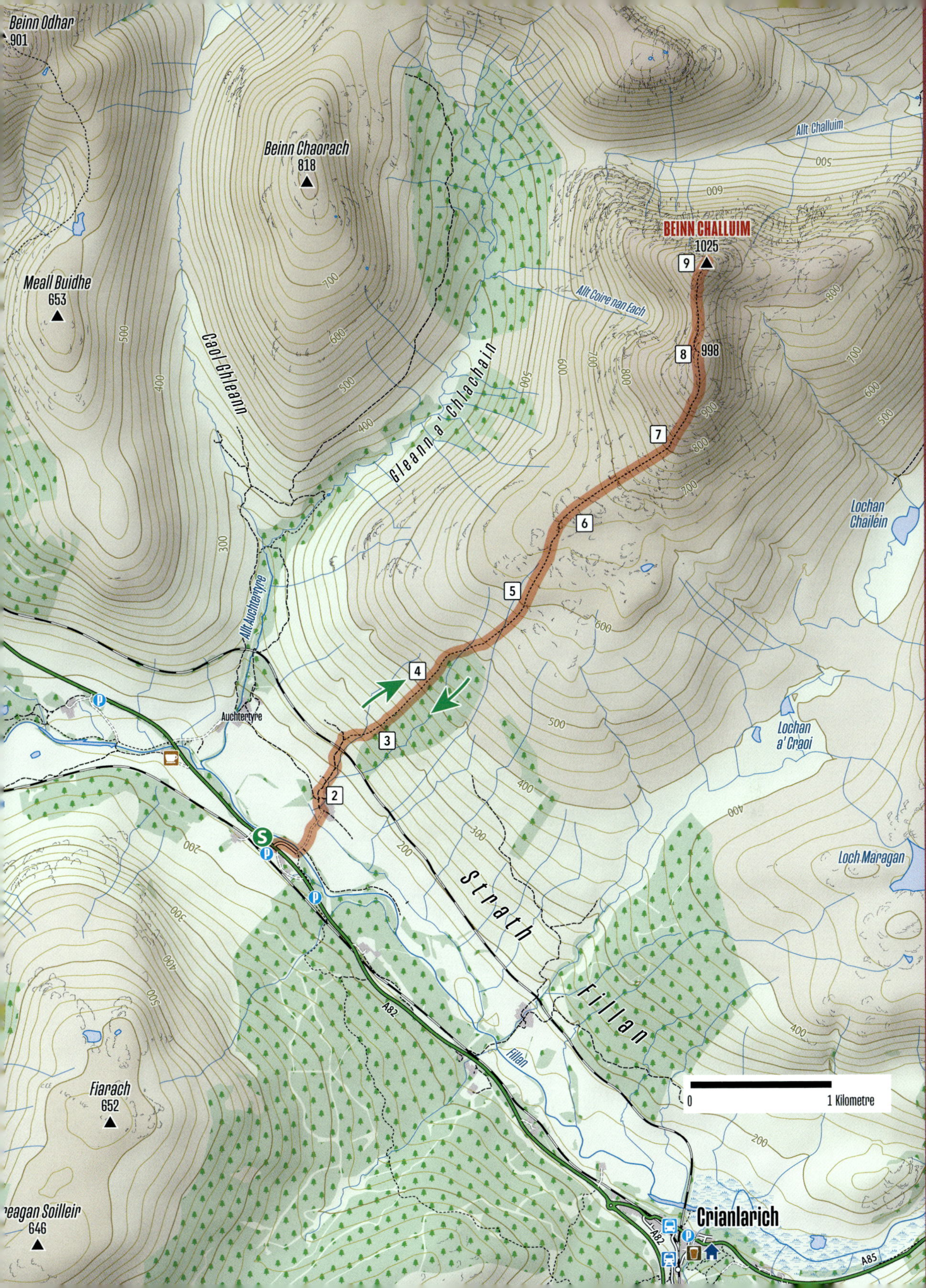

Beinn Odhar
901
Beinn Chaorach
818
Meall Buidhe
653
Caol Ghleann
Gleann a' Chlachain
Allt Challuim
BEINN CHALLUIM
1025
998
Allt Coire nan Each
Lochan Chailein
Lochan a' Craoi
Loch Maragan
Allt Auchtertyre
Auchtertyre
Strath Fillan
Fillan
A82
A85
Fiarach
652
eagan Soilleir
646
Crianlarich
0
1 Kilometre

AN CAISTEAL 995m
BEINN CHABHAIR 933m
BEINN A' CHROIN 942m

14 South-West Crianlarich Munros

HUB **Tyndrum – 10-minute drive**

OVERALL 1 2 3 **4** 5 TERRAIN 1 2 3 **4** 5
DISTANCE 1 2 **3** 4 5 NAVIGATION 1 2 3 **4** 5

DISTANCE **18.6km** ASCENT **1,645m**
TIMES **10:10hrs** (Walker) **8:55hrs** (Trekker)
6:45hrs (Fastpacker) **4:20hrs** (Trail runner)

START LOCATION **Parking area off the A82, south-west of Crianlarich**
START LATITUDE/LONGITUDE **56.3794, -4.6422**
START GRID REFERENCE **NN369239**
START W3W **///lorry.hurtles.common**

	PRONUNCIATION	NAME MEANING
An Caisteal	**'an cashtel'**	**the castle**
Beinn Chabhair	**'beyn chavair'**	**hill of the hawk**
Beinn a' Chroin	**'beyn a chroin'**	**hill of danger**

Introduction

Hardcore Munro-baggers will tick off this trio of summits as part of the Crianlarich Seven challenge – a gruelling 30-kilometre route with almost 3,500 metres of ascent. But if you don't fancy such a hectic day of relentless peak bagging, you can tick off these three summits in a far more sensible 18.6-kilometre route from the A82 lay-by to the south-west of Crianlarich. If even that is too much, Beinn Chabhair can be saved for another day, instead approaching it over boggy moorland from Inverarnan. But in the spirit of endurance set by the increasingly popular Crianlarich Seven challenge, I'd suggest hiking the full trio – you've got this.

Public Transport

If arriving using public transport, begin the walk at Crianlarich railway station which sits on the West Highland Line. Several bus services also stop in the village. From the station, walk south down Glenfalloch Road and then along the verge of the A82. This adds four kilometres to the total walk length.

Parking

Long, narrow parking area off the A82, south-west of Crianlarich, which offers free parking for over 20 cars. The nearest public toilets are in Crianlarich.

Escape Route

Retrace your steps after An Caisteal. Or complete the loop of An Caisteal and Beinn a' Chroin, leaving Beinn Chabhair for another day.

Refreshments

You'll find Artisan Cafe & Deli occupying the site of an old church on your way back to Tyndrum.

View towards the cloud-capped Bridge of Orchy Munros from An Caisteal.

Above Beinn Chabhair with Loch Lomond in the distance from An Caisteal. **Opposite** Walkers descending An Caisteal.

The Route

S Leave the parking area in Glen Falloch passing through a gate to the south and follow a gravel path towards the railway line. Pass through a tunnel under the railway line and cross a footbridge over the River Falloch. Continue uphill on a wide track with the river now on your left for just over 1km before passing through a gate.

2 Turn right straight after the gate following a soft grassy trail up the hillside. Walk south-west initially and then curve southwards to ascend the north ridge of Sròn Gharbh. For the last couple of hundred metres of vertical ascent on Sròn Gharbh the terrain becomes rockier but does not require scrambling. The top of Sròn Gharbh can be walked over or bypassed to the right.

3 From here you can appreciate the two neighbouring pairs of Munros – Ben More and Stob Binnein (page 79), and Cruach Ardrain and Beinn Tulaichean (page 75). Walk south-south-east on the path, descending slightly to point 687m on the OS map before continuing uphill towards An Caisteal via Twistin Hill. Weave left and right following an eroded path with an apt amount of twisting.

4 A couple of short sections of straightforward scrambling are required at NN377197, providing a little flavour of what is to come. The ground levels out on the final approach to An Caisteal's summit.

5 After taking in the view to Beinn a' Chroin and Beinn Chabhair, both of which look tantalisingly close but are in fact a decent slog away, leave the summit to the south-south-west following an eroded path which curves and weaves to the south-east towards Bealach Buidhe. The increasingly rocky ridge offers up a handful of big steps before dropping to the bealach at NN381185.

6 Many people choose to hike Beinn Chabhair separately to An Caisteal and Beinn a' Chroin but it's so close to An Caisteal that I've always seen the value in adding it on. (If you wish to hike this Munro separately, skip ahead in the instructions – to part way through **8** – and continue straight on to begin the ascent of Beinn a' Chroin which will reduce this walk to 14.8km.) If, like me, you can't resist the temptation, turn right to walk south-west for the first 150m before following a faint trail west towards the wide, lumpy saddle. As walking all seven Crianlarich Munros in a single day has become a popular endurance challenge, the route for this out-and-back has become clearer and clearer in recent years. As the ground gets wetter the trail becomes temporarily indistinct in places, but it becomes more obvious again as you approach the bealach.

7 After passing point 619m on the OS map the trail leads uphill and soon runs alongside a small stream. As the path becomes indistinct again, aim to pass in

Beinn a' Chroin, An Caisteal and Beinn Chabhair from the ascent of An Caisteal.

between two tiny lochans– marked on the OS map at 760m in elevation. Continue south-west uphill over grassy terrain avoiding the crags to reach a brief section of flatter ground on the 800m contour. The gradient steepens here so pass a section of bouldery scree and continue up over grassy ground, passing to the left of a prominent rocky outcrop. From here, keep left of the crags in front of you, ensuring that you top out just south of the summit. Keeping left of the crags also ensures a slightly gentler angle.

8 From the top, enjoy the view back across the valley to An Caisteal and south down Loch Lomond before retracing your steps back to Bealach Buidhe (**6**). At the bealach, turn right to begin your ascent of Beinn a' Chroin. Initially, the path leads to the right of the craggy face which looms in front of you before weaving left to the foot of a short scrambling section which requires a couple of awkward large steps up and can be tricky in the wet. There are plenty of handholds to assist. Once on the long summit plateau there are a number of lumps and bumps to pass over, the highest of which is at NN387185, atop a large smooth domed rock.

9 Walking along the top of this mountain on a sunny day is one of my favourite stretches as there are expansive views to the Bridge of Orchy Munros (page 87) to the north and Ben Lomond (page 17) to the south. Continue along the lengthy summit ridge and drop to the bealach at 877m; there is one rocky step towards the bottom which requires care but otherwise the trail is clear and straightforward.

10 From the bealach it's worth taking the right-hand fork and ascending the slightly smaller east top of Beinn a' Chroin for the view across to Cruach Ardrain and Beinn Tulaichean (page 75). From here, walk northwards following a path which is clear at first but becomes indistinct as it crosses wet ground with a lovely view of the River Falloch and many of its feeder streams as it curves hypnotically through the valley.

11 Towards the bottom the rocky path becomes a grassy, and sometimes very wet, path. On the 400m contour line, cross two streams one after the other, just above the point that they combine. Follow the grassy path down Coire Earb until it meets the main track, which follows the River Falloch for just over 2km back to the start.

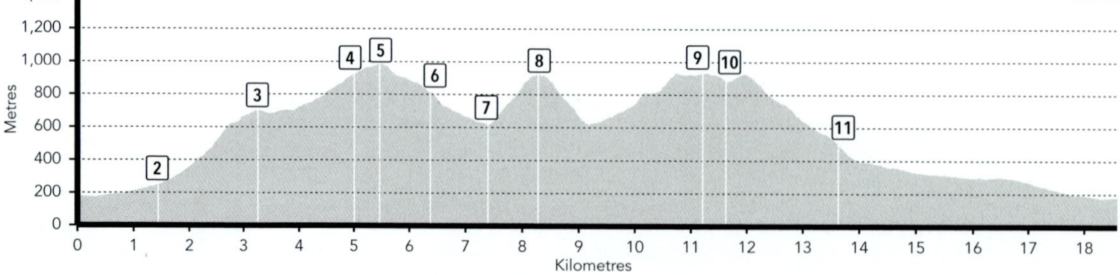

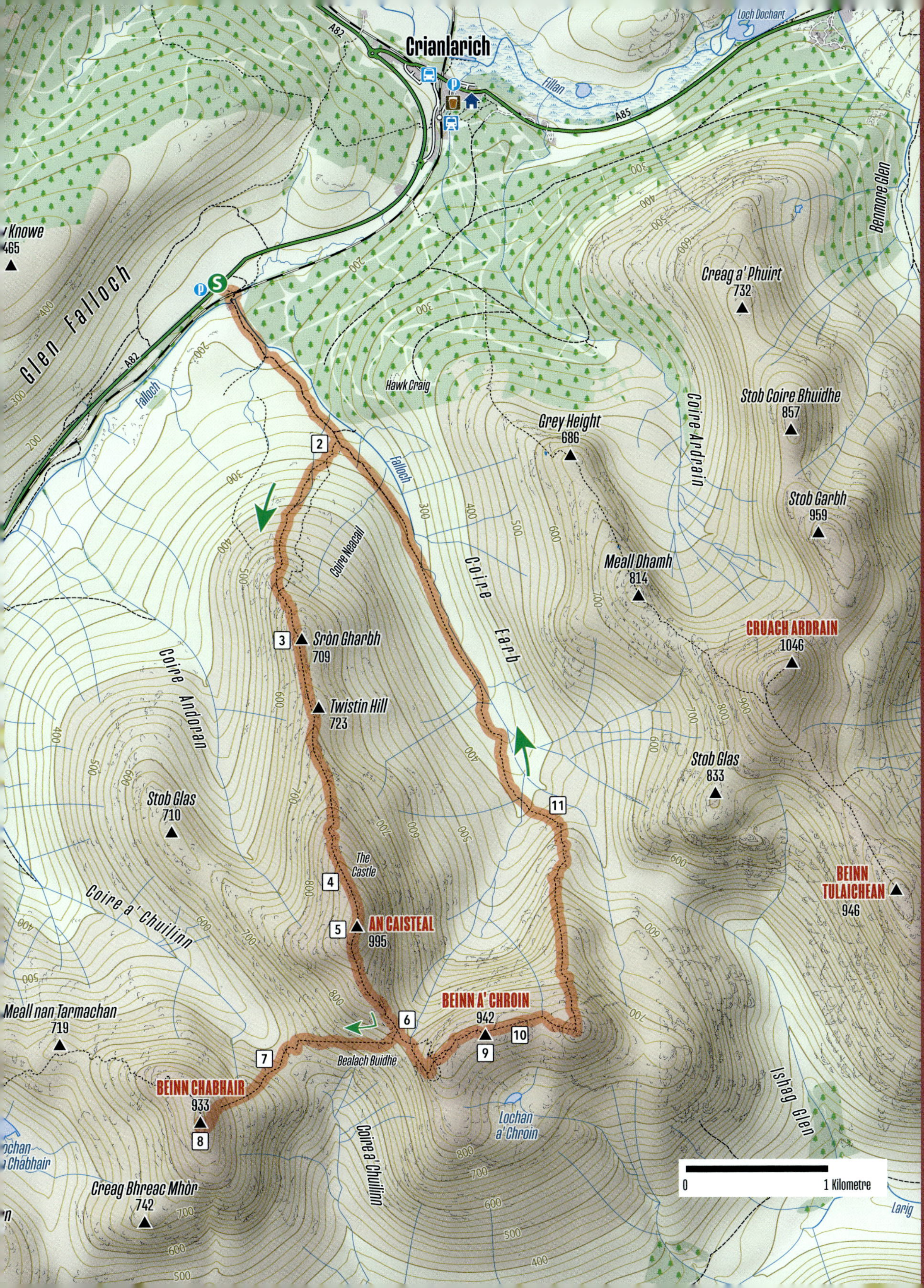

Crianlarich
Loch Dochart
Fillan
A82
A85
Benmore Glen
Knowe
465
Glen Falloch
A82
Falloch
Creag a' Phuirt
732
Hawk Craig
Grey Height
686
Coire Ardrain
Stob Coire Bhuidhe
857
Stob Garbh
959
Falloch
Coire Neacail
Coire Earb
Meall Dhamh
814
CRUACH ARDRAIN
1046
Sròn Gharbh
709
Twistin Hill
723
Coire Andoran
Stob Glas
710
Stob Glas
833
The Castle
BEINN TULAICHEAN
946
Coire a' Chuilinn
AN CAISTEAL
995
BEINN A' CHROIN
942
Meall nan Tarmachan
719
Bealach Buidhe
BEINN CHABHAIR
933
Ishag Glen
Lochan a' Chroin
Coire a' Chuilinn
Creag Bhreac Mhòr
742
0
1 Kilometre
Larig

15 Middle Crianlarich Munros

HUB **Tyndrum – 10-minute drive**

OVERALL 1 2 **3** 4 5 TERRAIN 1 2 **3** 4 5
DISTANCE 1 2 **3** 4 5 NAVIGATION 1 **2** 3 4 5

DISTANCE **16km** ASCENT **1,335m**
TIMES **8:35hrs** (Walker) **7:25hrs** (Trekker)
5:35hrs (Fastpacker) **3:35hrs** (Trail runner)

START LOCATION **Parking area off the A82, south-west of Crianlarich**
START LATITUDE/LONGITUDE **56.3794, -4.6422**
START GRID REFERENCE **NN369239**
START W3W **///lorry.hurtles.common**

	PRONUNCIATION	NAME MEANING
Cruach Ardrain	**'cru-uch arrdren'**	**high stacked heap**
Beinn Tulaichean	**'beyn toolikan'**	**hill of the hillocks**

Introduction

A thoroughly enjoyable outing in the Crianlarich area, this route feels a little like two-for-the-price-of-one. Starting from the A82 to the north, most of the effort is over once you've climbed Cruach Ardrain – a high bealach means the out-and-back to its smaller neighbour Beinn Tulaichean comes easier than expected. Cruach Ardrain is steep and knobbly on all sides and delivers great views to the pointed twins of Ben More and Stob Binnein.

Public Transport

If arriving using public transport, begin the walk at Crianlarich railway station which sits on the West Highland Line. Several bus services also stop in the village. It's possible to walk south from the train station up through the plantation to reach Grey Height and join the route instructions from there. The route remains a similar length.

Parking

Long narrow parking area off the A82, south-west of Crianlarich, which offers free parking for over 20 cars. The nearest public toilets are in Crianlarich.

Escape Route

None.

Refreshments

You'll find Artisan Cafe & Deli occupying the site of an old church on your way back to Tyndrum.

View from the bridge over the River Fillan to Cruach Ardrain (see route 13).

The Route

S Leave the parking area in Glen Falloch passing through a gate to the south and follow a gravel path towards the railway line. Pass through a tunnel under the railway line and cross a footbridge over the River Falloch. Continue uphill on a wide track with the river now on your left for just over 1km before passing through a gate and taking a left-hand turn when you are parallel with the edge of the plantation on the other side of the valley.

2 Drop down to the River Falloch and cross it using the new footbridge. From here, proceed north-east uphill over wet, grassy terrain to walk alongside a deer fence which separates the open hillside from the forestry plantation.

3 Follow the fence until the trail veers right, away from the plantation to ascend the north-west ridge of Grey Height (686m). The path is well-walked and is rocky in places but not technically difficult. On the 640m contour the trail passes to the east of a tiny lochan. Look out for a sharp left turn in the trail just after this. Continue from here to the summit of Grey Height.

4 The ground flattens here and undulates along a wide ridge heading south-south-east. Rocky outcrops pepper the ridge, as do small sections of boggy ground in the dips in the flattest ground. After reaching Meall Dhamh (814m), follow the path downhill for a short distance. The ground here is eroded in parts and there is one 'bad step' over slick rock, but this is avoidable using cut steps to the right without requiring scrambling. Continue along the ridge. The path here is clear as it veers around the westernmost flanks of Cruach Ardrain. There is higher ground to your left and lower ground down to your right.

Pausing on the descent of Cruach Ardrain overlooking Crianlarich.

5 At NN407209 turn left at the cairn to make the final part of the ascent of Cruach Ardrain. Just when you think you've reached the top, you must descend briefly before rising to the north-eastern top and the highest point of the Munro. My favourite view from here is to the twin peaks of Ben More and Stob Binnein (page 79), a fine pair of Munros separated from Cruach Ardrain by Benmore Glen to the north and Inverlochlarig Glen to the south. I have linked these two pairs of Munros in the past but it's a tricky undertaking given how low the bealach lies and the abundance of craggy ground to navigate around.

6 Retrace your steps to NN407209 (**5**) and continue south-east towards Beinn Tulaichean down the ridge following an eroded trail. Ignore the left-hand turn marked by a cairn which leads into Inverlochlarig Glen and continue descending to the lowest point of the saddle at 825m.

7 From the saddle, ascend to the summit of Beinn Tulaichean. This is undemanding walking at a more favourable gradient than the ascent up Cruach Ardrain.

8 Enjoy panoramic views of the surrounding mountains before retracing your steps back along the ridge to the cairned junction at NN407209 (**5**). From here, descend to the start point using the path of ascent.

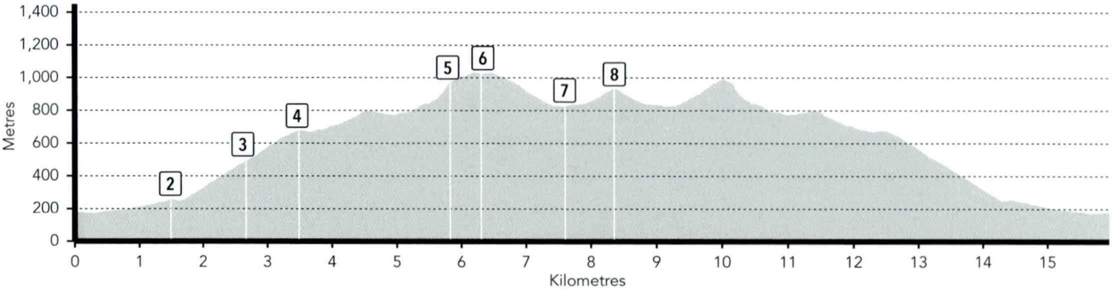

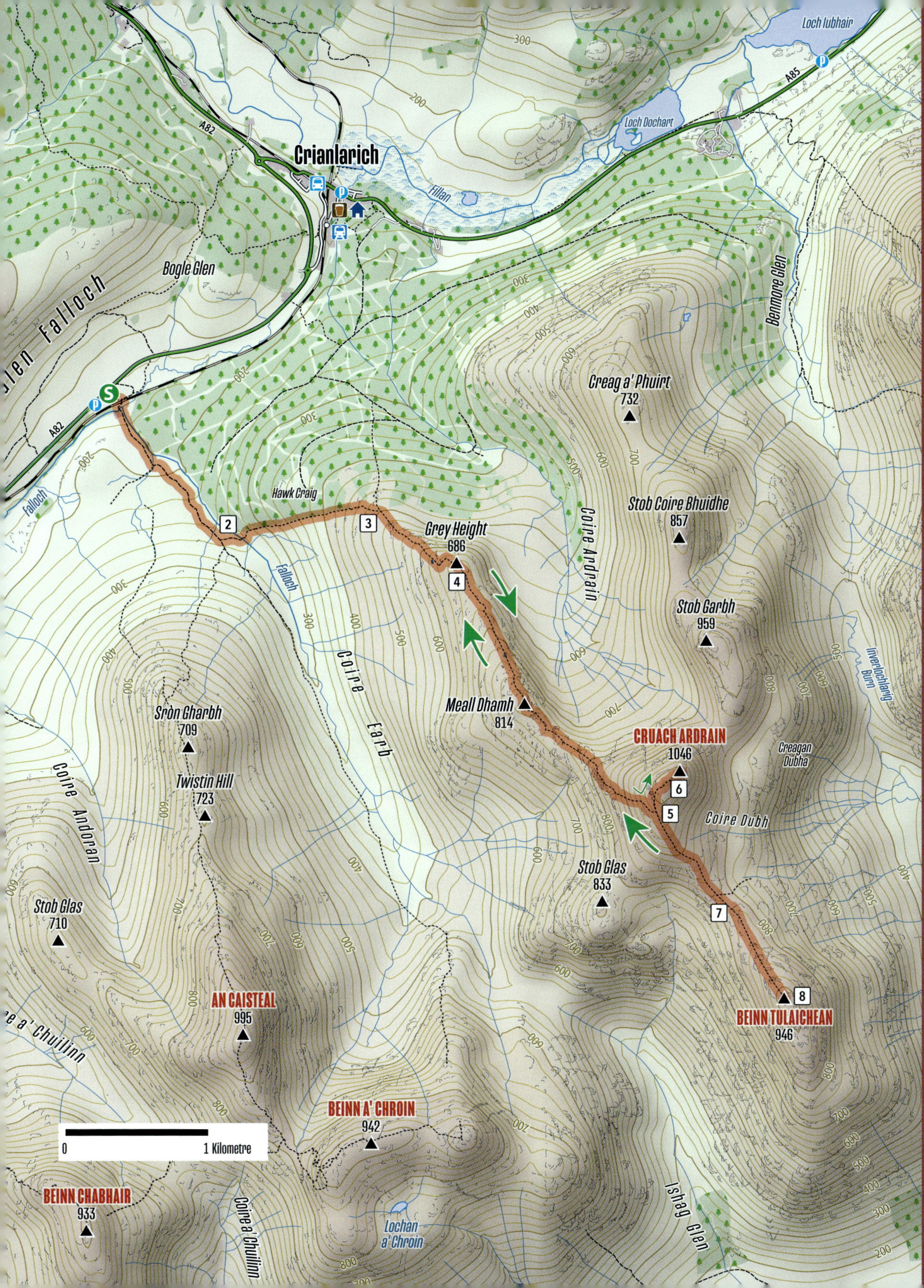

Crianlarich
Loch Iubhair
A85
Loch Dochart
Fillan
A82
Bogle Glen
Glen Falloch
Benmore Glen
Creag a' Phuirt
732
Hawk Craig
Coire Ardrain
Stob Coire Bhuidhe
857
Grey Height
686
Stob Garbh
959
Falloch
Inverhaggernie Burn
Coire Earb
Meall Dhamh
814
CRUACH ARDRAIN
1046
Creagan Dubha
Coire Dubh
Sròn Gharbh
709
Twistin Hill
723
Coire Andoran
Stob Glas
833
Stob Glas
710
AN CAISTEAL
995
BEINN TULAICHEAN
946
BEINN A' CHROIN
942
0
1 Kilometre
BEINN CHABHAIR
933
Coire a' Chuilinn
Lochan a' Chroin
Ishag Glen

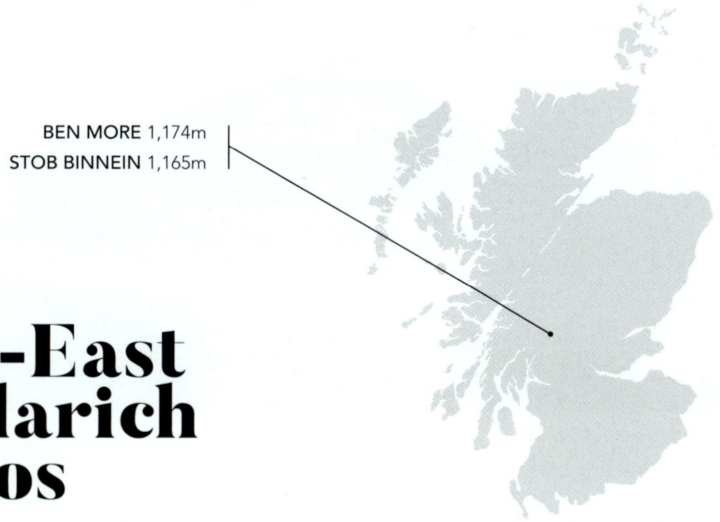

16 South-East Crianlarich Munros

HUB **Tyndrum – 10-minute drive**

OVERALL 1 **2** 3 4 5 TERRAIN 1 2 **3** 4 5
DISTANCE 1 **2** 3 4 5 NAVIGATION 1 **2** 3 4 5

DISTANCE **11.8km** ASCENT **1,340m**
TIMES **7:45hrs** (Walker) **6:45hrs** (Trekker)
5:05hrs (Fastpacker) **3:10hrs** (Trail runner)

START LOCATION **Lay-by on the A85, north-east of Benmore Farm**
START LATITUDE/LONGITUDE **56.4027, -4.5611**
START GRID REFERENCE **NN420263**
START W3W **///harnessed.scramble.harps**

	PRONUNCIATION	NAME MEANING
Ben More	**'ben more'**	**big hill**
Stob Binnein	**'stob bin-yen'**	**peak of the anvil**

Introduction

Soaring above the Crianlarich skies like two cresting waves, with shapely lines tapering beautifully to their distinctive summits, Ben More and Stob Binnein are handsome mountains. But they are not for the faint-hearted. If you're feeling sluggish and fancy a gentle hill bimble, do not opt for this route. The ascents are unrelenting, a series of lung-busting, energy-sapping slogs that seem to go on forever – it's a type-2 kind of sadistic fun. But persevere to the top and, as with much of life, you'll experience that wonderful truism: the harder the climb, the better the view.

Public Transport

Buses and trains stop in Crianlarich. Use the path which runs parallel to the A85 between Inverardran and Portnellan beside Loch Dochart to get to the start of the route. This adds 10 kilometres to the overall walk length.

Parking

Free parking for around 12 cars in the lay-by on the A85 north-east of Benmore Farm on the southern edge of Loch Iubhair. The nearest toilet facilities are in Crianlarich.

Escape Route

No escape routes; retrace your steps after the first Munro and return another day for the second.

Refreshments

Stop at the Artisan Cafe & Deli at Tom na Croiche on your return to Tyndrum – it occupies the site of an old church.

View from Beinn Challuim's south summit towards Ben More and Stob Binnein (see route 13).

View to Ben More from Stob Binnein.

The Route

S The walk begins in Glen Dochart with a plod south-west for 760m along the verge of the A85 to the path east of Benmore Farm at NN414259. A fingerpost marks the start of the path which is a left turn off the road. Pass through a gate and turn left to walk uphill on a wide gravel track. After 150m take the right-hand fork in the track to continue weaving steadily uphill on gravel. Pass through another gate.

2 After the gate the track splits once more. Make a mental note that this is where the descent track will bring you back to later on. But for now, take the stone-pitched steps up from a fingerpost which reads *Ben More path*. The steep pull from 300m to 1,174m now begins. Never has a mountain personified the phrase 'the mountain is my gym' like Ben More. This lung-busting, calf-crunching, snack-demolishing section of trail is pure mountain hiking exercise. I recommend staying hydrated, taking regular stops to refuel and making sure you turn around frequently to enjoy the views over Loch Iubhair.

3 The stone-pitched path assists you in making good progress at the beginning of the route and feels a tad like an outdoor step machine. After a few hundred metres, the stone pitching is less consistent, but in poor visibility, a clear line of erosion can be followed in a south-easterly direction up the mountainside. At NN429247, the trail starts to weave through some large crags before topping out.

4 The summit is marked by a trig pillar on slabby rock. Enjoy the view south-west from here across to the rugged profile of neighbour Cruach Ardrain (page 75). I have linked these mountains together in the past but it's a tough endeavour and should only be attempted by experienced navigators and long-distance mountain walkers. To the west, beyond Crianlarich, is fellow giant of the Southern Highlands, Ben Lui (page 57). Leave the summit of Ben More following the path to the south-south-east. It is rocky at first with one particular obstacle to overcome close to the start of the descent which requires a short downclimb. Hillwalkers who don't have rock climbing experience will prefer to avoid this by using the clear bypass path to the left. The remainder of the descent down the centre of the south shoulder of Ben More to Bealach-eadar-dha Bheinn is uncomplicated.

5 From the lowest point on the bealach, at 862m, begin the ascent of Stob Binnein, continuing to follow the eroded line forged by walkers before you. It leads south-south-east up Stob Binnein and offers no obstacles. Once on the summit plateau, the highest point is marked by a cairn. The summit of Stob Binnein is flatter than Ben More's which is noticeable when looking at the pair from the north.

6 Retrace your steps back to Bealach-eadar-dha Bheinn (**5**) where a small cairn marks the start of the descent north-west along the south-western slopes of Ben More. The descent trail is well walked and the land escapes into the valley from right to left at first. The path steepens in parts and is boggy in places but never technical.

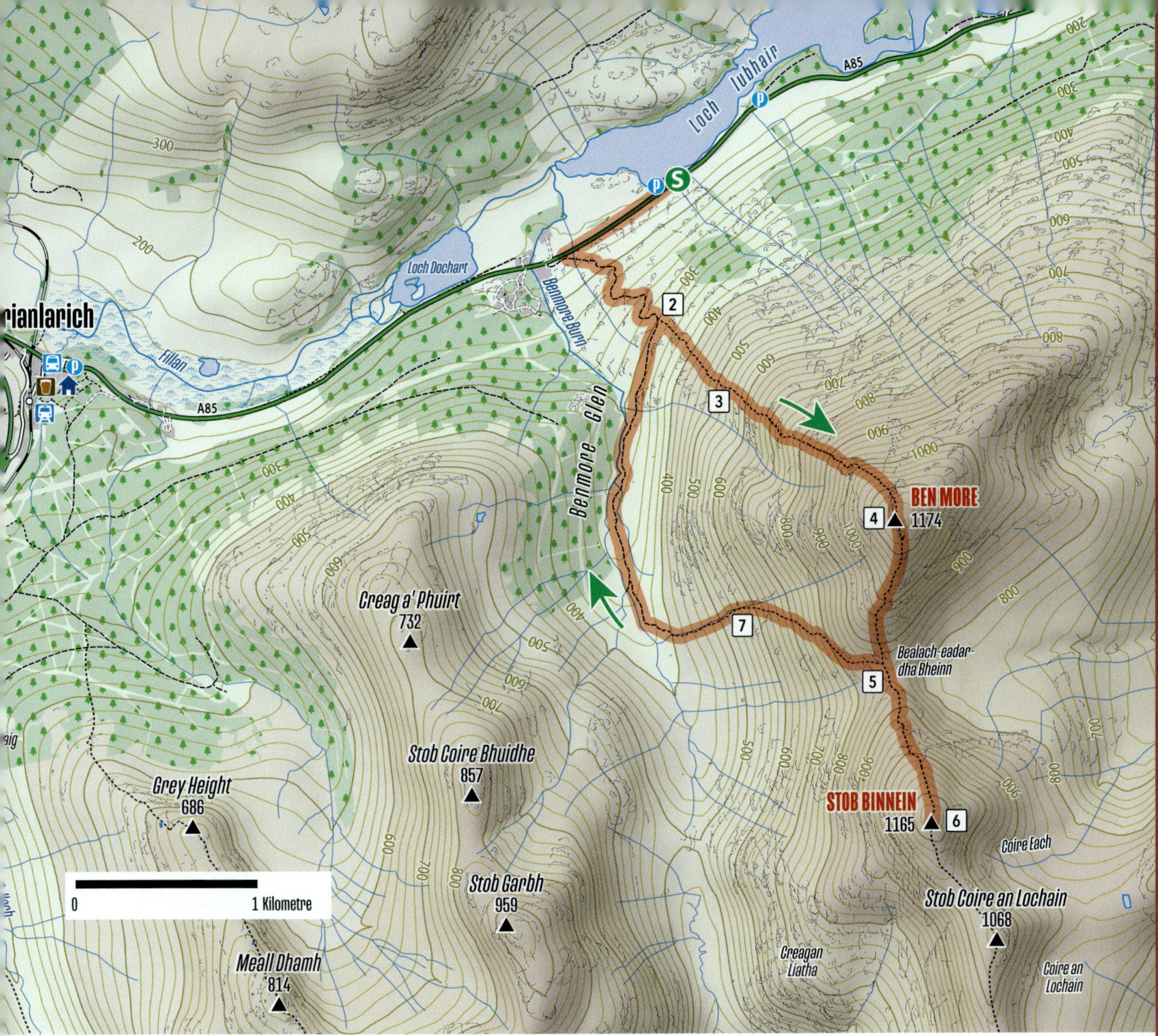

7 On the 600m contour, the path meets a stream and veers south-west to follow it almost to the valley floor. Keep the stream on your right until you're a mere 50m from the track. Cross to the other side of the stream here to reach the track in the bottom of Benmore Glen. Follow the track north as it undulates beside the Benmore Burn. The track leads to the gate you came through close to the start of the walk. Retrace your steps from here back to the start.

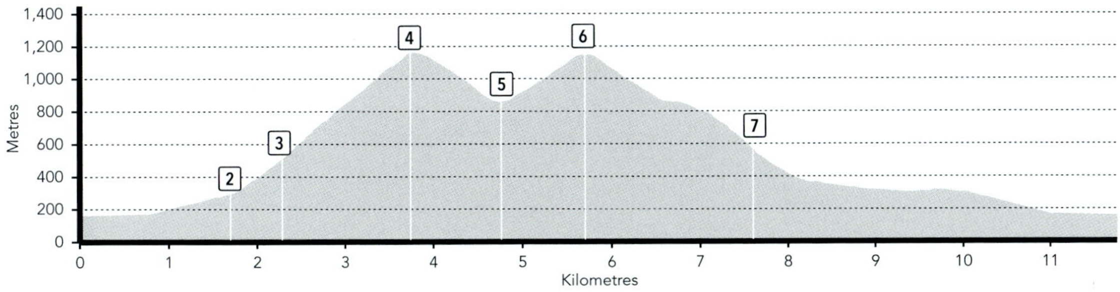

17 Beinn Mhanach

HUB **Tyndrum – 5-minute drive**

OVERALL 1 2 3 **(4)** 5 TERRAIN 1 2 3 4 **(5)** (Water)
DISTANCE 1 2 3 **(4)** 5 NAVIGATION 1 2 **(3)** 4 5

DISTANCE **21.9km** ASCENT **885m**
TIMES **7:55hrs** (Walker) **6:50hrs** (Trekker)
5:05hrs (Fastpacker) **3:30hrs** (Trail runner)

START LOCATION **Lay-by on the A82, north of Tyndrum**
START LATITUDE/LONGITUDE **56.4605, -4.7145**
START GRID REFERENCE **NN328331**
START W3W **///passively.spurring.fallback**

	PRONUNCIATION	NAME MEANING
Beinn Mhanach	**'beyn vannack'**	**monks' hill**

Introduction

Dumpy, pudding-shaped Beinn Mhanach is often dismissed as a forgettable Munro. It's certainly never made a top 10 list and, for most Munro-baggers, it's more of a 'get-it-done-and-never-return' kind of mission, rather than something to write home about. But, if you're in the right mood, you might find an understated charm in Beinn Mhanach. Located to the east of Bridge of Orchy, the recommended out-and-back route follows the West Highland Way from the A82 to Auch and then veers north-east up a lonely valley with several river crossings, before a steep, grassy slog to the summit.

Beinn Mhanach from Beinn an Dothaidh (see route 18).

Public Transport

The route can be lengthened to start at Bridge of Orchy railway station, which sits on the West Highland Line. Buses stop outside the Bridge of Orchy Hotel. Simply follow the West Highland Way south from Bridge of Orchy until you reach the bridge over the Allt Kinglass. Pick up the route description from here. This lengthens the route to 26.3km.

Parking

Free parking for around eight cars in a deep lay-by on the A82, halfway between Tyndrum and Auch. The nearest public toilets are in Tyndrum.

MTB-ability

The first and last 9.3 kilometres of the route can be mountain biked.

Escape Route

None.

Refreshments

A trip to the Highlands is never complete without a visit to The Green Welly Stop in Tyndrum. It has a restaurant, a snack bar, an outdoor store and a gift shop.

The Route

S Leave the lay-by to the north and walk along a track beside the railway. The West Highland Way soon merges in from the right to join the firm gravel track which runs parallel to the Allt Coire Chailein down to your left. Beinn Dòrain (page 87) is directly in front as you walk north along the track. In May, flowering gorse bushes leave a coconutty scent in the air.

2 Pass through a metal gate and turn right at the crossroads. Follow the estate track past some farm buildings and walk north-east into the Gleann Achadh-innis Chailein. Ford a stream with a concrete base just beyond the farm. After 500m walk under a railway viaduct, pass over a cattle grid and continue into the valley alongside the Allt Kinglass. Layers of bedrock protrude from the river when it's not in spate. Continue along the track with Beinn Mhanach now visible and getting closer with every step.

3 Cross the Allt Kinglass using another ford at NN341373. When it's not in spate you can get across the river with dry feet. With the river now on your right-hand side, there are another few shallow fords to cross on the next section of trail, the largest of which has a series of boulder stepping stones to assist your crossing.

4 Pass Ais-an t-Sìthean where there is a farm building close to the trail. The trail eases to the right as it curves eastwards, crossing a couple of minor fords over the Allt a' Chuirn which present no difficulty. Ignore the track turn-off on your left which leads around the side of Beinn a' Chuirn – this is an alternative line of ascent but not the one I prefer.

5 The track ascends gently before reaching an established bridge at a weir in the Allt a' Chùirn. A cairn here marks the start of the ascent, immediately after the bridge on the left. From here it's a big pull uphill to the summit of Beinn Mhanach. You'll gain over 500m in vertical ascent over steep, grassy and mossy terrain. Follow a faint path at the beginning which leads towards a line of fence posts which go in a north-easterly direction. Loch Lyon is visible to your right on the ascent. The fence posts can be handrailed north over steep grassy terrain to the 700m contour.

Descending Beinn Mhanach with Loch Lyon in the distance.

6 At the 700m contour the fence posts all but entirely collapse beside a rock. Take the opportunity here to peel off to the right and walk north-eastwards over pathless ground to the summit. Beinn Sheasgarnaich (page 109) dominates the view to the south-east over Loch Lyon, but I love to sit beside the summit cairn and enjoy the views to the Bridge of Orchy Munros (page 87) and the Achallader Munros (page 91) to the north and west. Back in 2019 I walked Beinn Mhanach, Beinn a' Chreachain, Beinn Achaladair, Beinn an Dothaidh and Beinn Dòrain together in a single route, wild camping along the way. While it *is* possible to link these five Munros, it's a big undertaking with a fair degree of pathless walking and it amounts to a total route of nearly 40km with over 2,000m of ascent.

7 Retrace your steps back to the start, being sure to zigzag over the steepest ground to be kind to your knees.

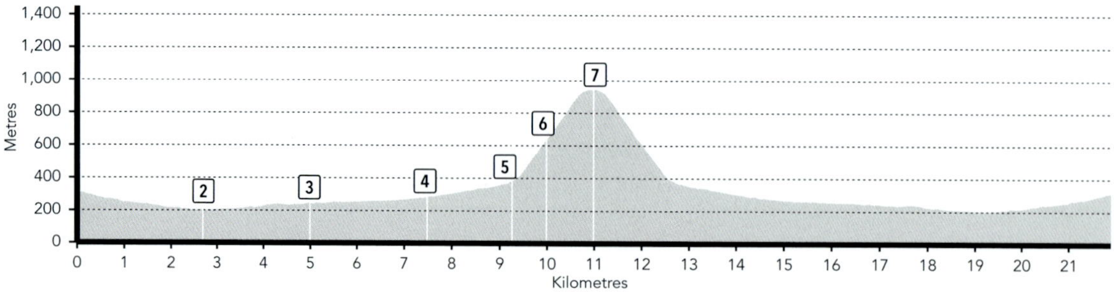

BEINN AN DOTHAIDH
1004
Coire a' Ghabhalach
Beinn a' Chuirn
923
BEINN MHANACH
953
Meall Tionail
728
Meall Garbh
994
BEINN DÒRAIN
1076
Gleann Achadh-innis Chailein
Allt Kinglass
Srath Tarabhan
Loch Lyon
Beinn nam Fuaran
806
Abhainn Ghlas
Beinn a' Chaisteil
886
Glen Coralan
Auch
A82
Cam Chreag
884
Beinn Odhar
901
Beinn Chaorach
818
0
1 Kilometre

BEINN DÒRAIN 1,076m
BEINN AN DOTHAIDH 1,004m

18 Bridge of Orchy Munros

HUB **Tyndrum – 10-minute drive**

OVERALL 1 **(2)** 3 4 5   TERRAIN 1 2 **(3)** 4 5
DISTANCE 1 **(2)** 3 4 5   NAVIGATION 1 **(2)** 3 4 5

DISTANCE **14km**   ASCENT **1,220m**
TIMES **7:45hrs** (Walker)   **6:40hrs** (Trekker)
5:00hrs (Fastpacker)   **3:15hrs** (Trail runner)

START LOCATION **Walkers' car park, next to Bridge of Orchy Hotel, A82**
START LATITUDE/LONGITUDE **56.5172, -4.7689**
START GRID REFERENCE **NN297395**
START W3W **///decades.prouder.spare**

	PRONUNCIATION	NAME MEANING
Beinn Dòrain	**'beyn dooren'**	**hill of the small stream**
Beinn an Dothaidh	**'beyn yen dooree'**	**hill of the scorching**

Introduction

Beinn Dòrain is an aesthetic mountain – beautifully conical, with the smooth lines and shapely curves of a volcanic triangle. It draws the eye from the A82, near Tyndrum, as if the mountains are calling – and you really must go. This walk is a premier outing, with loads of character throughout, as well as excellent views across the expanse of Rannoch Moor. Starting from Bridge of Orchy, it's difficult to make a circular route of these twin peaks, but they can easily be combined in a T-shaped route with an out-and-back approach. The location of Bridge of Orchy railway station also makes this walk perfect for a car-free outing.

Public Transport

Begin the walk at Bridge of Orchy railway station which is a short distance into the route. The railway station sits on the West Highland Line. Buses stop outside the Bridge of Orchy Hotel.

Parking

Free parking for at least 30 cars in the walkers' car park next to the Bridge of Orchy Hotel on the A82. The nearest public toilets are in Tyndrum.

Escape Route

No escape routes; retrace your steps after the first Munro and return another day for the second.

Refreshments

Pop into the Bridge of Orchy Hotel for lunch or dinner after your walk. Book a table to avoid disappointment.

The cairn marking the saddle between Beinn Dòrain and Beinn an Dothaidh with the south-western crags of the latter shrouded in fog.

A walker pauses to enjoy the view to the distant Crianlarich Munros from the ascent of Beinn an Dothaidh.

The Route

S Leave the car park and walk north alongside the A82 briefly before crossing the road and turning right up a minor road signposted to *Bridge of Orchy railway station*. Pass under the railway using the tunnel next to the station and ascend a set of stairs to reach a gravel path. Pass through a gate on to a track. Walk straight on here, heading north-east. Pass to the left of a communications mast and when the trail splits take the wider right-hand fork over eroded ground uphill which leads towards the bealach between the two Munros.

2 The wide, eroded path heads east-north-east uphill with grassy terrain on either side and crosses a couple of minor streams which present no difficulty. As the ground steepens, so does a ravine containing the Allt Coire an Dòthaidh on your left. Keep following the clear stony path which winds up through Coire an Dòthaidh right up to the top of the pass, steepening in parts with shifting stones and loose scree beneath your feet in places.

3 At the top of the pass, beside a cairn at 744m, you are faced with a choice. Turn right for Beinn Dòrain or left for Beinn an Dothaidh. Both will require you to return to the same point before you can summit the next Munro. These route instructions will tackle Beinn Dòrain first as it's the longest out-and-back and the higher Munro of the two. Turn right to follow the Beinn Dòrain path to the south-south-west. The start of the broad ridge is over gently sloping slabs which present no technical difficulty. The slabs give way to grassy plains and an eroded path leads past a small lochan.

4 When the path splits at NN323387, take the left-hand fork which leads south-south-east on to the crest of a rib of rock which forms the spine of Beinn Dòrain. While ascending along the ridge, be sure to look back over your left-hand shoulder to admire Beinn an Dothaidh, Beinn Achaladair, Beinn a' Chreachain (page 91) and Beinn Mhanach (page 83), nestled closely together in a tight diamond shape. Stick to the crest of the broad ridge all the way to Carn Sasunnaich, the top of which is marked by a large cairn. From here, drop slightly before making your final short ascent to the summit of Beinn Dòrain.

5 To add variety to the return route, you can continue straight on past the summit to follow the lower path which leads across the western flanks of Beinn Dòrain back to the grassy plains but there is no real benefit in doing this as the best 360-degree views can be appreciated by retracing your steps along the crest of the ridge. Return to the top of the pass at 744m (**3**), enjoying the view across the Black Mount to the Munro pairings of Stob Ghabhar and Stob a' Choire Odhair (page 95), alongside Creise and Meall a' Bhùiridh (page 173) over Loch Tulla. From the bealach take the stony path which leads north-east under crags.

6 Soon, the eroded trail hits a short section of bouldery ground but when you reach NN327401 the terrain becomes soft and grassy once more. A short section of peat hags can make navigation challenging for a brief moment but a path leading northwards can soon be picked up. This path circles to the right around Coire Reidh on the 925m contour, avoiding the worst of the boggy ground. Follow the path as it turns north-east once more to gain the edge from which Beinn an Dothaidh's three tops enclose Coire Reidh. Turn left to walk along the edge

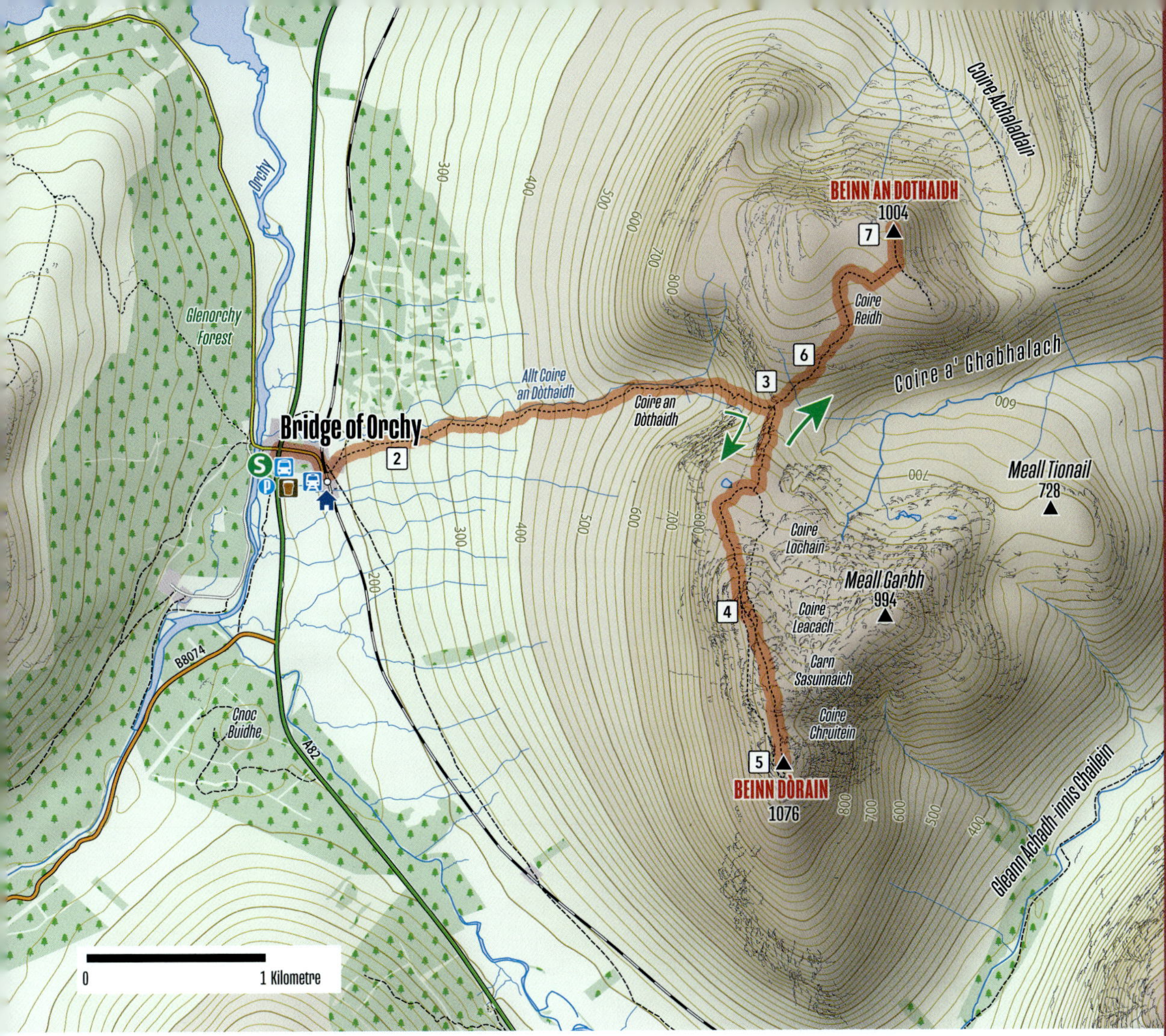

on a steady final climb north to the summit, which is the middle of three tops.

7 The view from the summit is in two contrasting halves with the flatlands of Rannoch Moor to the north and the abundance of mountainous terrain to the south including the long row of Crianlarich Munros (pages 69, 75 and 79) and Beinn Dòrain, with Ben Lui (page 57) looming in the distance. Retrace your steps back to the start.

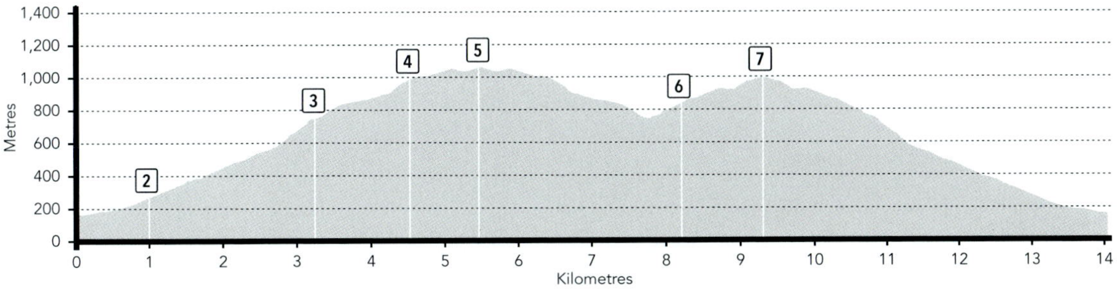

19 Achallader Munros

HUB **Tyndrum – 15-minute drive**

OVERALL 1 2 3 (4) 5 TERRAIN 1 2 3 4 (5) (Water)
DISTANCE 1 2 3 (4) 5 NAVIGATION 1 2 3 (4) 5

DISTANCE **21.8km** ASCENT **1,265m**
TIMES **9:25hrs** (Walker) **8:05hrs** (Trekker)
6:05hrs (Fastpacker) **4:05hrs** (Trail runner)

START LOCATION **Car park just off the A82 along the track to Achallader Farm**
START LATITUDE/LONGITUDE **56.5555, -4.7464**
START GRID REFERENCE **NN312437**
START W3W **///signs.trio.brambles**

	PRONUNCIATION	NAME MEANING
Beinn a' Chreachain	**'beyn akreken'**	**hill of the rocks**
Beinn Achaladair	**'beyn akoolader'**	**hill of the field of hard water**

Introduction
Handsomely chiselled, with rugged northern cliff faces, these two shapely Munros tower with superiority above the boggy flatlands of Rannoch Moor, serving up a first-class day of high-level ridge walking. Located to the east of Loch Tulla, forming the northern branch of the Beinn Dòrain range, the only issue with this walk is the tricky river crossing of the Water of Tulla, which should not be attempted when the river is in spate. The route is best walked clockwise, with the river crossing at the beginning – if it's impassable, it's better to realise early on, rather than at the end of a long day.

Public Transport
Not recommended, as it entails a five-kilometre walk along the verge of the busy A82 from Bridge of Orchy at the start and end of the walk.

Parking
Free parking for around 20 cars just off the A82 along the access track to Achallader Farm. Note that there used to be a car park at the farm itself, but parking here is no longer permitted. The nearest toilets are in Tyndrum.

Escape Route
No escape routes once committed to the broad ridge between the two Munros. It's too craggy, steep and pathless to descend early.

Refreshments
Pop into the Bridge of Orchy Hotel for lunch or dinner after your walk. Book a table to avoid disappointment.

Beinn Achaladair from Meall Buidhe.

The Route

S Leave the car park to the north-east along a track. Ignore the fingerpost that reads *To hill*, as this will be your route of descent. Instead, continue straight ahead past a communications mast. At the crossroads go straight on and then pass through a gate at Achallader Farm and head north-east along a track with a drystone wall on your right-hand side. Beyond said wall looms Beinn Achaladair, your second Munro of the day.

2 Pass through a gate and ford the Allt Ur. It is usually possible to keep dry feet here if the river is not in spate and you pick your path carefully. Minor stream crossings follow as you walk along the southern bank of the Water of Tulla before crossing to the northern side using a large footbridge. Pass through a gate and continue walking north-east along the track with the river now on your right.

3 Cross the bridge over a small stream at NN354470 and immediately turn right on a path through heather to the edge of the river and find a safe space to cross. When the water levels are low, you can hop over boulders and maintain dry feet, but after recent rainfall the river may require a careful wade. Walk south-east to pick up a path which leads up between two copses of woodland which are each fenced in. Continue on the boggy path to the railway tunnel and cross under the railway line. Continue walking south-east with a deer fence on your right and the Allt Coire an Lochain on your left.

4 At NN368449, just before the Lochan a' Chreachain comes into view, cross the stream and head east across soft grassy ground up the side of Coire an Lochain to gain the saddle between point 961m on the OS map and the main summit of Beinn a' Chreachain. Prepare for some steep, grassy and pathless walking. A faint path begins towards the col where the ground is drier. Arrive at the lowest point of the saddle and follow the path up to the Munro summit. Be sure to pause and enjoy the wonderful view over Rannoch Moor. On a clear day the profile of Ben Nevis is unmissable. From the summit, which is marked by a large cairn, the view to Beinn Achaladair and the glimpse as to the route ahead along the ridge are stunning.

5 From the summit, retrace your steps for 150m before heading west down an eroded path. Walk up and over Meall Buidhe which offers phenomenal views on both sides. The terrain is gently undulating, and the ridge is broad and unproblematic. Drop from Meall Buidhe south-west to the bealach at 813m.

6 From the bealach, ascend steeply for the final short pull to the summit of Beinn Achaladair. It is rocky in places and requires some big steps but doesn't constitute scrambling. Arrive at the first top and then walk along to the second, which is a couple of metres higher.

7 Follow the grassy ridge southwards to start your descent. On your left is Beinn Mhanach (page 83). It is possible to link these Munros, but I find it more pleasant to walk them separately. The ridge undulates slightly and passes just to the right of a minor top before descending to the bealach below Beinn an Dothaidh.

8 At NN340406, at a cairn, start your descent north-north-west into the corrie. The trail is an eroded line that sometimes becomes indistinct over small patches of boggy and rocky ground, but it largely follows the left-hand side of the Allt Coire Achaladair.

9 Towards the end of the route, the path follows the line of a deer fence before crossing a bridge over the railway line. At a fingerpost, fork left to finish the walk along a wet, grassy path back to the start point.

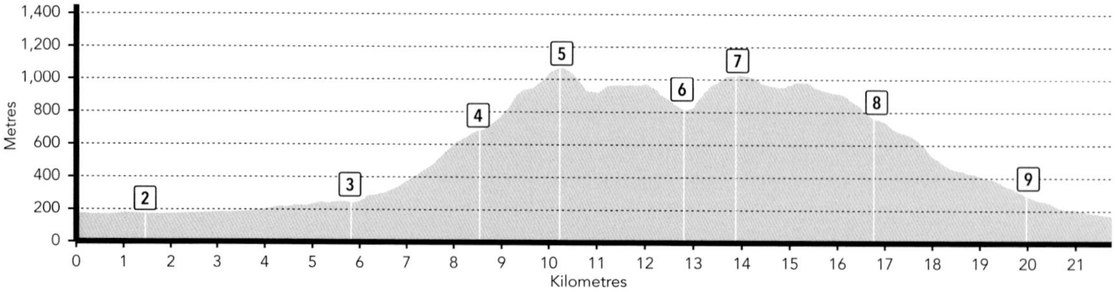

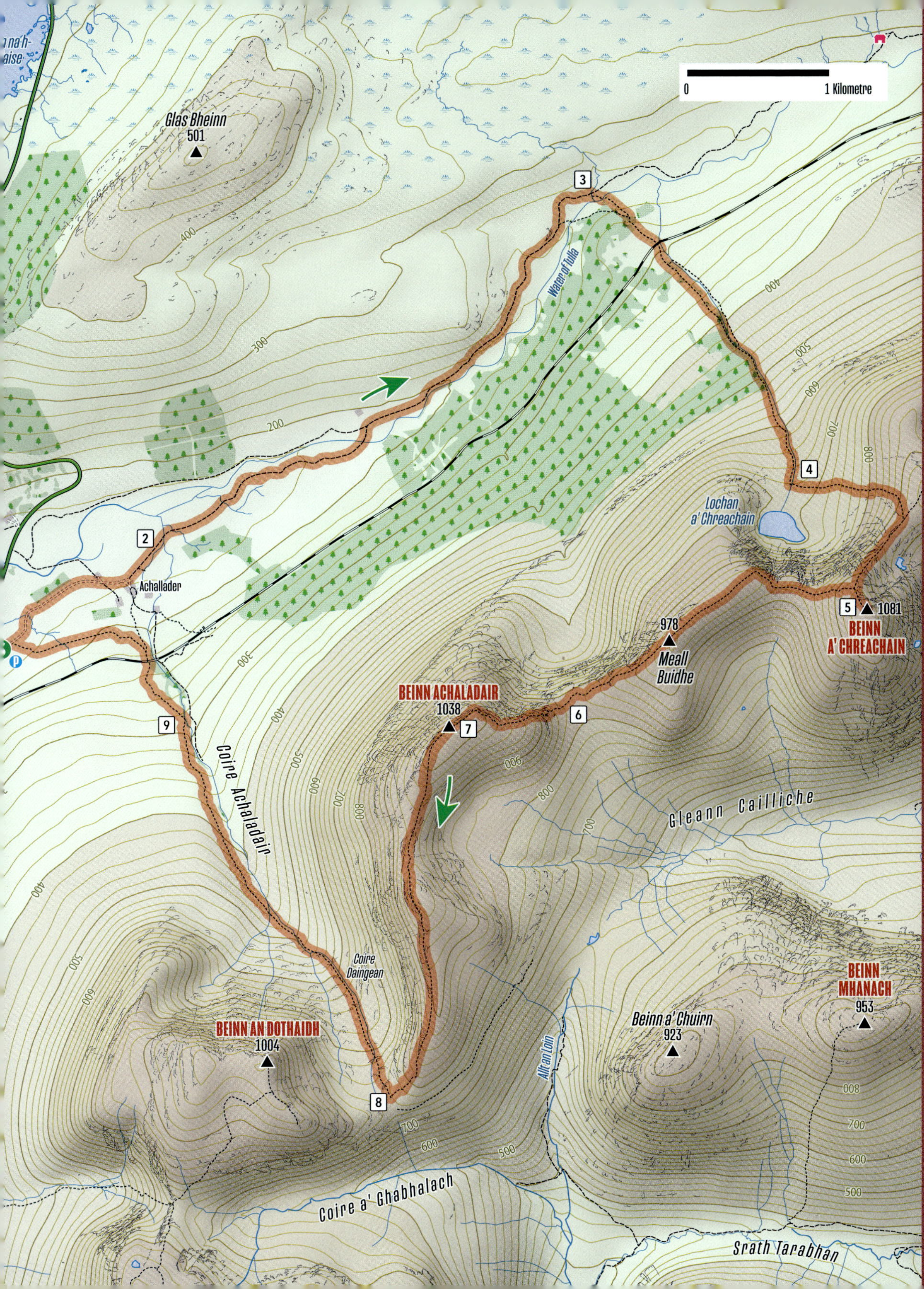

0
1 Kilometre
Glas Bheinn
501
Water of Tulla
Achallader
Lochan a' Chreachain
BEINN A' CHREACHAIN
1081
978
Meall Buidhe
BEINN ACHALADAIR
1038
Coire Achaladair
Gleann Cailliche
Coire Daingean
BEINN AN DOTHAIDH
1004
Allt an Loin
Beinn a' Chuirn
923
BEINN MHANACH
953
Coire a' Ghabhalach
Srath Tarabhan

STOB A' CHOIRE ODHAIR 945m
STOB GHABHAR 1,090m

20 South Black Mount Munros

HUB **Tyndrum – 20-minute drive**

OVERALL 1 2 3 **4** 5 TERRAIN 1 2 3 **4** 5
DISTANCE 1 2 **3** 4 5 NAVIGATION 1 2 3 **4** 5

DISTANCE **16.3km** ASCENT **1,250m**
TIMES **8:20hrs** (Walker) **7:10hrs** (Trekker)
5:25hrs (Fastpacker) **3:30hrs** (Trail runner)

START LOCATION **Victoria Bridge car park, Inveroran**
START LATITUDE/LONGITUDE **56.5368, -4.8141**
START GRID REFERENCE **NN270418**
START W3W **///strongman.backup.facelift**

	PRONUNCIATION	NAME MEANING
Stob a' Choire Odhair	**'stob a corriya oar'**	**peak of the dun-coloured corrie**
Stob Ghabhar	**'stob goo-er'**	**goat peak**

Introduction

'A stonking day in the hills' – full of adrenaline and gratitude after a dreamy, sun-drenched research outing up these twin Munros with a couple of friends, this is how I described my experience of this route at the time. Two of the four Black Mount Munros (see page 173 for the northern two), Stob a' Choire Odhair and Stob Ghabhar make a fine expedition for the discerning hillwalker – and touch wood you'll get the same bluebird conditions as I enjoyed. Rounded Stob a' Choire Odhair is less memorable, but Stob Ghabhar is bigger, gnarlier and more dramatic, and in-between the two there are excitingly narrow ridges above dark corries.

Public Transport

The route can be lengthened to start at Bridge of Orchy railway station, which sits on the West Highland Line. Buses stop outside the Bridge of Orchy Hotel. Simply follow the West Highland Way north from Bridge of Orchy to the road next to the car park. Pick up the route description from here. This lengthens the route to 26.5km.

Parking

Free parking for around 18 vehicles at the Victoria Bridge car park in Inveroran. The nearest public toilets are in Tyndrum.

MTB-ability

The first and last 2.1 kilometres of this route can be mountain biked.

Escape Route

From the bealach between the two Munros at point 668m on OS maps, you can descend south-east through Coire Toaig. Allt Toaig can then be followed back to the start.

Approaching the summit of Stob a' Choire Odhair with Stob Ghabhar behind.

Refreshments

The Inveroran Hotel has an on-site shop which sells breakfast rolls and stocks sandwiches and snacks. Non-residents are welcome for dinner and booking is required.

The Route

S Leave the car park to the north and join the West Highland Way along a tarmac road. Pass over Victoria Bridge to the other side of the Abhainn Shira. Take a left-hand turn in front of Forest Lodge on a track that leads west with the river on your left. Pass Clashgour Hut – this climbing hut remarkably used to be a primary school for local children – and turn right on to a grassy path just before you reach a bridge over the Allt Toaig. Walk northwards with the Allt Toaig on your left in the direction of the two Munros – Stob Ghabhar has a tendency to look pretty fearsome from all angles.

2 The path is boggy in parts until it starts to gain some height. At NN251439 make a note of the cairn on your left which marks the descent path which you'll use later today, but for now walk north-north-east uphill with the Allt Toaig on your left.

3 At NN253446, immediately following a bouldery crossing of the Allt Caolain Duibh, leave the track and turn right to head up the south ridge of Stob a' Choire Odhair. The path is an eroded line up the hillside, steep and stony in places, but it soon starts to zigzag which makes for slightly easier walking. For the final 200m of vertical ascent of Stob a' Choire Odhair the terrain is rockier, and the path can be tricky to spot in poor weather. Head northwards to gain the highest ground and then curve around to the north-east until you reach the stony summit which is marked by a cairn.

4 Views from Stob a' Choire Odhair stretch over the blanket moorland of Rannoch Moor. The Bridge of Orchy mountains (page 87) are prominent over Loch Tulla, and Creise and Meall a' Bhùiridh (page 173) – the northern Black Mount Munros – can be appreciated to the north. The view west to the bulk of Stob Ghabhar is magnificent. From the summit follow the path west-south-west as it leads down over stony ground to the bealach marked 668m on OS maps. While it's steep in places, it's not technical.

Stob a' Choire Odhair from the ascent of Stob Ghabhar.

5 From the bealach, the path begins to gradually ascend in a westerly direction before you're soon looking down on Coirein Lochain. Follow the zigzag path which leads south to the top of Couloir Buttress. The ground is steep and bouldery in places and may require you to put your hands down to assist your ascent at times, but it is not a graded scramble. Top out and enjoy the view to Bidean nam Bian (page 163) and Ben Nevis (page 195) on a clear day.

6 Drop down on to the ridge which appears narrow at first, but it is never particularly exposed. Stick to the eroded path and the crest of the ridge and you'll find it's a hugely enjoyable section of the walk. The ridge soon widens and the path curves around to the north to allow you to make your final ascent to the summit of Stob Ghabhar which is marked by a cairn. Pause and take in the view to the Ben Starav group (page 145) and the Glen Coe mountains from the summit.

7 Retrace your steps for 400m to a cairn at NN232452. Here, take the right-hand fork heading south off the ridge and on to the descent path. Descend along the eroded line which soon runs alongside a row of iron fence posts down the south ridge of Stob Ghabhar. The terrain on the descent is a mixture of grassy, rocky and boggy in places but never presents any technical difficulty in spite of its

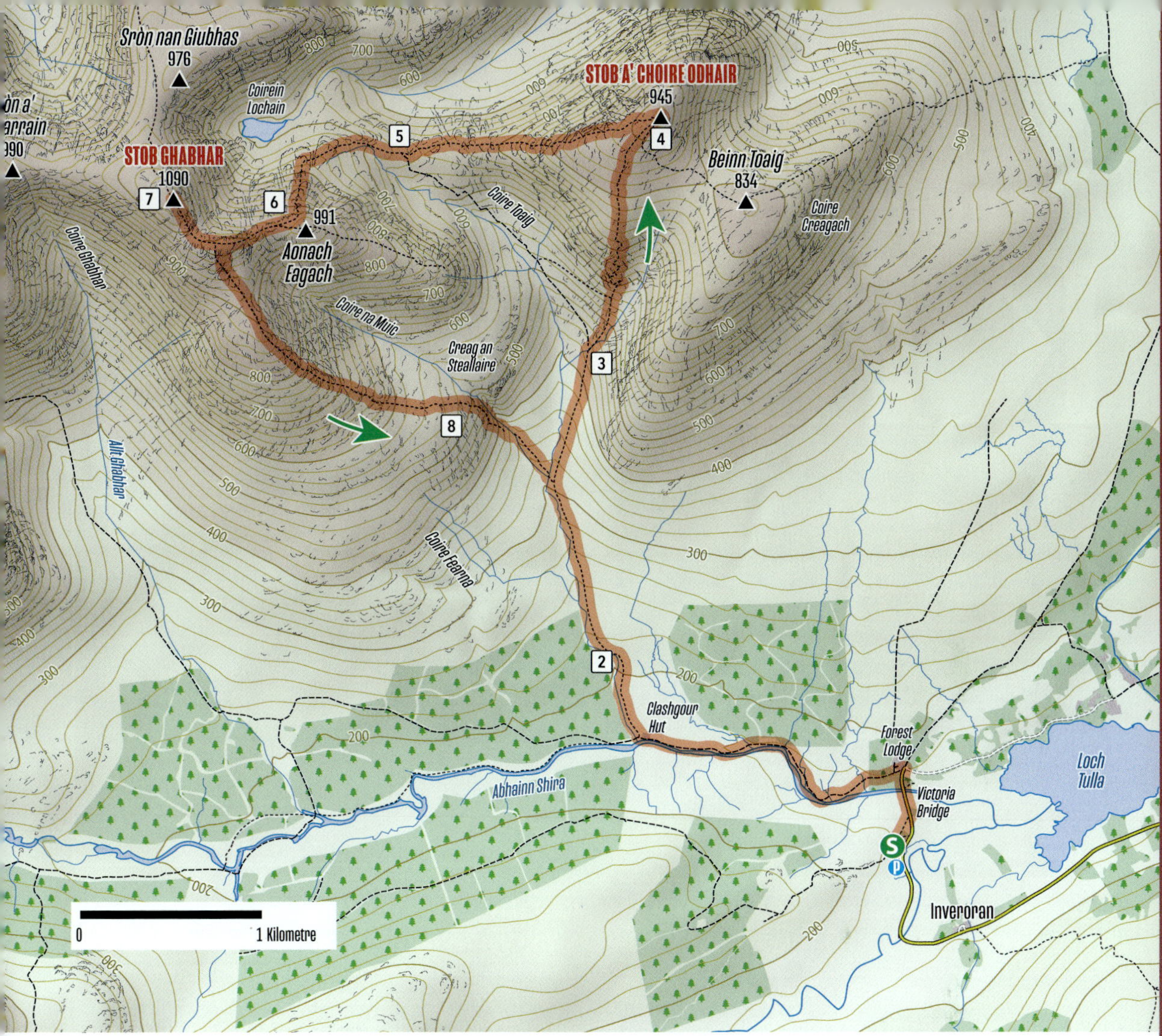

steepness in parts. The path curves south-east and then east towards Creag an Steallaire.

8 The path steepens at the point you're alongside the Allt Coire na Muic, which is the stream on your left. Cross the stream at NN249441 and continue your descent. The path leads down to another crossing point, this time over the Allt Toaig. A short climb up the other side of the small ravine leads you back to the route of initial ascent. Retrace your steps from here back to the start.

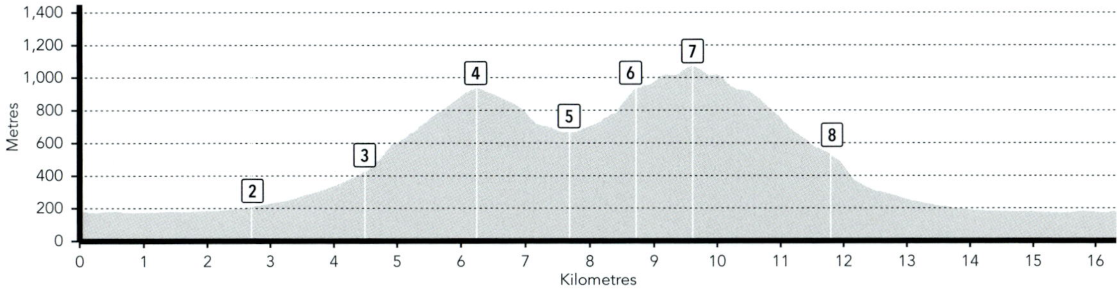

HIGHLAND HUB 5

KILLIN

The fifth of the Highland hubs is Killin, a small village beside the mesmerising Falls of Dochart. To the east you'll find Loch Tay, one of the deepest freshwater lochs in Scotland.

Killin has a supermarket for all your resupply needs plus a good selection of pubs, restaurants and cafes, especially for the size of the village, and Killin Outdoor Centre is on hand for any outdoor-kit-related calamities.

During my first Munro round I spent many days waiting out wet and stormy weather in Killin, keeping an eagle eye on the mountain forecast for the Ben Lawers range. Walks like the North-East Ben Lawers Range demand good weather as An Stùc is one of the steepest Munros in the collection with lots of eroded, loose terrain. Good visibility and dry ground are highly recommended.

As with Comrie, Killin had a railway station until the 1960s; today the only way to reach the hub using public transport is by bus. From the south, the easiest way to reach Killin is to change in Stirling and Callander. The C60 and S60 buses operate between Callander and Killin.

When on a budget and travelling by car, I like to set up my tent just outside Killin at Cruachan Farm Caravan and Camping Park. Staying on this small, friendly site means you get access to the facilities in the holiday park next door included in the very affordable price.

WHERE TO STAY

£	**££**	**£££**
Cruachan Farm Caravan and Camping Park	Killin Hotel	The Courie Inn

TRAVEL TIMES

21 Auchessan Munros	**15 minutes**
22 Glen Lochay Munros	**20 minutes**
23 Meall Ghaordaidh	**10 minutes**
24 Meall nan Tarmachan	**15 minutes**
25 Ben Lawers	**20 minutes**
26 North-East Ben Lawers Range	**15 minutes**
27 Loch an Daimh Munros	**45 minutes**
28 Glen Lyon Munros	**35 minutes**

Rest day ideas

- Visit the Scottish Crannog Centre on Loch Tay and explore the history of Iron Age crannogs – free-standing wooden structures forming loch dwellings which were inhabited over 2,000 years ago.
- Hire a boat or stand-up paddleboard on Loch Tay to take a leisurely tootle around the loch.
- Visit the Fortingall Yew, which is possibly the oldest tree in the United Kingdom. Conservative estimates put the age of the tree between 2,000 and 3,000 years old, but Forestry and Land Scotland state that it could be as old as 5,000 years.

Opposite Ben More from the descent of Sgiath Chùil (route 21).
Overleaf View from Meall Buidhe over Rannoch Moor towards Buachaille Etive Mòr and the Glen Coe Munros (route 27).

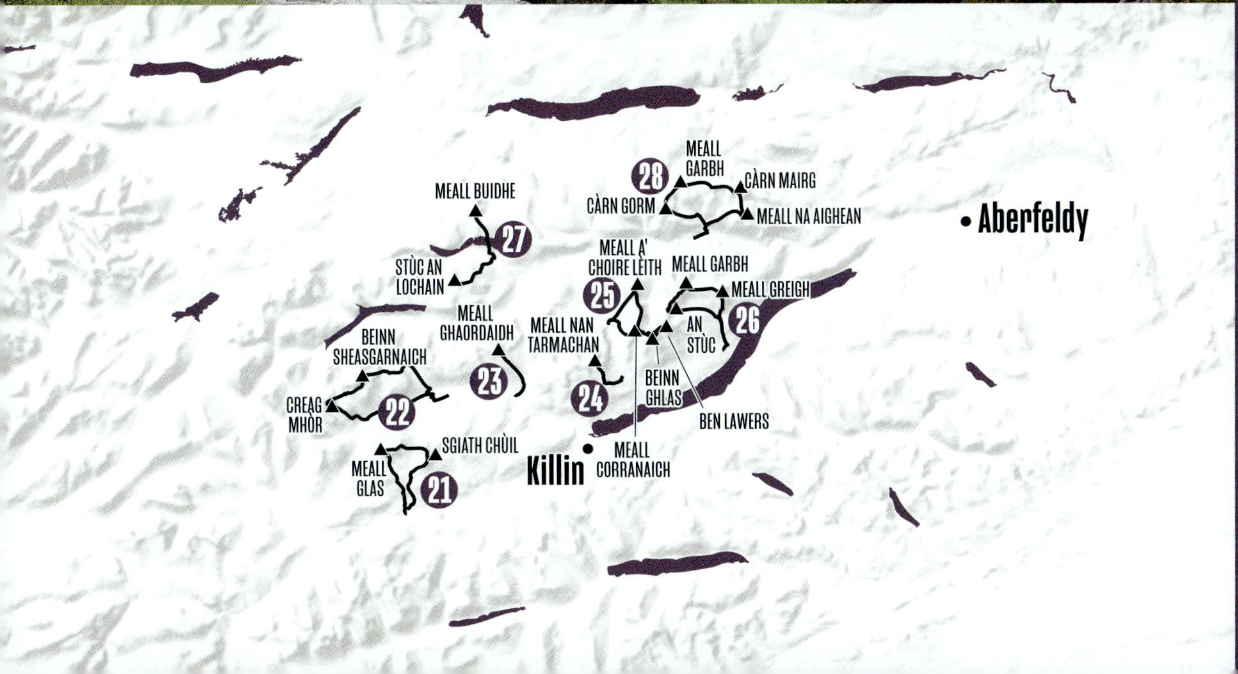
MEALL GARBH
28
CÀRN MAIRG
CÀRN GORM
MEALL NA AIGHEAN
Aberfeldy
MEALL BUIDHE
27
STÙC AN LOCHAIN
MEALL A' CHOIRE LÈITH
MEALL GARBH
MEALL GREIGH
25
26
AN STÙC
MEALL GHAORDAIDH
MEALL NAN TARMACHAN
BEINN SHEASGARNAICH
23
24
BEINN GHLAS
CREAG MHÒR
22
BEN LAWERS
SGIATH CHÙIL
MEALL CORRANAICH
Killin
MEALL GLAS
21

21 Auchessan Munros

HUB **Killin – 15-minute drive**

OVERALL 1 2 3 **4** 5 TERRAIN 1 2 3 **4** 5
DISTANCE 1 2 **3** 4 5 NAVIGATION 1 2 3 **4** 5

DISTANCE **15.7km** ASCENT **1,240m**
TIMES **8:05hrs** (Walker) **7:00hrs** (Trekker)
5:15hrs (Fastpacker) **3:25hrs** (Trail runner)

START LOCATION **Roadside parking at the top of the lane to Auchessan, just off the A85**
START LATITUDE/LONGITUDE **56.4148, -4.5178**
START GRID REFERENCE **NN447275**
START W3W **///added.coughed.perfumed**

	PRONUNCIATION	NAME MEANING
Meall Glas	**'meowl glass'**	**greenish-grey hill**
Sgiath Chùil	**'skeer hool'**	**back wing**

Introduction

A lofty bealach linking Munros, with minimal ascent and descent between the two, is worth its weight in gold – an energy-saving, morale-boosting geographical feature. Sadly, you do not get this between Meall Glas and Sgiath Chùil, the two Munros rising steeply above Glen Dochart, to the east of Crianlarich. There's a precipitous drop and re-ascent between the two summits – a cruel blow to tired legs – and there's steep, pathless, craggy ground thrown into the mix too. Good navigation skills are essential. Power through it all, however, and both summits are superlative viewpoints. A particular favourite is the V-shaped notch on the horizon seen from Sgiath Chùil – an aesthetically appealing and strikingly distinctive silhouette – created by the harmonious coming together of Ben More and Stob Binnein's flanks.

Public Transport

There are no public transport services close to this walk.

Parking

Free roadside parking for around six cars at the top of the lane to Auchessan, just off the A85. Arrive early and tuck in as much as you can to allow farm traffic to pass. The nearest public toilets are in Crianlarich.

Escape Route

From the bealach between Beinn Cheathaich and Sgiath Chùil, follow the Allt Riobain over rough terrain to meet the descent path at the dam.

Refreshments

Treat yourself to a meal at the Falls of Dochart Inn, right across the road from the famous falls in Killin.

Sgiath Chùil from the approach to Meall Glas over a stream.

Above Beinn Challuim and Meall Glas from Sgiath Chùil. **Opposite** On the ridge between Meall Glas and Beinn Cheathaich, looking over to Ben More.

The Route

S Walk along the lane northwards to a bridge across the River Dochart and on to Auchessan. At NN447279, turn left to pass behind an old farmhouse and walk along a gravel track behind the building which turns to lead northwards beside a stream.

2 When the track splits at a bridge, fork right and continue uphill on a grassy path past a sheepfold. The path then runs on the left-hand side of a deer fence. In summer, the higher you climb the thicker the bracken gets as you walk the path corridor through the middle of what feels like a bracken jungle. When you exit the bracken, be sure to look behind you to the opposite side of Glen Dochart, where Ben More (page 79) rises theatrically from the valley floor.

3 At NN444290, cross a stile and follow the path as it curves sharply to the right over soft ground. Now on the open hillside, continue first to the north-east and then to the north-west. The path hugs the stream as it ascends the grassy hillside, which becomes boggy in places. The twin peaks of Sgiath Chùil and Sgiath Chrom dominate to your right as you ascend. Despite it being wet underfoot, the path is fairly clear as it leads north-north-west towards your first objective – Meall Glas.

4 Cross the Allt Glas at NN436307 and look up to the smattering of crags guarding the higher reaches of the Munro. Despite starting beside the busy A85, by this point in the walk the area has the feeling of being far more remote. Ascend a small rocky spur at 650m elevation before continuing uphill on grassy terrain. When the grass becomes rough, head north and then north-north-west passing through broken crags on Meall Glas Beag. Aim for two small lochans on the OS map at NN434318 and continue north-west from here over grassy terrain around small broken crags to the summit of Meall Glas.

5 The summit is marked by a cairn on a small outcrop, and you're rewarded with stunning views to Beinn Challuim (page 65), plus the lesser-visited Munros of Creag Mhòr and Beinn Sheasgarnaich (page 109). Walk east-north-east from the summit along the grassy plateau and pick up a path which drops down grassy and then stony terrain to the bealach. Follow the path as it passes to the left of a 908m knoll and up Beinn Cheathaich, which has a trig pillar.

6 From here, you'll realise you were lulled into a false sense of security with the pleasantly undulating first part of the connection between the two Munros as you now need to carefully descend steep grassy

Sgiath Chùil from the summit of Meall Glas.

slopes littered with boulders and crags into a bowl of peat hags between Beinn Cheathaich and Sgiath Chùil, losing much of the height you'd previously gained. My preference is to head south-east and skirt carefully around the crags before passing to the south of two small lochans in the Lairig a' Churain at NN449324 where a path can be picked up. The path weaves through the boggy basin to the foot of the next ascent at NN456323.

7 Fuel your body ready for the steep grassy climb eastwards up to the ridge between Meall a' Churain and Sgiath Chùil. Top out at NN461323 after a gruelling mountain workout. Walk south, passing to the right of an unnamed knoll, before summitting Sgiath Chùil whose summit is also marked by a cairn on an outcrop. The view across to Ben More and Stob Binnein (page 79) from here is incredible.

8 Leave to the east-south-east at first following a grassy trail above the crags with a drop to your right initially. The path weaves over dry grassy ground southwards and skirts to the west of Sgiath Chrom, before it leads down into the valley below to the south-west.

9 The path leads to a dam. Cross the stream just below the dam and descend alongside the stream. Cross a bridge at a confluence of two streams and continue descending past some lovely pools. At NN451290, pass over a stile in the deer fence and continue following the stream downhill. The path becomes very muddy as it descends between the stream and the deer fence, and in summer it can also feel pretty wild and overgrown. The path merges on to a track which leads through numerous gates to a farmyard. Pass through the farmyard and walk past the houses at Auchessan. Retrace your steps to the start.

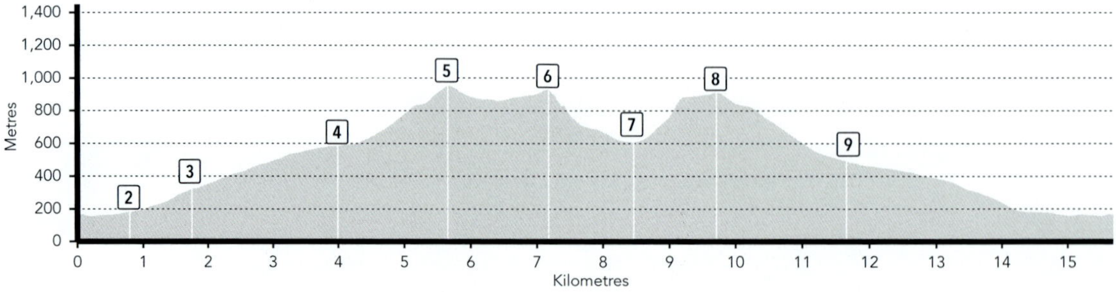

577
Creag an t-Searraich
Lochay
Beinn Cheathaich
937
Lairig a' Churain
Meall a' Churain
917
Meall Eòghainn
809
MEALL GLAS
959
5
6
7
Coire na Saobhaidhe
8
SGIATH CHÙIL
921
853
Sgiath Chrom
nan Imirean
849
Allt Coire nam Mòine
Allt Riobain
9
4
Beinn nan Clach
Creag nan Uan
3
Allt Riobain
Creag Iubhair
465
Eilean Garbh
2
Auchessan
A85
Dochart
Loch Iubhair
0
1 Kilometre
Stob an t-Suidhe
563

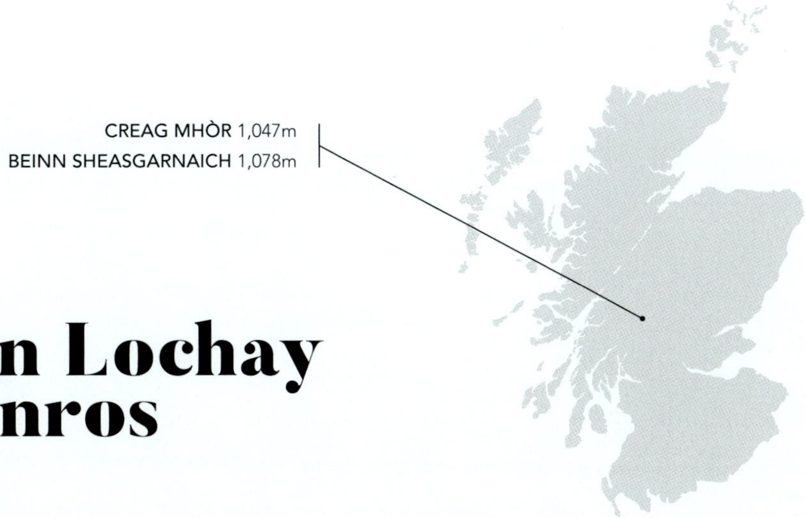

22 Glen Lochay Munros

HUB **Killin – 20-minute drive**

OVERALL 1 2 3 4 **5** TERRAIN 1 2 3 **4** 5
DISTANCE 1 2 3 4 **5** NAVIGATION 1 2 3 4 **5**

DISTANCE **25km** ASCENT **1,445m**
TIMES **10:50hrs** (Walker) **9:20hrs** (Trekker)
7:00hrs (Fastpacker) **4:40hrs** (Trail runner)

START LOCATION **Walkers' car park on the Glen Lochay Road, 1km east of Kenknock**
START LATITUDE/LONGITUDE **56.4990, -4.4758**
START GRID REFERENCE **NN476368**
START W3W **///united.slowness.musically**

	PRONUNCIATION	NAME MEANING
Creag Mhòr	**'craig voor'**	**big rock**
Beinn Sheasgarnaich	**'beyn shiskarneek'**	**sheltering mountain**

Introduction

This is not a route for the faint-hearted or navigationally challenged. There is loads of pathless terrain to negotiate and you'll need to competently pick your own lines through craggy ground. But, at the same time, it's not too daunting for the experienced hillwalker. Located to the south of Loch Lyon and to the north of Glen Lochay, Creag Mhòr and Beinn Sheasgarnaich are best approached from the Glen Lochay Road. There's a remote, secluded feel to the area. During a recent hike I didn't see a single other walker. The only sign of human activity was the hydroelectric pipeline on both sides of the valley – a prominent feature in an otherwise wild place.

Public Transport

There are no public transport services close to this walk.

Parking

Free parking for around 10 cars in the walkers' car park on the Glen Lochay Road, one kilometre east of Kenknock, beside the River Lochay. The nearest public toilets are in Killin.

MTB-ability

The first and last 2.3 kilometres are cyclable.

Escape Route

No escape routes; retrace your steps after the first Munro and return another day for the second.

Refreshments

Visit the small and welcoming Shutters cafe in Killin with its roaring fire and delicious cakes.

View to Creag Mhòr from the summit of Beinn Sheasgarnaich.

Above View down Glen Lochay from the ascent of Sròn nan Eun en route to Creag Mhòr. **Opposite** Beinn Challuim from the approach track to Creag Mhòr.

The Route

S Leave the car park and walk west on a tarmac road into Glen Lochay for around 1km, passing below a farm, to reach a crossroads. Turn right at this junction and pass through a gate on a tarmac road, heading uphill. After 1km of switchbacks the track forks. Take the left fork on to a gravel track, noting that the right-hand option is your descent route later.

2 On the gravel track cross over the pipeline and follow the track as it traverses below Creagan Fhear-chair. The Munros of Sgiath Chùil and Meall Glas (page 103) are across the glen on your left. After 5km cross the bridge over the Allt Bad a' Mhàim; leave the track immediately afterwards, taking another track to the right. This track soon dwindles to a faint grassy trail. At this point walk westwards uphill over rough ground to a gate in the deer fence at NN412351. Pass through the gate and walk alongside the fence as it curves up the hillside. The purpose of this is to avoid the steep crags at the foot of Sròn nan Eun.

3 Veer away from the fence after 250m, ascending between the crags beside a stream in a south-westerly direction. Make your ascent over soft grass, avoiding the rocks and crags which pepper the hillside. Aim for point NN406351 where a path can be picked up. This path runs along the left-hand side of the long nose of Sròn nan Eun.

4 Creag Mhòr rises up ahead but first you must descend slightly to Bealach Slaidearan. Continue from here following the trail up the left-hand side of the wide shoulder. As the terrain becomes rockier, the trail moves to the right-hand side, and an awesome view to the bulk of Beinn Sheasgarnaich commands attention. Loch Lyon can also be spotted down in the valley to the north. The path continues to snake up to the summit of Creag Mhòr without technical difficulty.

5 From the summit, Beinn Challuim (page 65) looks as if it's guarding the head of Glen Lochay. Walk west off the summit of Creag Mhòr, avoiding the craggy north-east face. Make your way over grassy and stony terrain on a gently descending gradient before curving northwards towards the bealach between Creag Mhòr and Meall Tionail. It's not necessary to walk all the way to the bealach; once clear of Creag Mhòr's crags you can curve underneath them walking east-north-east down into the bowl of peat hags at the foot of Sròn Tairbh.

Ben More and Cruach Ardrain from the summit of Beinn Sheasgarnaich.

6 After crossing the peat hags to the left of Lochan na Baintighearna, which is marked on the OS map but is barely distinguishable in real life, walk uphill over grassy terrain. A faint path weaves left and right as it heads north-east up the ridge and you're likely to stumble across it once you're free of the peaty ground.

7 Once you've topped out on Sròn Tairbh you can see the undulating route swinging from right to left towards the summit of Beinn Sheasgarnaich via Stob an Fhir-Bhogha. You can either go over Stob an Fhir-Bhogha or, as on the map and in the GPX file, you can skirt underneath it to the left, avoiding a touch of extra ascent. The faint path here fizzles out as the ground becomes softer, so it's necessary to make a beeline for NN412374 to pick up the path again as it leads northwards. On the final approach to Beinn Sheasgarnaich, which is frequently spelt Beinn Heasgarnich, look to your right over the top of Meall Ghaordaidh (page 115) to the Ben Lawers range (pages 123 and 127). Ben More (page 79) dominates to the south – this pyramidal Munro will always stand out among its proximate hills.

8 After enjoying the views from Beinn Sheasgarnaich, it's time to start the challenging, pathless descent to the private Glen Lochay to Glen Lyon road, which is visible for most of the descent. It evokes the feeling of being so close and yet so far for much of the journey. From the summit cairn, walk eastwards into Coire Bàn Mòr over soft grassy terrain keeping to the right of the crags and to the left of an unnamed lochan. Pick your way down the grassy slopes curving east-north-east towards the Allt Tarsuinn.

9 The Allt Tarsuinn can be crossed at NN433384 to the north of Lochan Achlarich. The stream parallel to it, which is fed by Lochan Achlarich, can be crossed at NN438385, albeit with some boggy ground to contend with between the two. Crossing the two burns becomes more difficult if you attempt it further downstream.

10 From here, walk uphill slightly in an easterly direction to gain the north ridge of the unnamed hill marked 779m on OS maps. Walk north-north-east down the ridge to meet a deer fence at its corner at NN444391. From here, walk east to meet the road which can be followed back to the start.

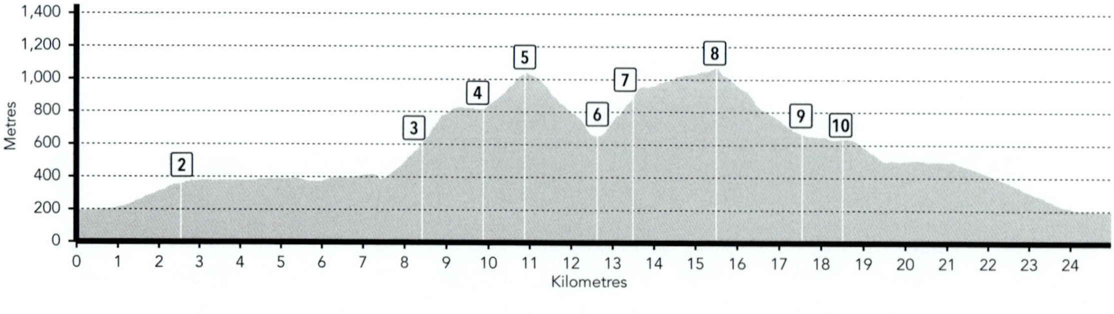

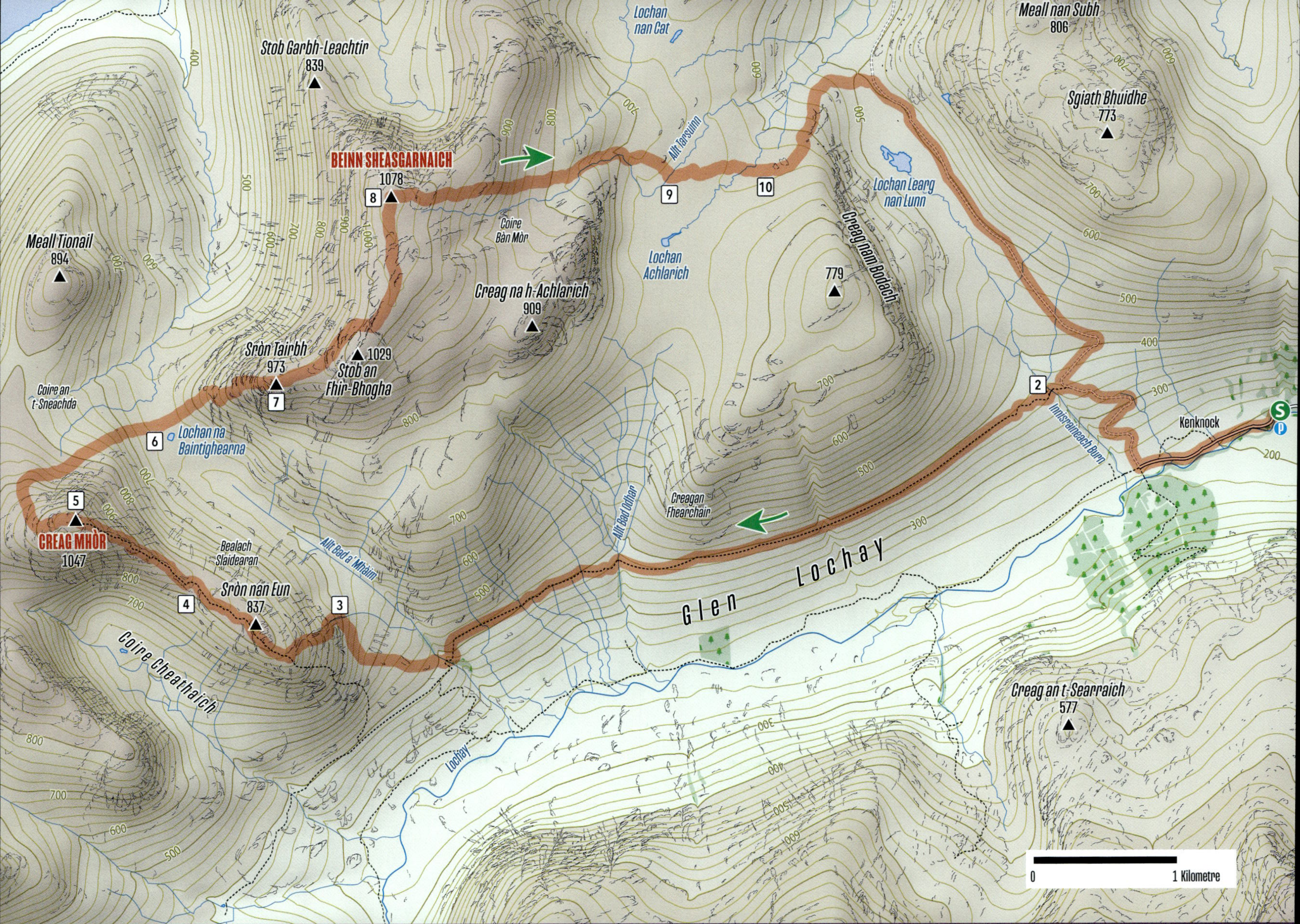

Meall nan Subh
806
Sgiath Bhuidhe
773
Lochan nan Cat
Stob Garbh-Leachtir
839
BEINN SHEASGARNAICH
1078
Allt Tarsuinn
Lochan Learg nan Lunn
Creag nam Bodach
779
Coire Bàn Mòr
Lochan Achlarich
Creag na h-Achlarich
909
Meall Tionail
894
Sròn Tairbh
973
Stob an Fhir-Bhogha
1029
Coire an t-Sneachda
Lochan na Baintighearna
CREAG MHÒR
1047
Bealach Slaidearan
Allt Bad a' Mhaim
Allt Bad Odhar
Creagan Fhearchair
Sròn nan Eun
837
Innisraineach Burn
Kenknock
Glen Lochay
Lochay
Coire Cheathaich
Creag an t-Searraich
577
0
1 Kilometre

23 Meall Ghaordaidh

HUB **Killin – 10-minute drive**

OVERALL 1 ② 3 4 5 TERRAIN 1 ② 3 4 5
DISTANCE ① 2 3 4 5 NAVIGATION 1 ② 3 4 5

DISTANCE **9.4km** ASCENT **895m**
TIMES **5:30hrs** (Walker) **4:45hrs** (Trekker)
3:35hrs (Fastpacker) **2:15hrs** (Trail runner)

START LOCATION **Roadside parking on the Glen Lochay Road at Tòrr Bharan**
START LATITUDE/LONGITUDE **56.4959, -4.3954**
START GRID REFERENCE **NN526363**
START W3W **///rests.circus.shorten**

	PRONUNCIATION	NAME MEANING
Meall Ghaordaidh	**'meowl ghurdie'**	**hill of the shoulder**

Introduction
'Meall Ghaordie' to his friends, this rounded Munro in the Breadalbane hills sits in solitary contentment away from its neighbours. As such, a multi-Munro-bagging route is off the cards for all except the most creative adventurers. This humble out-and-back route from Glen Lochay can be achieved in half a day, leaving ample time for some low-level sightseeing around Killin in the afternoon.

Public Transport
Killin is served by local buses – the road walk into Glen Lochay adds 12 kilometres to the total route length.

Parking
Free roadside parking for around four cars on a wide bend in the road west of the bridge over the Allt Dhùin Croisg on the Glen Lochay Road. Arrive early to avoid disappointment. The nearest public toilets are in Killin.

Escape Route
None.

Refreshments
Book a table at the Courie Inn, Killin. In Scottish Gaelic, 'courie' means 'to nestle or snuggle'. Throw in some good food and fine whisky and you've got yourself an ideal venue after a wet and windy Munro day.

Walker begins her ascent of Meall Ghaordaidh with Creag Mhòr behind her.

Ominous clouds above Beinn Cheathaich on the descent of Meall Ghaordaidh.

Glen Lochay from the ascent of Meall Ghaordaidh.

The Route

S Walk for 100 metres south-east along the road until you reach a large sign which reads *Access for Meall Ghaordie 1,040m.* Pass through the gate here and walk north-east along a grassy track which leads to a gate in the corner of a field. Pass through this and continue up the track passing mature sycamore and ash trees to the sound of the Allt Dhùin Croisg on your right.

2 At the top of the next field, go over the ladder stile and walk north. But not before pausing to take in the view back to Creag Mhòr, a 719m hill across the valley with patchy wooded lower slopes speckled with crags. In the summer months, shoulder-height bracken will flank both sides of the track. After around 700m, take the footpath on the left of the track which is marked by a cairn; there is also a pole in the bracken on the hillside which draws the eye.

3 The well-walked path leads to the right of the pole over grassy, wet and peaty terrain up the open hillside approaching Meall Ghaordaidh via its south-eastern slopes. On the ascent, be sure to turn around and enjoy the view out of Glen Lochay, nicely framed by Meall Dhùin Croisg. At your feet you might spot cloudberry from July and onwards through the summer months. They love the acidic, waterlogged soil that you'll no doubt curse at least once on your walk as you slip and slide over the peaty ground en route to the summit.

4 Around 1km from the top, the ground becomes firmer as the path leads up and around a series

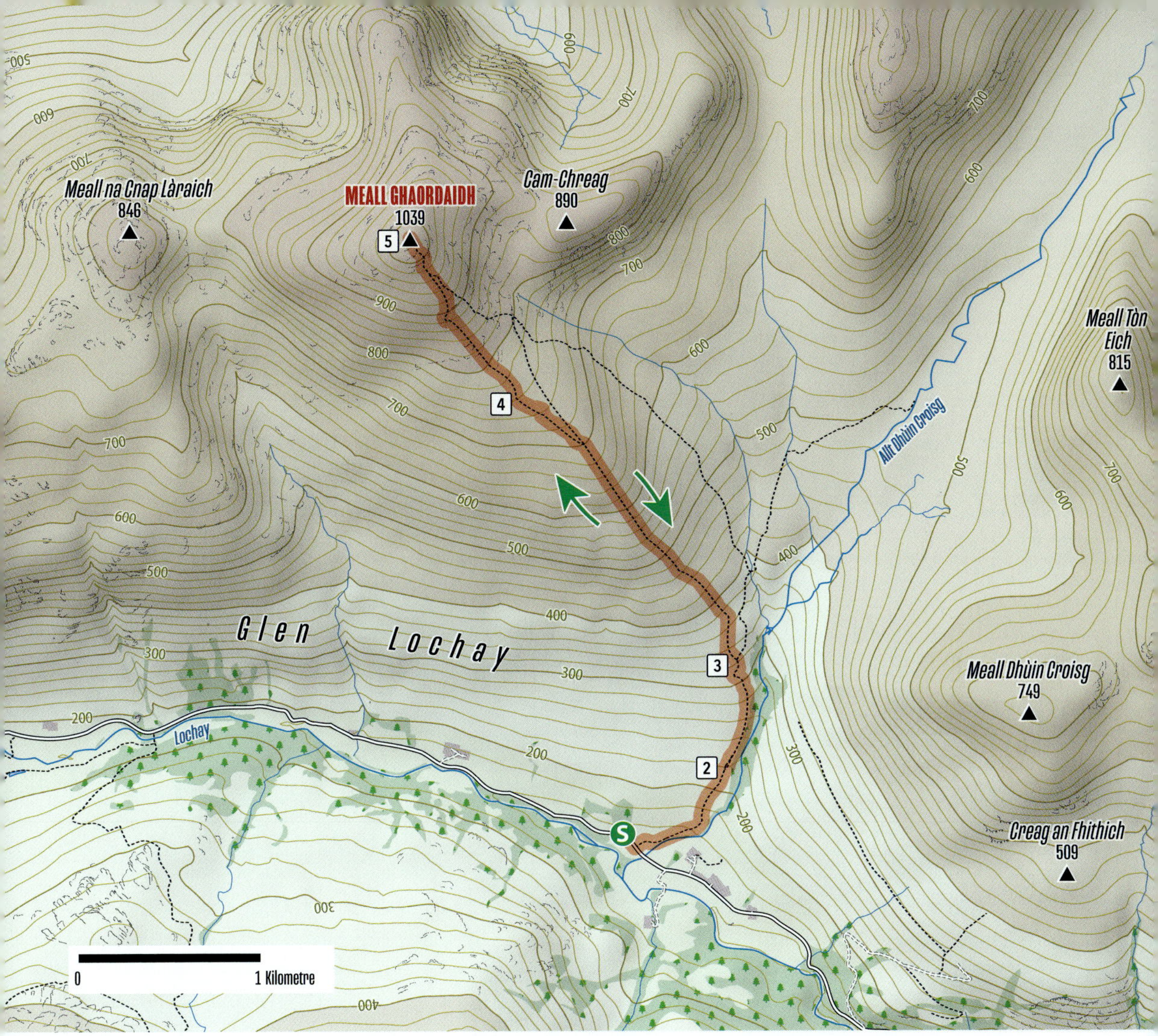

of small crags. At around 950m elevation, I saw a common frog during my last walk up this Munro – it's not unusual for frogs to find a home in what appears to be a harsh upland environment.

5 At the summit of Meall Ghaordaidh, you'll find a trig pillar inside a wind shelter and a view east to Ben Lawers (page 123). Descend via the route of ascent and appreciate the view west to Beinn Challuim (page 65), Beinn Sheasgarnaich and Creag Mhòr (page 109) at the end of the glen.

24 Meall nan Tarmachan

HUB **Killin – 15-minute drive**

OVERALL **1** 2 3 4 5 TERRAIN **1** 2 3 4 5
DISTANCE **1** 2 3 4 5 NAVIGATION **1** 2 3 4 5

DISTANCE **6.9km** ASCENT **650m**
TIMES **4:00hrs** (Walker) **3:30hrs** (Trekker)
2:35hrs (Fastpacker) **1:40hrs** (Trail runner)

START LOCATION **Car park south of Lochan na Làirige**
START LATITUDE/LONGITUDE **56.5155, -4.2708**
START GRID REFERENCE **NN603382**
START W3W **///cuff.remodel.remission**

	PRONUNCIATION	NAME MEANING
Meall nan Tarmachan	**'meowl nan tarmakan'**	**hill of the ptarmigan**

Introduction

Meall nan Tarmachan is wonderfully versatile. Fancy a simple half-day outing on good paths? No problem – as shown here, from the high car park near Lochan na Làirige, it's a straightforward out-and-back bimble to the summit of Meall nan Tarmachan, a beginner route which Mum, Dad, Grandma, Grandpa, the kids and your four-legged friends can all enjoy together. Or would you prefer a more intrepid, adrenaline-inducing option? Meall nan Tarmachan has got you covered. The gnarly, rocky ridge extending south-west from the summit is an exhilarating scramble up to grade 1.

Public Transport

Not recommended, as it entails walking along the busy A827 where there is very little verge.

Parking

Free parking for around 12 cars in a car park at the foot of Meall nan Tarmachan, south of the Lawers Dam (the dam at the end of Lochan na Làirige) via a rough track off the main mountain pass. The nearest public toilets are in Killin. If the parking area is full, there are some more spaces nearer the road. The mountain pass between Killin and the start point can be impassable following heavy snow or during icy conditions in winter.

Escape Route

None.

Refreshments

The Boathouse at Loch Tay welcomes non-residents and residents alike. Enjoy a drink or a bite over-looking the loch.

Meall Corranaich, Ben Lawers and Beinn Ghlas over Lochan na Làirige from Meall nan Tarmachan.

Meall Garbh and Beinn nan Eachan from the summit of Meall nan Tarmachan.

The Route

S Leave the car park to the south-west and walk along the track. Pass through a gate and remain on the track for another 300m to a crossroads; turn right here on to a footpath. Cross a small stream and continue on the well-maintained path through moorland.

2 Pass through the frame of an old gate and continue uphill. The stone-pitched path greatly assists your progress. The edges of the path are adorned with willow scrub, alpine lady's mantle, wild thyme and other floristically diverse plant life. John Muir, the Scots-born founder of the modern conservation movement, once described sheep as 'hoofed locusts' for their destructive effect on natural landscapes; conservation projects like the Tarmachan Habitat Restoration and Improvement Project have made great strides in reversing these effects in the Ben Lawers National Nature Reserve. The Loch Earn Munros of Ben Vorlich and Stùc a' Chroin (page 25) can be seen in the distance to the south.

3 At NN590377 the path turns sharply to the right and runs parallel to a fence. Walk northwards while admiring the pointy peak of Meall Garbh on your left. From here, it looks like Meall Garbh must be taller than Meall nan Tarmachan, but the angle is deceiving.

4 The path passes over an outlying south-east top on Meall nan Tarmachan where the view to Ben Lawers (page 123) over Lochan na Làirige is spectacular. From here, the path descends slightly to the bottom of a small dip. Cross a ladder stile and walk north-west to ascend the steeper stone-pitched path to gain the higher reaches of the Munro. At the very

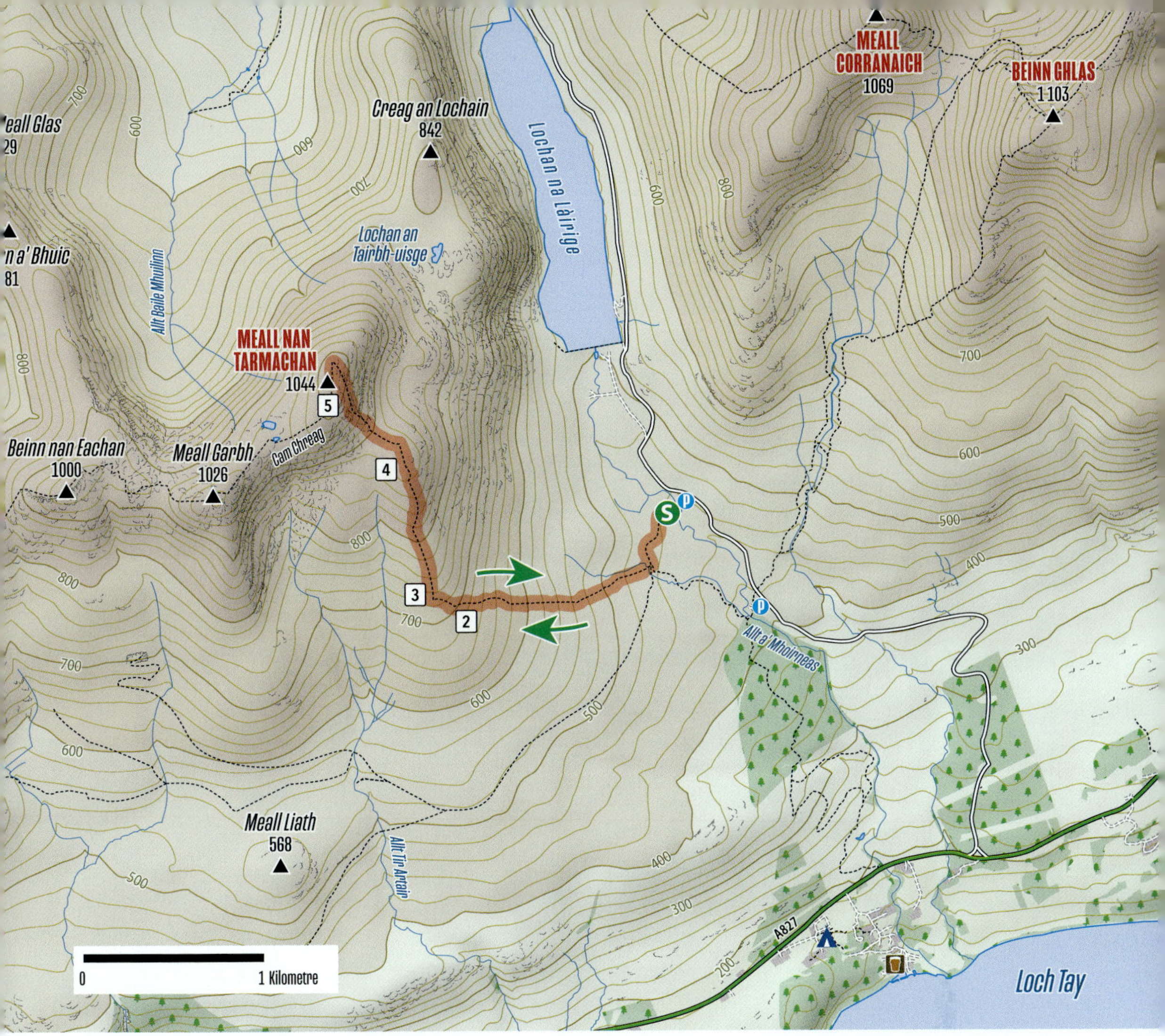

top, a sharp left-hand turn is required before the final short walk to the summit, which is marked by a cairn.

5 The rugged Tarmachan ridge extends south-west of the summit. Confident Munro-baggers who are comfortable with scrambling up to grade 1 with a touch of exposure may wish to continue along the ridge and finish with a rough, wet descent back to the bottom track. Many will prefer to delight in a leisurely half day of walking on good paths; if so, simply retrace your steps back to the start.

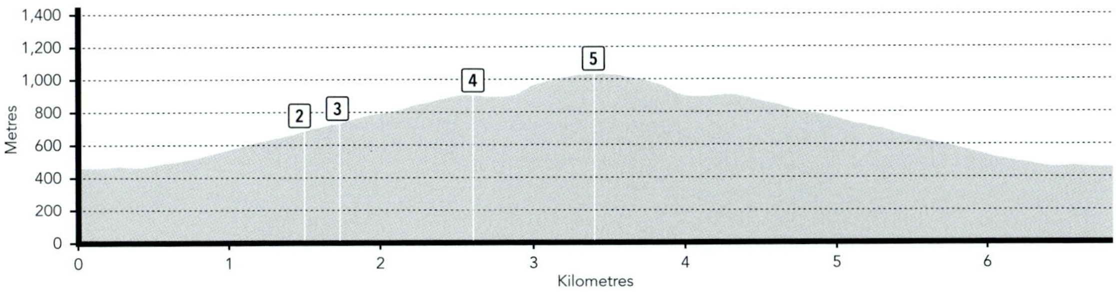

25 Ben Lawers

HUB **Killin – 20-minute drive**

OVERALL 1 2 **3** 4 5 TERRAIN 1 2 **3** 4 5
DISTANCE 1 2 **3** 4 5 NAVIGATION 1 2 **3** 4 5

DISTANCE **15.5km** ASCENT **1,420m**
TIMES **8:50hrs** (Walker) **7:35hrs** (Trekker)
5:45hrs (Fastpacker) **3:40hrs** (Trail runner)

START LOCATION **Small car park north of Lochan na Làirige**
START LATITUDE/LONGITUDE **56.5451, -4.2890**
START GRID REFERENCE **NN593415**
START W3W **///unpainted.dime.magnetic**

	PRONUNCIATION	NAME MEANING
Meall a' Choire Lèith	**'meowl a corriya lay'**	**hill of the grey corrie**
Meall Corranaich	**'meowl korraneech'**	**notched hill**
Beinn Ghlas	**'beyn glass'**	**grey-green mountain**
Ben Lawers	**'ben lorrs'**	**hill of the loud stream**

Introduction

The zenith of an expansive seven-Munro massif to the north of Loch Tay, Ben Lawers is the tenth highest Munro of all and the loftiest in the Southern Highlands. It's a favourite of impatient Munro box-tickers. Here it's possible to bag up to seven summits in a single linear route, with a car parked at each end, but it's a big day and the vehicle logistics are a faff.

A better option, perhaps, is this circular route, ticking off four of the Ben Lawers peaks – including the mighty 1,214-metre overlord – starting and finishing with a welcome leg-up from the car park, over 500 metres above sea level.

Public Transport

There are no public transport services close to this walk.

Parking

Free parking for six cars in the small car park north of Lochan na Làirige. The nearest public toilets can be found in Killin. The mountain pass between Killin and the start point can be impassable following heavy snow or during icy conditions in winter.

Escape Route

Consider picking up the descent instructions after Meall Corranaich and leaving Beinn Ghlas and Ben Lawers for a separate outing.

Refreshments

The Boathouse at Loch Tay welcomes non-residents and residents alike. Enjoy a drink or a bite over-looking the loch.

Beinn Ghlas from Ben Lawers.

The Route

S Leave the car park using a path through the heather to the north past a conical cairn. After 300m the path joins a more substantial one heading north-east at first which soon gives way to peat hags. At a fork in the trail at NN596418, choose the left-hand trail to begin your approach to Meall a' Choire Lèith, noting that the option straight ahead is your route of descent. The faint path leads north-east over flat, boggy ground before dropping gently downhill. The trail becomes indistinct over wet ground and crosses the Allt Gleann Dà-Eig at NN601424 just south of the intake.

2 From here, follow the trail north-east uphill. The higher you get, the more charming the view down the Gleann Dà-Eig becomes. Cross the stream at NN607432 and continue north-east up the steeper bilberry-lined trail up Meall a' Choire Lèith. On the upper reaches glance left towards the Munros of Stùc an Lochain and Meall Buidhe (page 133) which are made easier to identify by the Giorra Dam (at the end of Loch an Daimh) which splits them.

3 From the summit of Meall a' Choire Lèith look south-west to admire Meall nan Tarmachan (page 119) whose ridgelines look striking from this angle. Head south-east from the summit to walk along the edge above Coire Liath down to the bealach between Meall a' Choire Lèith and Meall Corranaich which is marked 775m on OS maps. Swift progress can be made on this gradual descent. The ascent is straightforward on the other side of the bealach too.

4 The path meets the same stream you crossed earlier as you approach Meall Corranaich along the western side of the spine of Meall Corranaich's northern ridge at first. At the top of the stream join the crest of the ridge and pass over point 1,018m. Parallel to this ridge, Ben Lawers and Beinn Ghlas can now be appreciated on the other side of Glen a' Chobhair.

5 At Meall Corranaich's summit cairn note the path to your right as this is the descent route for later. But for now, leave the summit following the path to the south-east. Beinn Ghlas is so close you can see any people who happen to be on the summit, but it first requires a steep, wiggly drop of around 200m to the bealach on a largely stone-pitched path.

Reascending Meall Corranaich with Ben Lawers behind.

6 This path meets the popular ascent trail for Ben Lawers but crosses this to continue straight ahead on the ascent of the grassy slopes of Beinn Ghlas. This path starts leading to the south-east but soon curves north-east and then south-east once more to ascend beside the edge of the small coire to the north of Beinn Ghlas's summit. From the top there's an alluring view to Ben Lawers, An Stùc and Meall Garbh (page 127) which immediately draws the eye, but the view down to Loch Tay is also superb.

7 Descend to the north-east, following a clear path down the gradual slope to the bealach at the foot of Ben Lawers. From the bealach walk up a stone-pitched path to the summit of Ben Lawers, which is marked by a trig pillar and what remains of an old view indicator. As this is the highest point in the historic county of Perthshire the views are expansive, and you certainly get the sensation of being on top of the world.

8 Retrace your steps from the summit of Ben Lawers to the bealach at the foot of Beinn Ghlas and follow the main path south-west for 1.2km to avoid reascending Beinn Ghlas. The same cannot be said for Meall Corranaich though which blocks the way between you and the car park. From the crossroads (**6**), reclimb Meall Corranaich and just before the

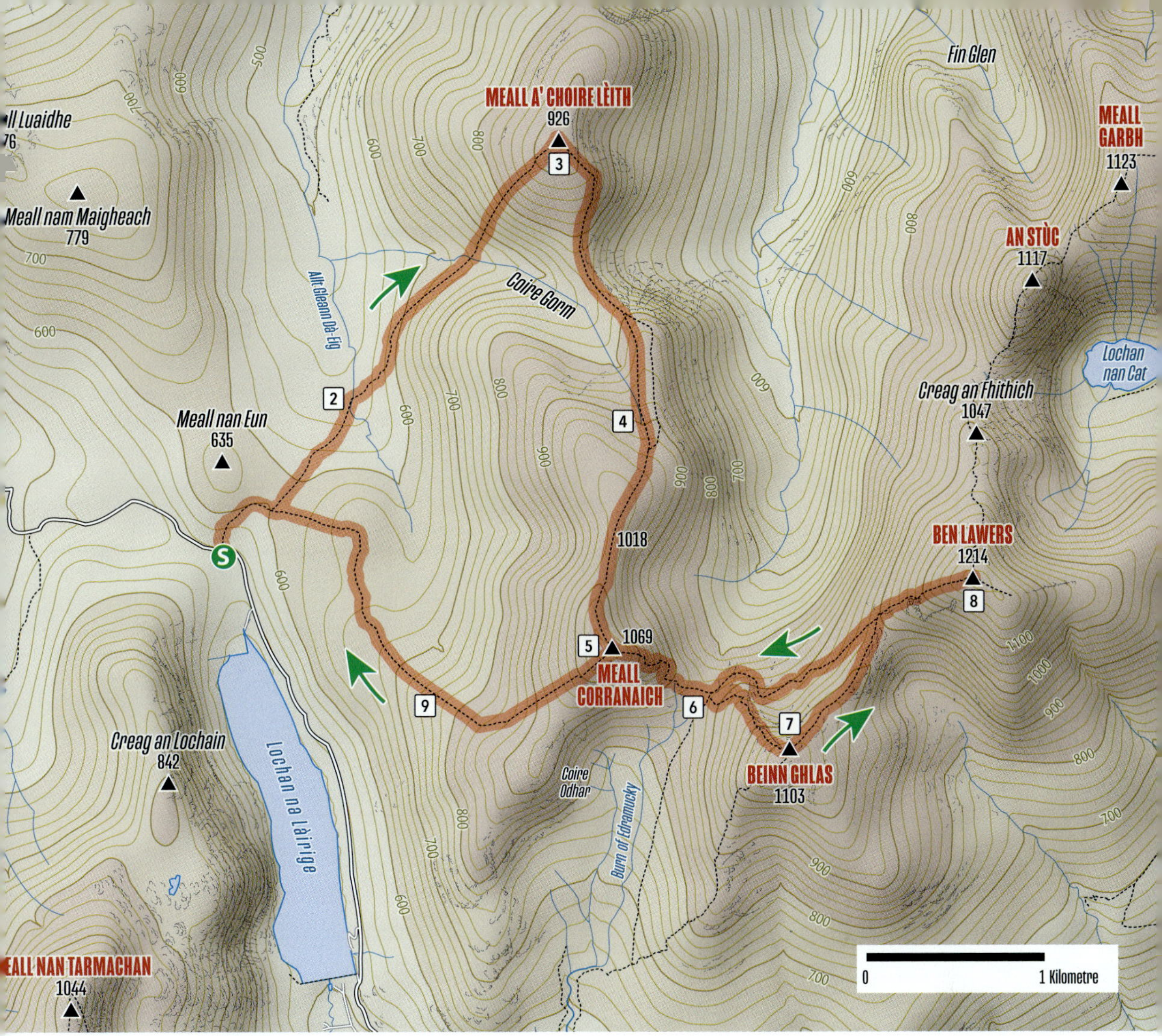

summit, turn left to walk alongside some old iron fence posts to the south-west. The clear path descends the shoulder but loops around to the right down towards the peat-filled bowl above the road.

9 When the path becomes wet and indistinct, use the trusty iron fence posts as a guide and cross the peat hags which make a return as the ground levels out. Meet up with the point you peeled off towards Meall a' Choire Lèith earlier in the day and retrace your steps to the car from there.

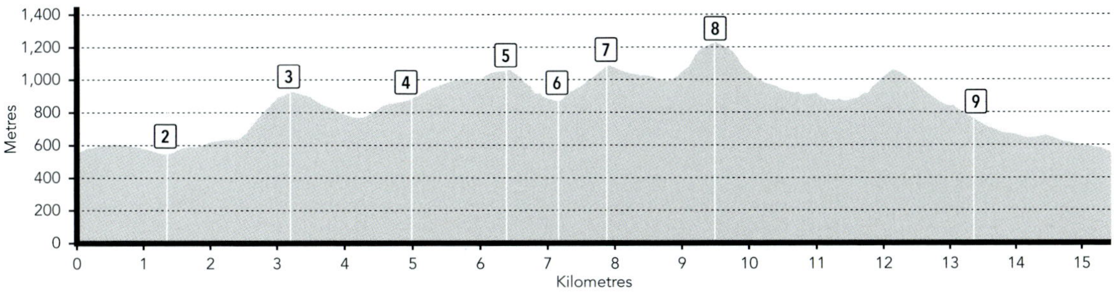

MEALL GREIGH 1,001m
MEALL GARBH 1,123m
AN STÙC 1,117m

26 North-East Ben Lawers Range

HUB **Killin – 15-minute drive**

OVERALL 1 2 3 **(4)** 5 TERRAIN 1 2 3 **(4)** 5
DISTANCE 1 2 **(3)** 4 5 NAVIGATION 1 2 3 **(4)** 5

DISTANCE **17.3km** ASCENT **1,270m**
TIMES **8:35hrs** (Walker) **7:25hrs** (Trekker)
5:35hrs (Fastpacker) **3:40hrs** (Trail runner)

START LOCATION **Walkers' car park, close to Ben Lawers Hotel**
START LATITUDE/LONGITUDE **56.5295, -4.1519**
START GRID REFERENCE **NN677395**
START W3W **///result.apes.scanning**

	PRONUNCIATION	NAME MEANING
Meall Greigh	**'meowl grey'**	**hill of the horse studs**
Meall Garbh	**'meowl garav'**	**rough hill**
An Stùc	**'an stook'**	**the peak**

Introduction
Fit Munro hikers often bag this trio as part of the so-called Ben Lawers Seven – a classic route for efficient peak-bagging – but they work just as well as a standalone walk. Forming the eastern stretch of the Ben Lawers massif, a hike from Loch Tay of Meall Greigh, Meall Garbh and An Stùc completes a delightful, anticlockwise circumnavigation of the Lochan nan Cat corrie. The crux of the route is An Stùc, a no-nonsense peak with steep terrain and rocky obstacles aplenty. It's not technically difficult scrambling, but it shouldn't be underestimated either. Give An Stùc your full focus, however, and it's an exciting element of this excellent walk, best enjoyed on a clear, sunny day.

Public Transport
A school bus stops outside the Ben Lawers Hotel twice a day on schooldays only. There are no other nearby public transport options.

Parking
In the summer months, follow the *Hill Walkers' Car Parking* sign beside the Ben Lawers Hotel – this leads to chargeable parking for around 10 cars in a makeshift car park off the A827. It may also be possible to park in the hotel car park for a fee. Public toilets can be found in Killin.

Escape Route
There are no pleasant alternative descent options beyond Meall Greigh.

Refreshments
The Ben Lawers Hotel is perfectly placed for you to grab something to eat or drink after your walk.

An Stùc from Meall Garbh.

Above View back towards Meall Garbh from An Stùc. **Opposite** Descending An Stùc towards Lochan nan Cat.

The Route

S Leave the car park and walk north up the A827 for 500m until you pass a small cluster of buildings, including the Horn Carver – a shop which sells hand-carved deer antler and cow horn. Take the track on the left after you pass the shop. This track leads behind the properties and ascends gradually north-north-west beside a stream. The track ends at Machuim where a series of signs guide walkers away from the farm buildings, through a gate and north-west through woodland above the Lawers Burn. The slim path can become overgrown in the summer months. The trail leads over a selection of small bridges and crosses two ladder stiles before exiting on to a grassy path on the open hillside. Meall Greigh and Sròn Mhòr are directly in front of you.

2 The path passes a series of old shielings for just over 500m before reaching a junction at NN675416. Fork right on to a grassy trail, noting that the other path is the descent route. At NN673420 the path meets another; fork right here and continue uphill heading north-north-east towards the top of Sròn Mhòr. Behind you the distant Munros of Ben Vorlich and Stùc a' Chroin (page 25) are visible across Loch Tay. Through the trained zoom of a camera, you can clearly see the well-walked path up the north ridge of Ben Vorlich. Meall Garbh and An Stùc are visible on your left as you ascend Sròn Mhòr, alongside Ben Lawers (page 123), the highest peak of the three.

3 The path bypasses left of the summit of Sròn Mhòr and continues northwards up the final 200m of vertical ascent to the top of Meall Greigh, without any technicality.

4 Leave the main 1,001m summit of Meall Greigh to the north-west dropping into a small dip before rising once more to pass over the north-west summit of Meall Greigh at 993m. From here, the wide triangular profile of Schiehallion (page 277) can be spotted to the north-east. Walk north-west and then west following the clear path which passes to the left of the first of a series of small humps and then alongside a fence on your right.

5 Reach the soggy saddle at the foot of Meall Garbh and continue uphill, handrailing the fence as it snakes up the hillside. Pause halfway to admire the view to the Munros of the Glen Lyon horseshoe (page 137) on your right. When the fence takes a sharp downhill right-hand turn continue south-south-west over uncomplicated terrain to the summit of Meall Garbh which is marked by a cairn. Meall Garbh proffers a positively frightening view of An Stùc which looks impenetrable from here, but don't be deterred, there is a way up the steep crags.

6 First, continue along the remainder of the airy summit ridge before dropping south-west to the bealach over mixed terrain which is rocky in some places and gravelly in others with one awkward step towards the bottom.

Ben Lawers and Beinn Ghlas from An Stùc.

7 From the 991m bealach, begin the short but steep ascent of An Stùc. The path starts on the left-hand side of the craggy face and zigzags up the mountain encountering loose ground and rocky obstacles along the way until you top out. Take care, take your time and the short climb is over before you know it. That said, it shouldn't be underestimated as many a Munro-bagger has reported having a 'wobble' here. An alternative approach from the bealach would be to walk below the crags and approach the summit from the north without the need for any scrambling.
8 The summit high point is the mound you arrive at just before the cairn. Leave the summit to the south-west down a grassy slope which becomes rockier as it bends southwards towards Bealach Dubh. Ben Lawers (page 123) looks equal parts imposing and inviting up ahead, but we'll leave that for another day.
9 At Bealach Dubh take the left-hand fork to follow a stream on your right-hand side. The path zigzags around craggy ground before it drops sharply for an awkward descent down a heavily eroded line which can be treacherous in the wet. Descend cautiously and arrive at the bottom at Lochan nan Cat.
10 Walk on the south side of both Lochan nan Cat and Lochan nan Uan following a faint path over very boggy ground. After passing both lochans, continue eastwards alongside a stream to a dam where an access track begins. Follow the track south-east for just short of 300m when you reach a path junction marked by a small cairn.
11 Fork left at the path junction on to a grassy path. Relish the drier ground as the path traverses high above the Lawers Burn before descending bracken-enclosed switchbacks to a bridge across the stream. Cross the bridge and follow the path into the valley, retracing your steps to the start.

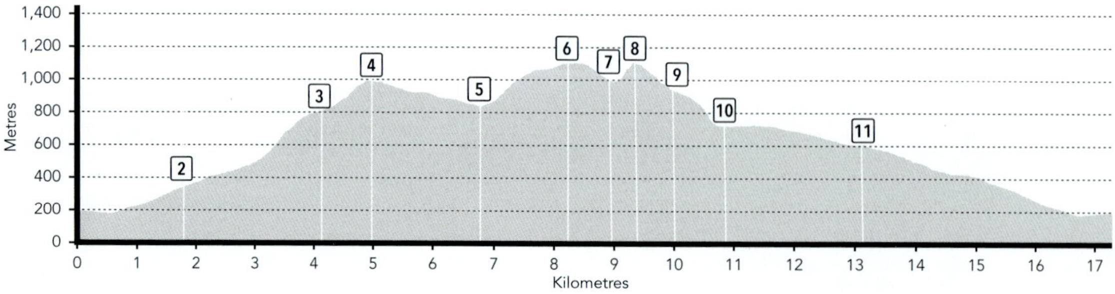

Creag Roro
Bealach Creag a' Bhannaich
Fin Glen
5
MEALL GREIGH
4
1001
6
MEALL GARBH
1123
7
AN STÙC
1117
8
Cat Gully
Creag a' Bhuic
Lochan nan Cat
Lochan nan Uan
3
Sròn Mhòr
805
Ravens Gully
9
10
11
Lawers Burn
1047
Creag an Fhithich
BEN LAWERS
1214
East Mealour
504
2
Dubh Eas
Lawers Burn
Machuim
Meall Odhar
568
Lawers
A827
Allt a' Choire Chireachain
Allt an Tuim Bhric
Loch Tay
0
1 Kilometre

27 Loch an Daimh Munros

HUB **Killin – 45-minute drive**

OVERALL 1 2 (3) 4 5 TERRAIN 1 (2) 3 4 5
DISTANCE 1 2 (3) 4 5 NAVIGATION 1 2 (3) 4 5

DISTANCE **17.4km** ASCENT **1,270m**
TIMES **8:35hrs** (Walker) **7:25hrs** (Trekker)
5:35hrs (Fastpacker) **3:40hrs** (Trail runner)

START LOCATION **Walkers' car park, east of Giorra Dam, Loch an Daimh**
START LATITUDE/LONGITUDE **56.5857, -4.4244**
START GRID REFERENCE **NN511463**
START W3W **///economies.dummy.blunders**

	PRONUNCIATION	NAME MEANING
Stùc an Lochain	**'stook an lochan'**	**peak of the little loch**
Meall Buidhe	**'meowl booyer'**	**yellow hill**

Introduction

'The one where you return to the start in the middle.' It sounds illogical, but it makes sense. These Munros are walked as two separate out-and-backs from the car park next to Giorra Dam, which holds back water from the Loch an Daimh reservoir. It's the epitome of a walk of two halves. Hike one Munro in the morning; hike the second in the afternoon; and return to your car in between. It's a bit like the Munro version of Groundhog Day, and it feels a little functional. But it's a novelty diving into your car to shelter from the wind and rain, turning the radio on and enjoying your cheese and pickle sandwiches in blissful safety.

Public Transport

There are no public transport services close to this walk.

Parking

Free parking for around 10 cars in a small walkers' car park, east of Giorra Dam and Loch an Daimh. The nearest public toilets are in Killin. A comprehensive visitor information board regarding the Meggernie and Lochs estates can be found beside the car park with detail around stalking, lambing and wildlife. The road to the dam involves passing through an unlocked estate gate which must be closed again after use. The mountain pass between Killin and the start point can be impassable following heavy snow or during icy conditions in winter.

Escape Route

None.

Refreshments

Rest your weary legs on one of the benches at the Glenlyon Tea Room overlooking the River Lyon, a stunning setting.

Walker approaching the summit of Meall Buidhe along the broad grassy ridge.

The Route

S Walk west along the road from the car park and cross the bridge which is parallel to the dam. The track leads up through woodland before exiting at the easternmost edge of Loch an Daimh reservoir atop the impressive Giorra Dam, 463 metres in length and 35 metres tall. The 'massive buttress' dam acts as storage for the Cashlie Power Station and forms part of the Breadalbane Hydroelectric Power Scheme which was constructed in the 1950s.

2 Follow the track south for 150m and take the left-hand fork, following an eroded trail up the hillside which leads west for 500m and then curves southwards. The trail is stony in parts, and then boggy as it coexists with a stream for a short distance.

3 Reach the ridge at NN501453; from here a set of iron fence posts guides the way up the mountainside to the west. Follow the eroded trail until the western end of Loch an Daimh comes into view. Here, the trail veers left, as do the fence posts, and the highest point on Stùc an Lochain comes into view on the other side of the corrie.

4 The trail traverses the southern rim of the bowl as it undulates mildly towards the final part of the ascent to the summit. The terrain steepens but remains non-technical. From Stùc an Lochain's summit, which is frequently spelt Stuchd an Lochain, enjoy the view down to the south-west to Loch Lyon and beyond to Beinn Sheasgarnaich (page 109). To the west are the Bridge of Orchy Munros (page 87) and to the north is your next objective, Meall Buidhe.

5 The easiest way there is to retrace your steps back to the start point, re-group and start the ascent of Meall Buidhe when you're ready. Despite being back where you started, the ascent of a second Munro is more appealing due to the fact that you're already above 400m at the car park. Most Munro-baggers will seek to walk these Munros on the same day, but if you're feeling tired, simply return for Meall Buidhe another day. From the car park, pass around the barrier and follow the gravel track north to begin the ascent of Meall Buidhe.

Reflections in Loch an Daimh from the Giorra Dam wall on the approach to Stùc an Lochain.

6 After 300m at the track junction, turn left and walk north-west along the track with a view of Giorra Dam and Loch an Daimh down to your left. A cairn on the track at NN507467 indicates that it's time to turn right. Follow the eroded trail northwards up the mountainside. The ground has a tendency to feel much softer and wetter than on Stùc an Lochain.

7 After nearly 1km of steady uphill progress, at NN506476, a faint trail leads off to the left to gain a spur. A clear trail leads northwards up the spur until a series of peat hags are reached over flatter ground. As the route crosses the peat hags, any sign of a path dwindles away but if you head north-north-west to NN504487, you'll pick the trail back up at a cairn which sits just below the saddle between Meall Buidhe's south top and Meall a' Phuill.

8 From here, head uphill on to the long summit ridge passing the south top of Meall Buidhe. A pleasant walk along the broad grassy ridge for just over 1km brings you to Meall Buidhe's highest point, with expansive views across Rannoch Moor to Glen Coe. Loch Ericht can also be seen to the north, with a distant thickset Ben Alder (page 259).

9 Retrace your steps back to the start point.

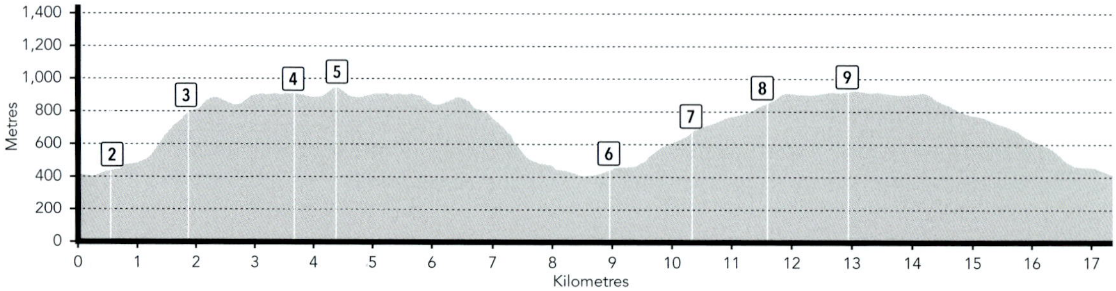

Garbh Mheall
912
MEALL BUIDHE
932
9
Glas Choire
Meall a' Phuill
878
917
8
824
Càrn Làirig Meachdainn
Cam Chreag
862
Cùl Lochan
7
Creagan nan Gobhar
629
Allt na Gealainn
Loch an Daimh
6
Druim nan Calum
S
P
2
Creag an Fheadain
888
3
Coire Bàn
STÙC AN LOCHAIN
960
Lochan nan Cat
5
Coire an Duich
4
Sròn Chona Choirein
927
Allt Ruighe Ghiuthas
Allt Conait
Ross
Wester Moar
0
1 Kilometre
Meall Dubh
770
Creag a' Chullaich
562
Glen Lyon
Lyon
An Grianan
864

CÀRN GORM 1,029m
MEALL GARBH 968m
CÀRN MAIRG 1,041m
MEALL NA AIGHEAN 981m

28 Glen Lyon Munros

HUB **Killin – 35-minute drive**

OVERALL 1 2 **(3)** 4 5 TERRAIN 1 2 **(3)** 4 5
DISTANCE 1 2 3 **(4)** 5 NAVIGATION 1 **(2)** 3 4 5

DISTANCE **20.3km** ASCENT **1,385m**
TIMES **9:40hrs** (Walker) **8:20hrs** (Trekker)
6:15hrs (Fastpacker) **4:05hrs** (Trail runner)

START LOCATION **Large lay-by off the Glen Lyon Road, 1.1km west of Invervar**
START LATITUDE/LONGITUDE **56.6031, -4.1915**
START GRID REFERENCE **NN655478**
START W3W **///lifestyle.mass.sprinkle**

	PRONUNCIATION	NAME MEANING
Càrn Gorm	**'carn gorrom'**	**blue hill**
Meall Garbh	**'meowl garav'**	**rough hill**
Càrn Mairg	**'carn merrig'**	**rusty hill**
Meall na Aighean	**'meowl ne eye en'**	**hill of the heifers**

Introduction

There is something utterly satisfying about a 'proper' high-level, summit-bagging horseshoe route. The looping circularity of it all has an appealing logic and neatness, and when you climb high, you stay high, thus maximising your time on the sun-drenched ridges and airy bealachs – exactly where you want to be. This is precisely what you get with the Glen Lyon Munros, a beautifully undulating, circular round bagging four Munros and serving up delightful views across Loch Rannoch and the Ben Lawers massif. Located in a Site of Special Scientific Interest, Càrn Gorm and Meall Garbh are known for their unique montane flora and fauna, while Glen Lyon is commonly praised as 'Scotland's longest, loneliest and loveliest glen'. It's worth nothing that Meall na Aighean is called Creag Mhòr on some older maps.

Public Transport

There are no public transport services close to this walk.

Parking

Free parking for around 18 cars in the large lay-by off the Glen Lyon Road, 1.1 kilometres west of Invervar. There used to be a walkers' car park in Invervar itself, but parking here is no longer permitted. The nearest public toilets can be found in Kenmore.

Escape Route

From Meall a' Bhàrr, descend part way down the south ridge before branching left to reach a hut beside the Allt Coire a' Chearcaill. Cross the stream and follow a good path into the valley from here.

Refreshments

Pop into the Ben Lawers Hotel for a drink or a meal on your way back to Killin.

Descending Meall na Aighean.

The Route

S Walk eastwards for 1.1km through Glen Lyon on the minor road towards Invervar. Turn left to pass through a gate at a fingerpost which reads *Lint Mill 150 yards*. Follow the footpath north alongside the Invervar Burn on your left. Pass what remains of a late-18th-century lint mill, a quirky, cylindrical building with a conical roof, before the path sweeps right to cross a deer fence using either the stile or the gate. Continue northwards, passing through another gate and walk through a deforested area uphill on a gravel path through the deadwood to the sound of cascading water from the Invervar Burn.

2 Pass through a gate to join a grassy track – the leftmost of two tracks – and continue north beside a deer fence. Pass through a wider gate across the track before journeying down and crossing a disconcertingly bouncy footbridge following a fingerpost which reads *Route to Càrn Gorm*. Walk north-west on the path which now runs along the left-hand side of the Invervar Burn, with a fenced plantation on your left.

3 Around 200m after the bridge the path zigzags sharply to the plantation corner before stretching south-west across the open hillside, away from the Invervar Burn, to ascend Càrn Gorm. The route follows a long-eroded line up the Munro, with grasses and blaeberry bushes lining the heathland either side of the path. The route here is soft and boggy in parts as it gradually veers west and then north-west.

4 At a fork in the path at around 720m in elevation, continue north-west along the eroded trail, slightly to the right of the crest of the shoulder. Towards the top, the route gets steeper as higher ground is gained but, as Munros go, it's a surprisingly gentle ascent to the high point of 1,029m. The summit is marked by a small cairn on a rocky outcrop beside the path. The prize for the most enticing view goes to the Ben Lawers Munros (pages 123 and 127) with the crags of An Stùc looking particularly formidable from here.

5 Continue along the summit ridge, passing a fallen trig pillar before descending north along the eroded trail to the bealach at 851m. From here, you might wish to detour to An Sgòrr (a Munro Top), or (as is shown on the map and in the GPX file) continue north-east on the main path to the saddle between An Sgòrr and Meall Garbh. From here, ascend to the summit of Meall Garbh over increasingly rocky terrain with a few old iron fence posts to assist navigation in poor weather. Here, you'll find a cairn comprised largely of rusty iron fence posts.

6 Descend eastwards over loose rocky terrain to a bealach at 864m before the trail undulates alongside old fence posts towards another Munro Top – Meall a' Bhàrr. When you reach the most northerly point on the summit ridge of Meall a' Bhàrr at NN666518, walk south-south-east along to the highest point at 1,004m.

7 From the summit of Meall a' Bhàrr, walk east along comfortable grassy terrain where you can make speedy progress to a cairn at 999m before dropping to a bealach and starting your ascent of Càrn Mairg. Skirt to the right under the 1,001m north-west top and, as you reach the summit plateau, make a sharp left turn past huge quartz boulders to reach the large summit cairn which marks the highest point of the Glen Lyon horseshoe. This spot offers arguably the best vantage point north-east to the 'Matterhorn of Perthshire' – Schiehallion (page 277).

8 My preferred line of descent here is down to the south-east initially for around 75 metres and then forking left. An eroded trail leads to a small boulder field, avoiding the crags. This is the trickiest patch of terrain on the walk, although it's short-lived. At the base of the boulder field, cross the low wall and pick up a grassy trail heading south-east and then south. Those collecting Munro Tops, of which there are an abundance along this horseshoe, will wish to pop up Meall Liath before continuing.

9 Descend over grassy terrain to the boggy bealach beneath Meall na Aighean, your fourth and final Munro of the day. Continue southwards to ascend the other side and, when you're above 900m elevation, the path veers to the left to approach the outcrop of slanted rock atop Meall na Aighean from west to east. Meall na Aighean used to carry the name Creag Mhòr on old maps and you'll sometimes still come across it being referred to by that name.

10 Leave the summit initially by the route of ascent until the trail splits after 150 metres at NN693496. Take the left fork which leads over the south-west top of Meall na Aighean. From here, head north-west

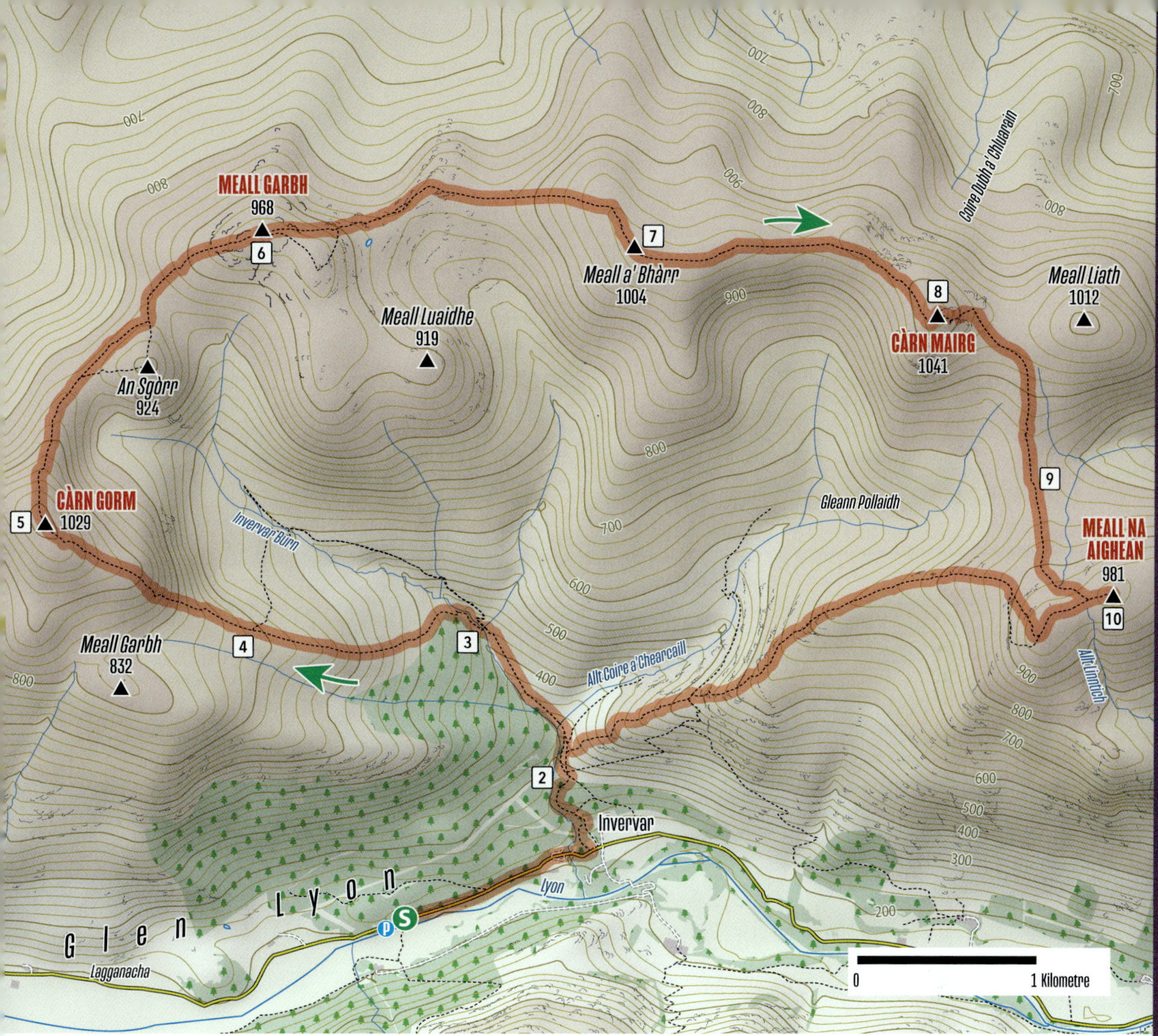

over soft ground to pick up the descent path down the western ridge of Meall na Aighean which runs parallel to the Allt Coire a' Chearcaill in the valley to your right. Enjoy the fine view down Glen Lyon as the eroded path steepens and winds through heather. Pass through a gate at the bottom of the hill and walk over grassy terrain to the track in the valley bottom. From here, retrace your steps to the start.

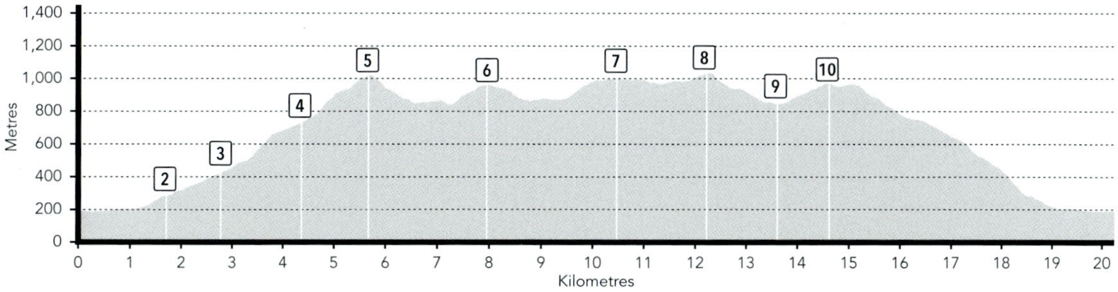

HIGHLAND HUB 6
GLENCOE

The sixth of the Highland hubs is Glencoe, a village at the mouth of the famous glen which was formed by a glacier during the last ice age.

Glencoe has a small supermarket which is handy for resupplying and limited places to eat out, but the Glencoe Gathering serves food and drink which will be wolfed down after a day in the hills. The Red Squirrel Campsite is perfectly placed for the Munros in this collection but can get busy. I've known it be noisy late at night in the summer months, so if that's not your scene and you do want to camp, I recommend Blackwater Hostel and Campsite in Kinlochleven for the Munros in the northern half of the hub, and Glencoe Mountain Resort for the southern walks. These are quieter sites, but you may have further to travel to reach certain Munros or to access facilities.

Glen Coe itself is something to be marvelled at. I would argue that the A-road which cuts through the glen is one of Scotand's most impressive routes to drive, especially early in the morning before the coaches arrive. The mountains with their fierce beauty are simply towering over you from both sides and after winding through Rannoch Moor, it's a real contrast in scenery.

Direct bus services link Glasgow and Glencoe in under three hours. It's also possible to travel by train as far as Bridge of Orchy or Fort William and catch a local bus service to Glencoe from each of those places.

There are some thrilling walks from this hub, many on challenging varied terrain, so while Glencoe might be calling your name, it's worth saving these walks until you've built up some experience, maybe on the softer, more rounded Munros in the Braemar hub or the well-walked routes in Arrochar.

WHERE TO STAY

£	££	£££
Red Squirrel Campsite	Glencoe Youth Hostel	Clachaig Inn

TRAVEL TIMES

29 Ben Starav Three	**45 minutes**
30 South-East Glen Etive Munros	**45 minutes**
31 North-West Glen Etive Munros	**45 minutes**
32 Beinn a' Bheithir	**10 minutes**
33 Bidean nam Bian	**10 minutes**
34 Buachaille Etive Mòr and Buachaille Etive Beag	**15 minutes**
35 North Black Mount Munros	**20 minutes**
36 Sgorr nam Fiannaidh	**1km walk**
37 Meall Dearg	**5 minutes**
38 East Mamores	**15 minutes**

Rest day ideas

- Follow one of the waymarked Glencoe Lochan trails. These are short, low-level walks amid redwood trees with stunning mountain views.
- Catch a game of shinty, Scotland's community sport. Shinty is played with sticks and a ball and is most similar to hockey but there are plenty of variances in play. There is a shinty club at neighbouring Ballachulish.
- Pop into the Glencoe Visitor Centre – watch a short film about the history of the area in their cinema space, check out their exhibitions and learn about their conservation work.

Opposite View down to the Pap of Glencoe and Loch Leven from the approach to Sgùrr Eilde Mòr (route 38).
Overleaf Bidean nam Bian from the ascent of Stob Coir' an Albannaich (route 30).

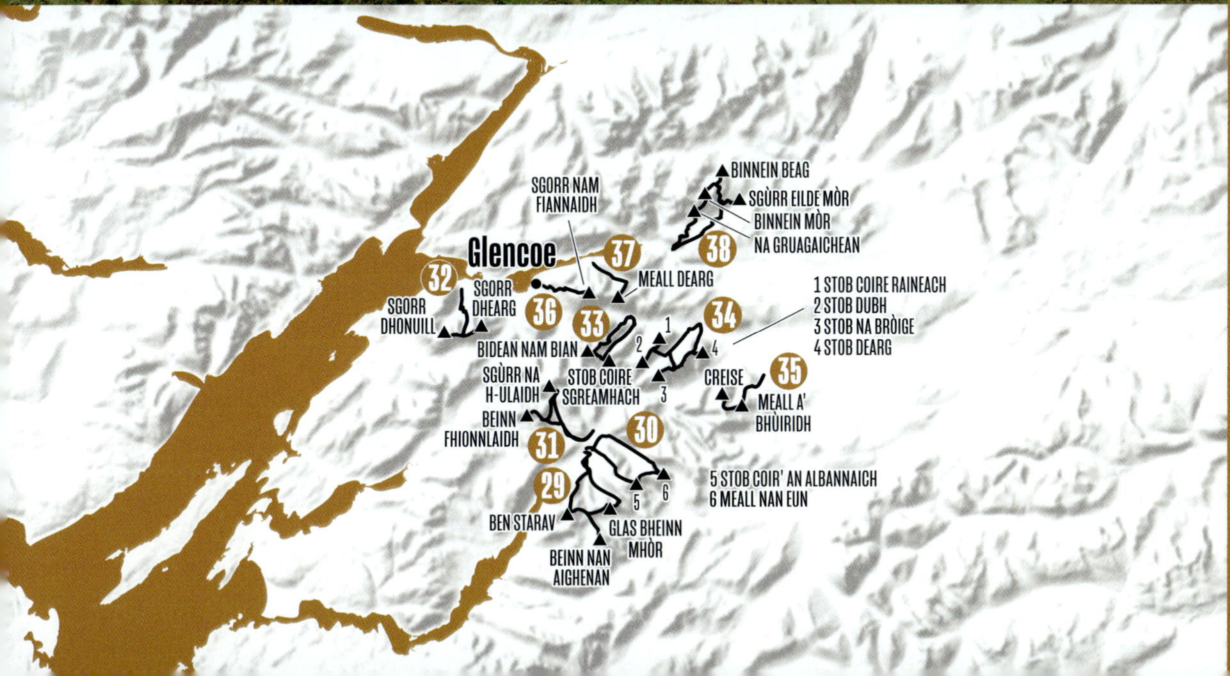

Glencoe
SGORR NAM FIANNAIDH
BINNEIN BEAG
SGÙRR EILDE MÒR
BINNEIN MÒR
NA GRUAGAICHEAN
MEALL DEARG
SGORR DHONUILL
SGORR DHEARG
BIDEAN NAM BIAN
STOB COIRE SGREAMHACH
SGÙRR NA H-ULAIDH
BEINN FHIONNLAIDH
1 STOB COIRE RAINEACH
2 STOB DUBH
3 STOB NA BRÒIGE
4 STOB DEARG
CREISE
MEALL A' BHÙIRIDH
5 STOB COIR' AN ALBANNAICH
6 MEALL NAN EUN
BEN STARAV
GLAS BHEINN MHÒR
BEINN NAN AIGHENAN
29
30
31
32
33
34
35
36
37
38

29 Ben Starav Three

HUB **Glencoe – 45-minute drive**

OVERALL 1 2 3 **4** 5 TERRAIN 1 2 3 **4** 5
DISTANCE 1 2 3 **4** 5 NAVIGATION 1 2 **3** 4 5

DISTANCE **20.6km** ASCENT **1,965m**
TIMES **12:00hrs** (Walker) **10:25hrs** (Trekker)
7:50hrs (Fastpacker) **5:00hrs** (Trail runner)

START LOCATION **Parking area, Glen Etive Road, opposite track to Coileitir**
START LATITUDE/LONGITUDE **56.5765, -5.0351**
START GRID REFERENCE **NN136468**
START W3W **///brilliant.loft.sulk**

	PRONUNCIATION	NAME MEANING
Ben Starav	**'ben starav'**	**the hill of the rustling noise**
Beinn nan Aighenan	**'beyn nan eye-uh-nen'**	**hill of the hinds**
Glas Bheinn Mhòr	**'glass vane vore'**	**big grey hill**

Introduction
Imposingly rugged and brutally built, there is nothing easy about this Glen Etive adventure. It's a challenging route with a bit of everything: a remote approach, secret waterfalls, a lung-busting climb, sweeping ridgelines and a bit of boulder-hopping fun. There's a cheeky out-and-back to bag Beinn nan Aighenan – which could be dropped and walked on another day, if so desired – but the real highlight is Ben Starav. It's a pointy, boulder-strewn summit, a *real* mountain imbued with the tenacity and boldness of Scotland's wild places.

On the ascent of Ben Starav.

Public Transport
There are no public transport services close to this walk.

Parking
There is free parking for around 12 cars in the gravel parking area on the Glen Etive Road, just before the track that leads to Coileitir Cottage. The nearest public toilets are in Glencoe. There are also toilets at the Kingshouse Hotel and Glencoe Mountain Resort if you use the bar/cafe or pay for parking.

Escape Route
Descend north beside the Allt nam Meirleach from Bealachan Lochain Ghaineamhaich (**6**) back to the bottom track. Note: the estate sometimes requests that walkers avoid this escape route during stalking season. Observe the information board at the car park for details.

Refreshments
There are no cafes or pubs in Glen Etive, so it would be rude not to stop by the Boots Bar at the Clachaig Inn on your way back to Glencoe. Regularly making

Above Ascending Glas Bheinn Mhòr with Ben Starav behind. **Opposite** Rock formations on Stob Coire Dheirg with Glas Bheinn Mhòr in the distance.

the shortlist for The Great Outdoors Reader Awards (and winning), the Clachaig Inn is a hillwalking institution serving locally sourced meals and excellent ale.

The Route

S Leave the car park and walk east down the track opposite which is signposted to *Glen Ceitlein Cottage*. Cross the bridge over the River Etive and after 100m take the right-hand fork at a major track junction. Remain on the main track which forks right around Coileitir Cottage and follows the river.

2 Follow the track until you reach a bridge over the Allt Mheuran. Cross the bridge and walk south-east on a new footpath which leads to the foot of the long north-east branch of the north ridge of Ben Starav, passing the lower falls on the Allt Mheuran on the way. Ascend on a stony, dry and eroded path. There is low chance of navigational error on the early part of this ascent due to the increasing popularity of the mountain in recent years; it's simply a case of powering yourself to the top from near-enough sea level along the path heading south-west and then latterly south.

3 The terrain is rockier the higher you climb, and you might want to put a hand down on occasion to assist your ascent. The final 100m of vertical ascent is comprised of bouldery terrain. There is no mistaking the summit of Ben Starav as it's simply the point when you can climb no more and the expanse of Loch Etive stretches out in front of you. There is no plateau to walk across to gain the summit; one minute you're climbing steeply and the next minute you're touching the summit cairn.

4 From the summit, walk along the broad ridge to a south-east subsidiary top and then descend steeply down the ridge to the north-east over loose terrain bound for Stob Coire Dheirg. Most will choose to scramble along the crest of the ridge but there is a bypass path to the right for those who don't fancy the exposure. The routes soon merge and, when they do, ascend boulder scree to the top of Stob Coire Dheirg, a Munro Top.

5 Descend south-east and then east down the eroded path to Bealachan Lochain Ghaineamhaich, enjoying the sight of the rocky pinnacles to your left and passing a huge outcrop of quartzite which runs like a vein through the mountainside. Beinn nan Aighenan is in view throughout the descent and most Munro-baggers won't be able to resist branching off to it as an out-and-back side-quest. But if you do want to leave it for another day you can use the escape route as your route of ascent and descent (see page 145).

View from the car park of the route with Glas Bheinn Mhòr on the left and the long approach ridge for Ben Starav from centre to right.

6 Leave the ridge at NN139423 and take the path to the right which leads around the south-western flanks of Meall nan Trì Tighearnan down to a bealach at the foot of Beinn nan Aighenan. The path continues across the bealach and reveals the best route up the Munro. The terrain is rocky and there are a handful of big steps to take care on. The summit is marked by a cairn and has outstanding views across the Inner Hebrides.

7 Retrace your steps to the bealach at the foot of Beinn nan Aighenan. While it is possible to walk from here directly up to Meall nan Trì Tighearnan dodging the crags, it's a much less arduous approach to simply follow the path back to Bealachan Lochain Ghaineamhaich (**6**) first, then turn right and walk east up a straightforward path.

8 Pass over the summit and descend Meall nan Trì Tighearnan east to the foot of Glas Bheinn Mhòr, then walk north-east on your final ascent of the day. Be sure to look back towards Ben Starav plenty of times on the ascent – I think it's the best view of the day.

9 The summit of Glas Bheinn Mhòr is marked by a cairn on a small plateau. Sit here and enjoy a view across Glenetive Forest to Sgùrr na h-Ulaidh and Beinn Fhionnlaidh (page 155) and pick out the shapely outlines of Buachaille Etive Beag and Buachaille Etive Mòr (page 167). Begin your descent using the long east ridge. Just as you're getting used to the comfortable gradient of the first kilometre, the path turns north, then north-east, and steepens as it drops to the saddle between Glas Bheinn Mhòr and Stob Coir' an Albannaich.

10 Cross the stream in the bealach. Strong walkers may continue uphill from here to the summit of Stob Coir' an Albannaich and merge this route with the one on page 151, making for a tiring but rewarding day. Most will turn left and follow the path north-west into the valley. Keep the Allt Mheuran on your left for the duration of the descent but cross multiple other streams which rush downhill to join the main burn.

11 Towards the bottom of the route, the ground becomes boggier and overgrown in the summer months before you ultimately meet back up with the approach track. Retrace your steps to the start from here.

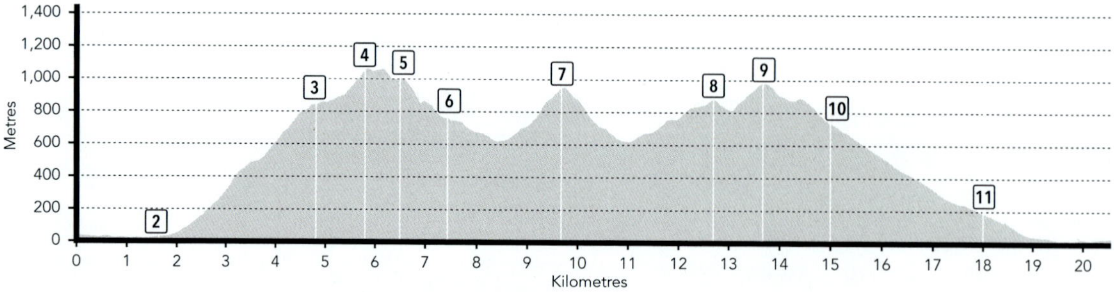

0
1 Kilometre
Glen Etive
Creag Dhubh
303
Loch Druimachoish
Allt Ceitharnan
Etive
Glen Ceitlein
Allt Ceitlein
Coileitir
Kinlochetive
Beinn Chaorach
850
Coire Glas
STOB COIR' AN ALBANNAICH
1044
Allt nam Meirleach
Allt Mheuran
Allt Choire Ghuibh
GLAS BHEINN MHÒR
997
BEN STARAV
1078
Stob Coire Dheirg
1020
Meall nan Tri Tighearnan
892
Bealach an Lochain Ghainmhich
Meall Cruidh
930
Stob an Duine Ruaidh
918
Coire na Caime
BEINN NAN AIGHENAN
960
Coire Hallater
Coire Ghiusachan

30 South-East Glen Etive Munros

HUB **Glencoe – 45-minute drive**

OVERALL 1 2 **(3)** 4 5 TERRAIN 1 2 **(3)** 4 5
DISTANCE 1 2 **(3)** 4 5 NAVIGATION 1 2 3 **(4)** 5

DISTANCE **15.7km** ASCENT **1,355m**
TIMES **8:35hrs** (Walker) **7:25hrs** (Trekker)
5:35hrs (Fastpacker) **3:35hrs** (Trail runner)

START LOCATION **Parking area, Glen Etive Road, opposite track to Coileitir**
START LATITUDE/LONGITUDE **56.5765, -5.0351**
START GRID REFERENCE **NN136468**
START W3W **///brilliant.loft.sulk**

	PRONUNCIATION	NAME MEANING
Stob Coir' an Albannaich	**'stob core an albanaik'**	**peak of the corrie of the Scotsman**
Meall nan Eun	**'meowl nan ee-an'**	**hill of the birds**

Introduction

Wedged between Loch Etive and the Black Mount peaks, to the north-east of Ben Starav, the twin Munros of Stob Coir' an Albannaich and Meall nan Eun deliver another challenging slice of Glen Etive hillwalking adventure. There's lots of slabby rock here, a key geological feature of the area, and the humongous, exposed blocks of granite can be lethal in wet or icy conditions. But, on a good day, the walking is delightful and the descent of Meall nan Eun serves up eye-watering views across the Glen Coe skyline including to Buachaille Etive Mòr and Bidean nam Bian.

Public Transport

There are no public transport services close to this walk.

Parking

There is free parking for around 12 cars in the gravel parking area on the Glen Etive Road, just before the track that leads to Coileitir Cottage. The nearest public toilets are in Glencoe. There are also toilets at the Kingshouse Hotel and Glencoe Mountain Resort if you use the bar/cafe or pay for parking.

Escape Route

No escape routes; retrace your steps after the first Munro and return another day for the second.

Refreshments

There are no cafes or pubs in Glen Etive, so it would be rude not to stop by the Boots Bar at the Clachaig Inn on your way back to Glencoe.

The shimmering slabs of Stob Coir' an Albannaich from the descent of Meall nan Eun.

The Route

S Leave the car park and walk east down the track opposite which is signposted to *Glen Ceitlein Cottage*. Cross the bridge over the River Etive and after 100m take the left-hand fork at a major track junction. After 100m, at NN141467, there is a small gap in the woodland on the right. It's not easy to notice it from the track so keep an eye out. It passes through a gap in the deer fence, providing access to the bracken-filled and birch-dotted hillside. In the height of summer, you may have a fight on your hands against the bracken that grows to over head height. It is interspersed with grassy sections which provide blessed relief. There are some boulders and holes in the terrain so be careful with your footing as you walk south-east to reach the northern bank of the Eag an Uillt.

2 There is a break in the bottom half of the deer fence at NN145464. Pass under this and continue south-east uphill on the grassy open hillside. A view opens out down to Loch Etive to the south-west. Remaining on the northern side of the stream, follow a faint path which zigzags through the grass to gain higher ground, passing around the odd crag. The gradient is relentlessly steep, but the walking is non-technical. Pause to take a breath and enjoy the view behind you across Glen Etive to Beinn Fhionnlaidh and Sgùrr na h-Ulaidh (page 155). In the event that you lose the faint path in the long grass, don't stray too far from the stream and you'll be sure to pick it back up again.

3 From around 500m in elevation, loose gravel takes over as the path ascends a spur beside the stream, before then becoming a grassy pull once again. At NN152459, around 660m in elevation, cross the Eag an Uillt and continue south-east as the gradient begins levelling off towards the summit of Beinn Chaorach which is reached by navigating between spaced-out boulders on grass. Latterly, flat slabs are crossed to reach the top of Beinn Chaorach. Ben Starav (page 145) dominates the view to the south-west; it's also a great vantage point to observe the ridge which leads from Ben Starav to Glas Bheinn Mhòr.

4 Leave the summit over slabby rock to the south, admiring the view to Stob Coir' an Albannaich. Pass to the right of two small lochans and continue south before crossing a small stream and easing south-east and then east, keeping clear of the crags. The approach to Stob Coir' an Albannaich is straightforward over cropped grass. Gain the ridge and veer south-east along to the summit which is perched right on the edge of the crags above Coire Leacach. On a clear day you can see your objective, Meall nan Eun, which you'll get to via Meall Tarsuinn. The Black Mount Munro Stob Ghabhar is also visible beyond this (page 95).

5 Leave the summit to the east and descend the shoulder. As tempting as it may be to exit early to the left, the best line to the bealach below Meall Tarsuinn is around 500m down the ridge (NN174442) and it is marked by a small cairn. The path is well-worn here as you walk north towards the bealach, passing over some slanted slabby ground en route. Towards the very bottom there are a couple of big rocky steps to scramble down.

6 From the bealach, pass to the right of a collection of small pools, before curving to ascend Meall Tarsuinn to the north-east. A faint path indicates the preferred line. Be sure to glance behind you to marvel at the slabs coming off Stob Coir' an Albannaich. It really is a dazzling sight when the sunlight shimmers on the rock, especially after recent rain. From the top of Meall Tarsuinn, walk east-north-east down to the next bealach over another set of slabs before beginning your grassy and occasionally boggy ascent of Meall nan Eun. A cairn marks the summit on a wide grassy plateau.

7 Leave the summit to the north-west with the Glen Coe Munros filling the skyline dramatically, notably the undulating skyline of Buachaille Etive Mòr (page 167) and the immensity of Bidean nam Bian (page 163). The descent is gradual at first and follows a path down the shoulder before reaching a set of crags. Pick your way carefully down here. Once you reach around 550m in elevation the steep crags are behind you and a wet, grassy descent north-west into

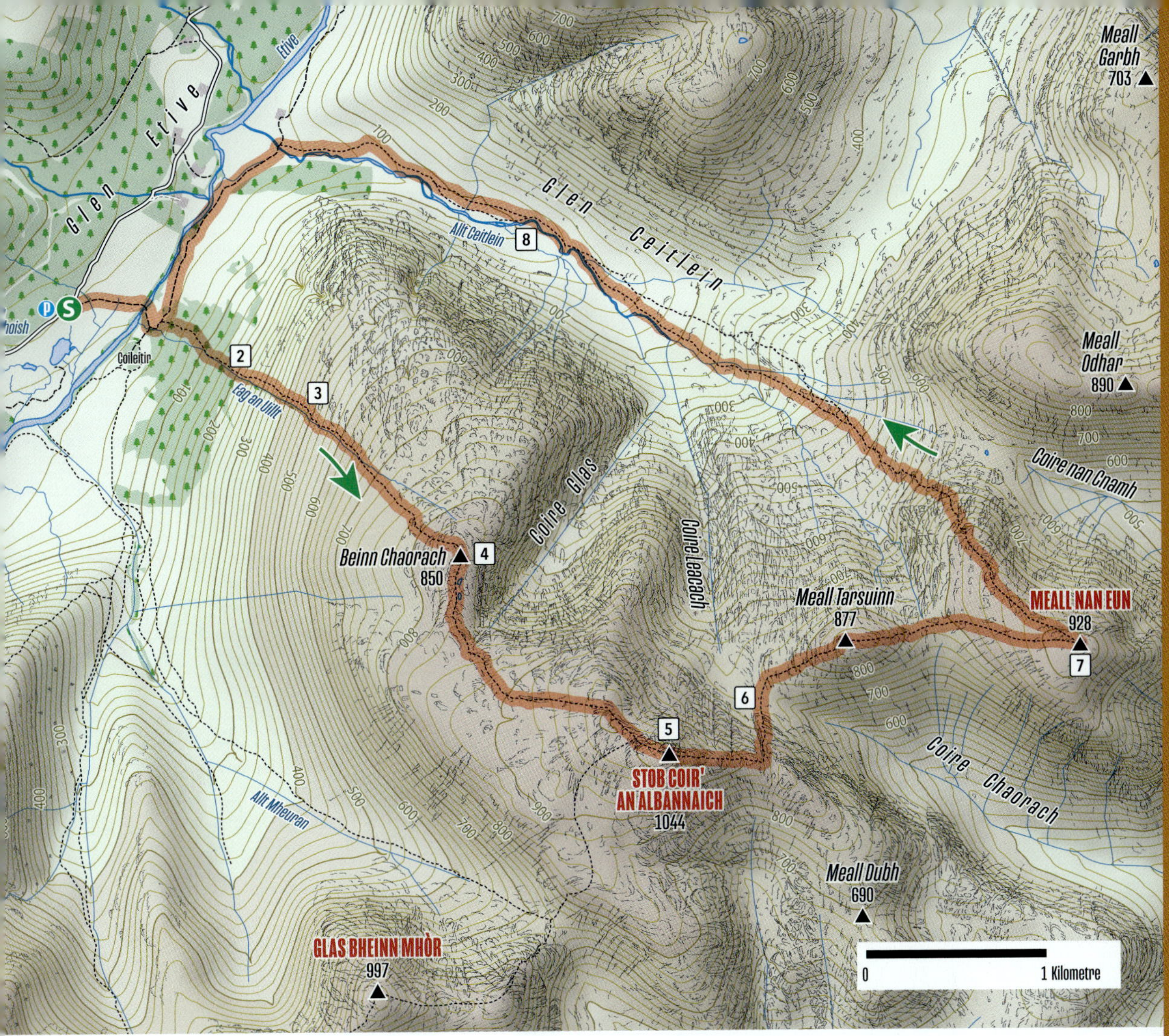

Glen Ceitlein commences. The path becomes clearer the closer you get to the hydro track in the basin.

8 The track begins at the weir – follow it out of the glen, passing through two gates along the way, both of which have ladder stiles beside them in the event that they are locked. At a T-junction in the track, turn left and cross a bridge over the Allt Ceitlein. Pass through a further two gates and follow the track back to the start.

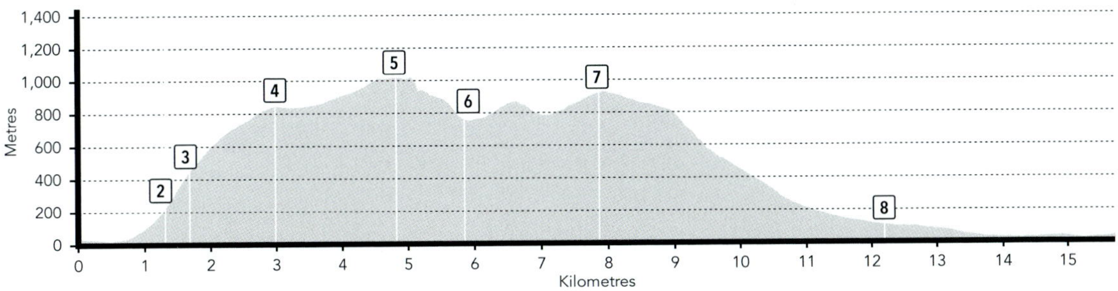

31 North-West Glen Etive Munros

HUB **Glencoe – 45-minute drive**

OVERALL 1 2 3 **4** 5 TERRAIN 1 2 3 **4** 5
DISTANCE 1 2 **3** 4 5 NAVIGATION 1 2 3 **4** 5

DISTANCE **18.8km** ASCENT **1,550m**
TIMES **10:00hrs** (Walker) **8:40hrs** (Trekker)
6:30hrs (Fastpacker) **4:10hrs** (Trail runner)

START LOCATION **Small parking area, Glen Etive Road, Invercharnan**
START LATITUDE/LONGITUDE **56.5911, -5.0256**
START GRID REFERENCE **NN143484**
START W3W **///headliner.ripen.festivity**

	PRONUNCIATION	NAME MEANING
Sgùrr na h-Ulaidh	**'skor na hooli'**	**peak of treasure**
Beinn Fhionnlaidh	**'beyn yun-lai'**	**Finlay's hill**

Introduction

Secreted away to the south-west behind the bulk of mighty Bidean nam Bian, it's as if shy Sgùrr na h-Ulaidh and Beinn Fhionnlaidh are happiest in the shadows, safely veiled from the Glen Coe madness. Here you'll see less of the Instagram hillwalking crowd. Instead, this is a place for the discerning hillwalker, who craves remote, pathless, tranquil hiking, rather than iconic photograph spots. For separate walks, Sgùrr na h-Ulaidh can be tackled from Glen Coe and Beinn Fhionnlaidh from Glen Creran, but my preferred route is to tick them off together from Invercharnan. This committing route features pathless terrain and just a touch of scrambling.

Public Transport

There are no public transport services close to this walk.

Parking

Free parking for around five cars in the small parking area on the Glen Etive Road near Invercharnan. Be sure to leave enough room for forestry vehicles. The nearest public toilets are in Glencoe village. There are also toilets at the Kingshouse Hotel and Glencoe Mountain Resort if you use the bar/cafe or pay for parking.

MTB-ability

The first and last 3.4 kilometres of this route can be cycled.

Escape Route

No escape routes; retrace your steps after the first Munro and return another day for the second.

View to Beinn Fhionnlaidh from the start of the descent of Sgùrr na h-Ulaidh.

Refreshments

There are no cafes or pubs in Glen Etive, so it would be rude not to stop by the Boots Bar at the Clachaig Inn on your way back to Glencoe.

The Route

S Cross the bridge south of the parking area and leave the road to the right when it bends sharply left. Walk south-west on a track and pass through a gate. Walk for 1km along the track as the view to Ben Starav (page 145) dominates up ahead. The track then curves north-west to run high above, but parallel to, the Allt nan Gaoirean up and out of what remains of the Glenetive Forest.

2 Ignoring all minor tracks that branch off the main track, continue to NN120490 where it abruptly ends. Cyclists can leave bikes here. Continue in the same direction, following a faint path over wet grassy ground. The trail leads over a handful of small streams and gradually ascends towards the saddle between Sgùrr na h-Ulaidh and Meall a' Bhuiridh, known as Bealach Clach nam Meirleach.

3 The saddle itself is shielded from view for much of the ascent by the western slopes of Meall a' Bhuiridh. Ease around these slopes before heading north and then north-east, picking up a line of old iron fence posts which climb up to the far north-western edge of the bealach below Sgùrr na h-Ulaidh's crags.

4 Follow the fence posts as they turn sharply left at NN117513. Follow a faint path as it weaves around crags, heading north-west up the hillside. One scramble move is required for which there are adequate hand and footholds, before the path faces south-west briefly to pass underneath a huge sheet of slanted rock at NN116513. Fifty metres later, at NN115513, a cairn marks the point that the ascent in a north-westerly direction continues, initially over rocky steps and then back on to grassy terrain. There is a stunning view down Loch Etive from here.

5 On the upper reaches of the mountain there are a few deep rocky steps which require some care, but it's mostly a straightforward ascent to the top with a fantastic view to Beinn Fhionnlaidh all the way up. Some of the iron fence posts have survived the altitude gain as the path continues to follow the same line. The posts are a great handrail in clag. A cairn marks the summit of the Munro, from which there are epic views to the Glen Coe giants that Sgùrr na h-Ulaidh is nestled behind including Bidean nam Bian (page 163).

Touching the summit cairn of Sgùrr na h-Ulaidh.

6 Retrace your steps to the Bealach Clach nam Meirleach and follow the fence posts down into Bealach Caol Creran at the foot of Beinn Fhionnlaidh.

7 Instead of tackling the crags, it's safer to head south from NN111505 and walk across the bowl. Walk over rough, pathless ground keeping below the crags mostly in the 450–475m contour range. After 1km, rise to 500m to reach a path at NN109492. Turn right on this path, which is also your route of descent, and walk alongside the steep gorge east of Meall nan Gobhar. The stony path begins to zigzag, and when the switchbacks end, the path leads north-west. The path peters out when it hits a patch of wet, grassy terrain, so aim for point NN102498 on the ridge at an old fence corner, where only the posts remain, at the bealach between the main summit and the eastern top of Beinn Fhionnlaidh.

8 From here, follow the path west-north-west to start the final 150m of vertical ascent that remains. The path, which is stony at first and then grassy, leads to the foot of two rocky obstacles close to

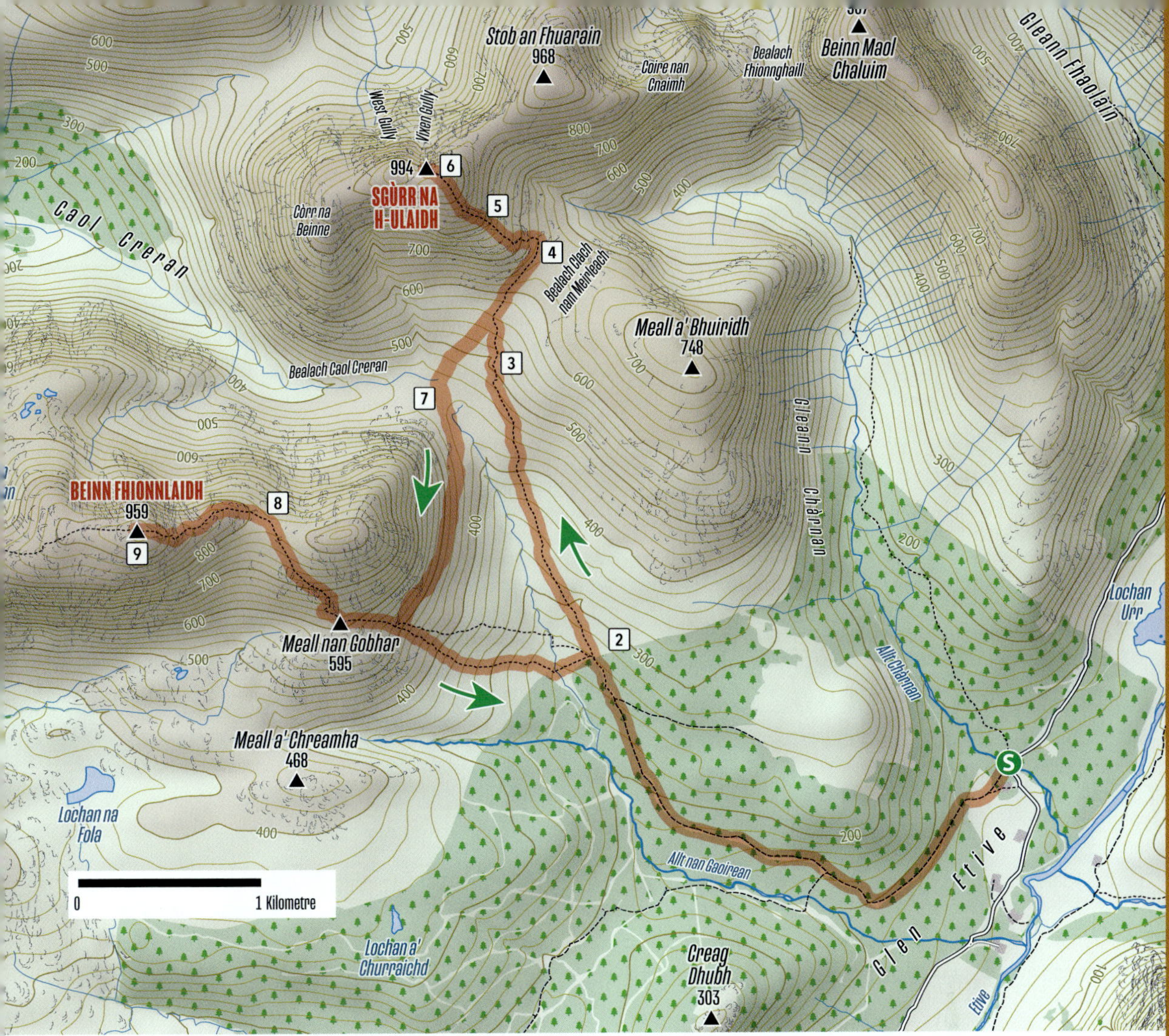

the summit. A couple of simple scramble moves with adequate hand and footholds, and you've made it to the summit plateau.

9 From the top, spot the nearby Munro Beinn Sgulaird (page 41) to the south-west. Retrace your steps on the descent path back to NN109492. From here, continue descending south-east and then east into the valley. The path leads down to a bridge over a stream and then up to the track (**2**), from which you can retrace your steps to the start.

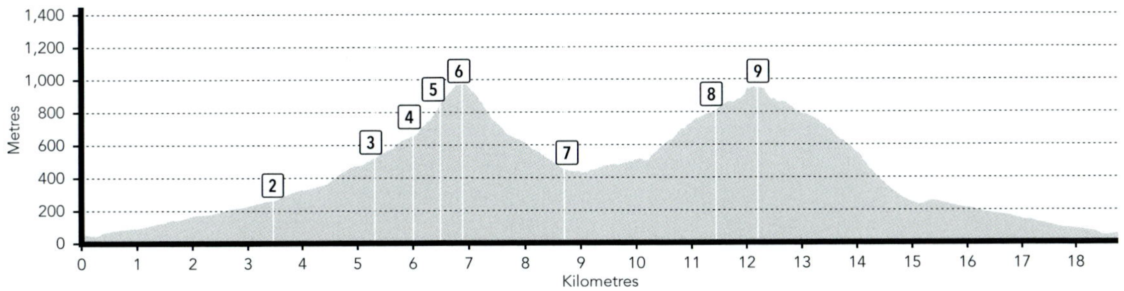

32 Beinn a' Bheithir

HUB **Glencoe – 10-minute drive**

OVERALL 1 2 **3** 4 5 TERRAIN 1 2 3 **4** 5
DISTANCE 1 **2** 3 4 5 NAVIGATION 1 2 3 **4** 5

DISTANCE **12.2km** ASCENT **1,260m**
TIMES **7:30hrs** (Walker) **6:30hrs** (Trekker)
4:55hrs (Fastpacker) **3:05hrs** (Trail runner)

START LOCATION **Glenachulish Forestry and Land Scotland car park, Glenachulish**
START LATITUDE/LONGITUDE **56.6809, -5.1899**
START GRID REFERENCE **NN046588**
START W3W **///foam.thrashing.tiny**

	PRONUNCIATION	NAME MEANING
Beinn a' Bheithir	**'beyn a veh-er'**	**hill of the thunderbolt**
Sgorr Dhonuill	**'sgor gnaw-nil'**	**Donald's peak**
Sgorr Dhearg	**'sgor yerreg'**	**red peak**

Introduction

Beinn a' Bheithir – which has two Munros, Sgorr Dhonuill and Sgorr Dhearg – might just be the unsung hero of the Glen Coe area. Tucked away to the west, detached from the main hustle 'n' bustle of the valley, here you get quieter walking – but still with a dusting of that Glen Coe magic. Sgorr Dhonuill offers exceptional views across Argyll, yet it's Sgorr Dhearg that truly steals the show – as a platform for gazing across Aonach Eagach, Bidean nam Bian and the rest of the Glen Coe big hitters, it packs a real punch. Please note the forestry tracks at the start of this route are subject to landscaping work, so look out for signage indicating any diversions.

Public Transport

The nearest bus stop served by Scottish Citylink buses from Glasgow is Ballachulish Bridge on the A82, 1.5 kilometres from the start point.

Parking

Free parking for around eight cars in the Glenachulish Forestry and Land Scotland car park. The nearest toilet facilities are in Ballachulish.

Escape Route

No escape routes; retrace your steps after the first Munro and return another day for the second.

Refreshments

The Laroch Restaurant and Bar in nearby Ballachulish is on your way back to Glencoe. It's a great place to go for food after your walk. I can always be swayed by a good sticky toffee pudding.

View of Sgorr Dhearg from the descent of Sgorr Dhonuill.

Looking down from the summit of Sgorr Dhearg over Loch Leven and Glencoe village.

The Route

S Leave the car park to the south using the track which leads uphill through woodland to a gate. Pass through the gate and continue on the track for just over 500m until you meet a wide track junction. Turn left here and cross a bridge over a burn before turning right immediately afterwards. This good footpath leads south up Gleann a' Chaolais (Glenachulish). After 1km the path meets a track. Cross the track and pick up a footpath on the other side. A fingerpost indicates this is the path for hillwalkers.

2 This path also exits on to a track just under 500m later – cross the track and continue straight on up a stone-pitched staircase. Here, the trees start to thin out and you're treated to glimpses of a view down to Loch Linnhe. This part of the trail reminds me of the hiking I've done in Canada with the added bonus that you don't have to be on alert for disturbing bears. The path exits out of the woodland on to the left-hand side of a corrie.

3 Reach a fence and follow the trail briefly downhill. Another fence joins from the right and the path follows this uphill once more. Navigate over some wet, boggy ground and then pass under the deer fence where there is a large gap at NN049560.

4 Continue ascending on the other side of the fence. Gain the flatter grassy plateau where a path leads to the base of steeper ground at the head of the coire. Walk south up the path which skirts around crags towards the bealach. The path splits just before the saddle: fork right to ascend Sgorr Dhonuill first. It may be the smaller of the two Munros but it's the more challenging climb.

5 There is a good path to follow from the bealach with a gorgeous view down into Gleann a' Chaolais. The ridge levels out before the final pull up to the summit of Sgorr Dhonuill which reminds me of a Walnut Whip from this angle. The view down the Firth of Lorn to the left of the Munro is incredible. While it looks steep and somewhat intimidating, there are no complicated scramble moves on the final bit of the ascent. The path starts on the left-hand side before coming across to the right where there is a touch of exposure thanks to a sheer drop down to your right. If you prefer, this can be avoided by keeping slightly to the left to scramble over boulders instead.

6 The summit of Sgorr Dhonuill offers exceptional views across Argyll. The view to Sgorr Dhearg is very impressive due its proximity to Sgorr Dhonuill – it's only 1.7km away as the crow flies. Retrace your steps to the bealach (**5**). Cross the bealach and ascend Sgorr Dhearg to the east and then north-east following an eroded trail. The terrain becomes stony and loose as you walk in shallow zigzags up the mountain.

7 The summit views are simply phenomenal as you look down on Glencoe village and across to Aonach Eagach (pages 177 and 181), Bidean nam Bian (page 163) and Ben Nevis (page 195). It's a real feast for the eyes with so many peaks and lochs to take in. Retrace your steps to the bealach (**5**), enjoying the view to Sgorr Dhonuill with Mull off to the left and head back down Gleann a' Chaolais via the ascent route.

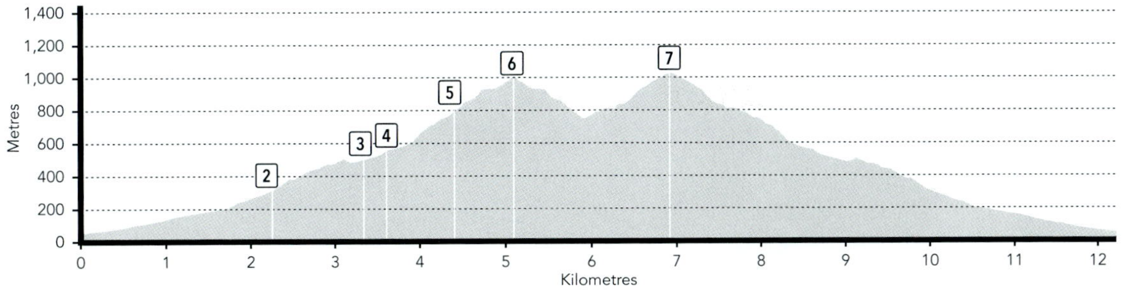

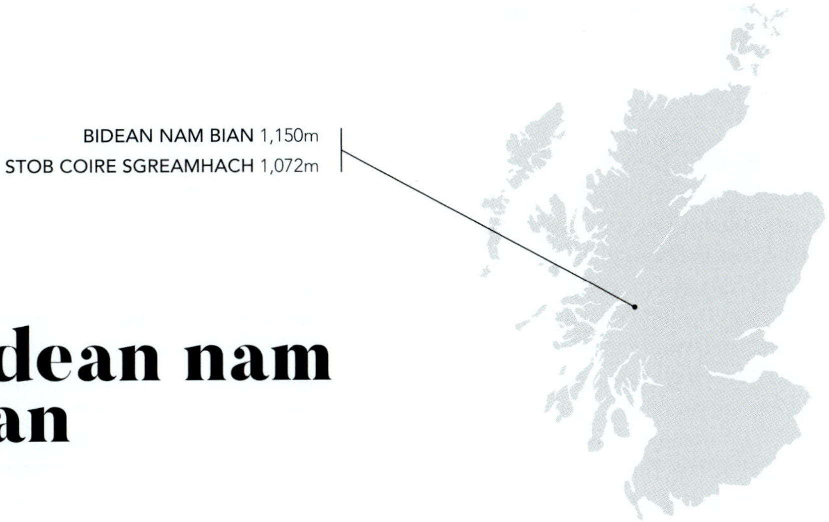

33 Bidean nam Bian

HUB **Glencoe – 10-minute drive**

OVERALL 1 2 **3** 4 5 TERRAIN 1 2 3 **4** 5
DISTANCE 1 **2** 3 4 5 NAVIGATION 1 2 3 **4** 5

DISTANCE **11.5km** ASCENT **1,455m**
TIMES **8:10hrs** (Walker) **7:05hrs** (Trekker)
5:20hrs (Fastpacker) **3:20hrs** (Trail runner)

START LOCATION **Three Sisters viewpoint and car park, A82**
START LATITUDE/LONGITUDE **56.6678, -4.9867**
START GRID REFERENCE **NN170568**
START W3W **///harp.makeup.tasteful**

	PRONUNCIATION	NAME MEANING
Bidean nam Bian	**'beedyan nam bee-an'**	**peak of the mountains**
Stob Coire Sgreamhach	**'stob corriya skray-ach'**	**peak of the dreadful corrie**

Introduction
The 'peak of the mountains' in Gaelic, Bidean nam Bian is the overlord of Glen Coe: a colossal fortress soaring 1,150 metres above Scotland's most dramatic valley. With its lower flanks closely guarded by the craggy Three Sisters, Bidean nam Bian – with its subsidiary Munro peak Stob Coire Sgreamhach – is a hard-won mountain, but the juice is worth the squeeze. The summit views of Aonach Eagach's barbaric spine are mesmerising, while the descent into the Lost Valley – a hidden, glacier-carved hollow used by 17th-century cattle rustlers – is wonderfully romantic.

Public Transport
The nearest bus stop is in Glencoe village. Scottish Citylink buses linking Glasgow with Fort William via the A82 will sometimes stop by request at the Three Sisters car park, but this is at the discretion of the driver and would leave an eight-kilometre walk back to Glencoe at the end of a tough mountain day if you don't manage to get the bus.

Parking
There is free parking for over 50 cars in the Three Sisters car park on the A82. The nearest toilet facilities are in Glencoe.

Escape Route
Descend from Bidean nam Bian to the bealach at the top of the Lost Valley (Coire Gabhail) and leave the out-and-back of Stob Coire Sgreamhach for another day.

Refreshments
It would be rude not to stop by the Boots Bar at the Clachaig Inn on your way back to Glencoe.

Bidean nam Bian from Stob Coire Sgreamhach.

The Route

S Leave the car park using the footpath which leads south-west downhill and turn right on to the bottom path. After 200m, turn left to head into the mountains. Cross the bridge over the River Coe and ascend on the well-maintained path. Ignore a grassy turn-off to the left and stay on the stone path as it leads between two of the Three Sisters – Geàrr Aonach and Aonach Dubh – crossing a couple of streams along the way.

2 The path hugs the left-hand side of a wide gully, the Coire nan Lochan. There are a handful of slabby rocks to traverse along the way, but the path is mostly clear and straightforward. Ascend to the sound of the constant flow of water beside you while the crags of Stob Coire nan Lochan loom up ahead. Remain on the left-hand side of the stream, ignoring paths to the right, to pass an impressive waterfall.

3 Arrive at a cluster of lochans in the upper corrie and walk east-south-east on a faint, wet, grassy path which leads to a short section of loose scree, up and on to the ridge. Turn right at a cairn at NN155550 and walk south-west up the ridge of Stob Coire nan Lochan with the conical Stob Coire Sgreamhach piercing skywards on your left. The length of Aonach Eagach (pages 177 and 181) can also be marvelled at from here and, if visibility allows, Ben Nevis (page 195) can be seen towering beyond that.

4 The ridge is rough and rocky and vertical ascent is gained quickly. Bidean nam Bian soon comes into view too. There is some scrambling required on the ridge, up to grade 1 in exposed positions, but with plenty of handholds and footholds. Much of the scrambling can be avoided by using a bypass path on the left. With a final couple of hoists, top out on to the summit of Stob Coire nan Lochan. On a hazy morning, the silhouetted views down the glen over Buachaille Etive Beag and Buachaille Etive Mòr (page 167) can't be beaten.

5 The view west to Beinn a' Bheithir with its twin Munro summits (page 159) can also be enjoyed before joining the boulder scree path to the bealach at the foot of Bidean nam Bian. From Stob Coire nan Lochan, the route ahead looks terrifying but, in reality, this short ascent is more straightforward than the one up Stob Coire nan Lochan. The path stays to the left of the steepest ground before topping out on the boulder-strewn summit of this fine Munro.

6 Walk south-east off the summit over rough terrain down the ridge, keeping left of the crest at first before undulating over rocky, scree-splattered alpine terrain. Views down the Lost Valley (Coire Gabhail) can be appreciated down to the left.

7 Pass the cairn in Bealach Dearg which marks the start of the descent at NN150536 and glance down the eroded channel that you'll be taking later. For now, continue uphill to the east on the path up Stob Coire Sgreamhach. The terrain continues to be rough and loose underfoot but there is no scrambling on this part of the route. As you've already put most of the ascent work in, the summit arrives sooner than expected.

8 After delighting at the views down the gnarly ridge of Beinn Fhada, the first of the Three Sisters, retrace your steps to NN150536 (**7**) and, after checking there's nobody coming up, make your way down the red scree chute. Take your time and use the walls of the chute to assist your descent. It's somewhat of a relief when stone-pitched steps appear. Follow the path as it zigzags steeply into the Lost Valley. The route gradient eases as the path hugs the left-hand side of the valley above the Allt Coire Gabhail before dropping into the bowl.

9 Walk across the flat bottom of the Coire Gabhail, aiming for the bottom right-hand corner where you can pick up a path which rises uphill briefly. The only obstacle from here is one of the Lost Valley's huge boulders which blocks your path. You can pass it either to the left or right, both involving a mild scramble. From here, drop to a crossing over the Allt Coire Gabhail using stepping stones.

10 Make your way down through a section of trees and boulders. Keep the stream on your right and ignore the high path which leads off to the left. You'll no doubt pass plenty of day trippers whose ultimate destination is the Lost Valley itself. You've earned

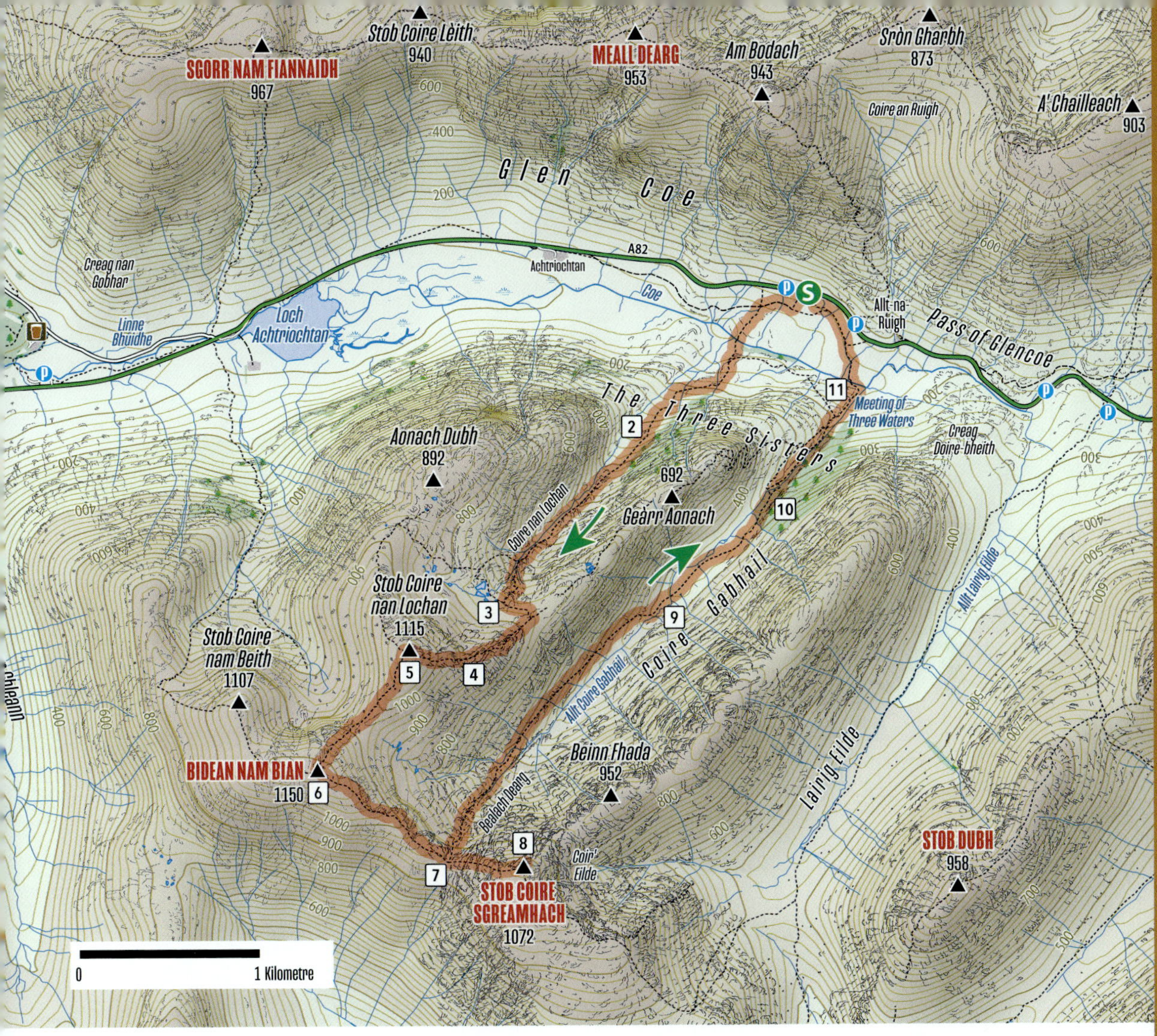

the right to feel a tad smug by this point. Descend a section of slabby rock, where safety chains reminiscent of those seen on popular trails in mountainous regions of the Alps and Pyrenees have been added.

11 Cross the bridge over the River Coe and ascend the metal staircase from where you can follow the path back to the start.

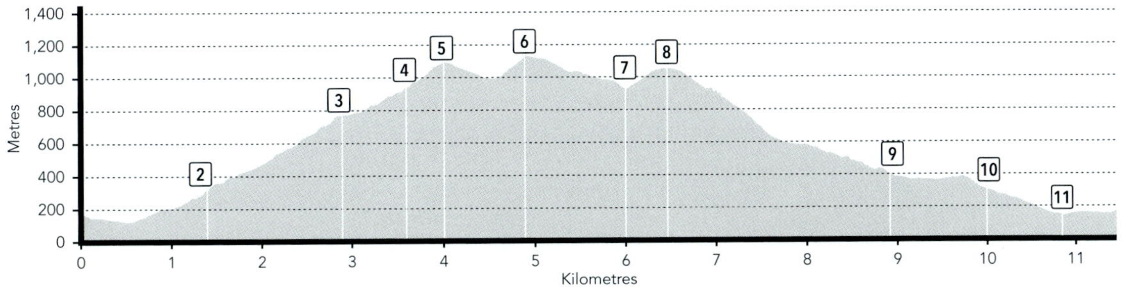

STOB DEARG 1,021m
STOB NA BRÒIGE 956m
STOB COIRE RAINEACH 925m
STOB DUBH 958m

34 Buachaille Etive Mòr & Buachaille Etive Beag

HUB **Glencoe – 15-minute drive**

OVERALL 1 2 3 ④ 5 TERRAIN 1 2 3 ④ 5
DISTANCE 1 2 3 ④ 5 NAVIGATION 1 2 ③ 4 5

DISTANCE **21km** ASCENT **2,045m**
TIMES **12:25hrs** (Walker) **10:45hrs** (Trekker)
8:05hrs (Fastpacker) **5:10hrs** (Trail runner)

START LOCATION **Lay-by opposite the Buachaille Etive Mòr car park, A82**
START LATITUDE/LONGITUDE **56.6646, -4.9058**
START GRID REFERENCE **NN220563**
START W3W **///trappings.bombshell.arrives**

	PRONUNCIATION	NAME MEANING
Buachaille Etive Mòr	**'booerchil etive more'**	**big herdsman of Etive**
Stob Dearg	**'stob jerreck'**	**red peak**
Stob na Bròige	**'stob na broykya'**	**peak of the shoe**
Buachaille Etive Beag	**'booerchil etive bake'**	**little herdsman of Etive**
Stob Coire Raineach	**'stob corriya ranyoch'**	**peak of the corrie of ferns**
Stob Dubh	**'stob doo'**	**dark peak**

Introduction

Buachaille Etive Mòr, or simply the Buachaille as it is affectionately known, is one of Scotland's most photographed mountains. It's a colossal sentry guarding the head of Glen Coe, soaring into the skies with a handsome pyramidal profile – like a child's drawing of a perfectly triangular mountain. At its base is the impossibly idyllic whitewashed cottage of Lagangarbh, star of a million and one Instagram posts. Next door is Etive Mòr's namesake, Buachaille Etive Beag, another epic mountain with excellent hillwalking ground. You may prefer to give each the full attention of a dedicated route, but bagging both together is more than possible – it makes for an iconic day in the hills. This route should not be attempted in winter due to the risk of avalanche.

Public Transport

Scottish Citylink buses linking Glasgow with Fort William stop at the bottom of the access road for the Glencoe Mountain Resort, at the junction with the A82. Buses will sometimes stop by request close to the start point at Altnafeadh but this is at the discretion of the driver and would leave a 5.8-kilometre walk along the West Highland Way to the bus stop at the end of a tough mountain day if you couldn't guarantee a pick-up from Altnafeadh for the return journey.

Parking

Free parking for around eight cars in the lay-by opposite the Buachaille Etive Mòr car park off the A82 at the head of the track to Lagangarbh (white climbing hut). The car park itself is very rough and

Buachaille Etive Mòr (Stob na Bròige) seen from the Glen Etive Road.

Above Buachaille Etive Beag and Buachaille Etive Mòr from Beinn Fhionnlaidh (see route 31). **Opposite** Stob na Bròige from the summit of Stob na Doire.

full of potholes, so I prefer to park in the lay-by. An alternative for parking is further west where the return route meets the A82 (NN213559, ///froze.shipwreck.simple). The nearest public toilets are in Glencoe. There are also toilets at the nearby Kingshouse Hotel and Glencoe Mountain Resort if you use the bar/cafe or pay for parking.

Escape Route

Consider these Munros in pairs. To retire after the two Munros that comprise Buachaille Etive Mòr, simply turn right in the Lairig Gartain and return to the start. Return another day for Buachaille Etive Beag.

Refreshments

White Corries Cafe sits within the Glencoe Mountain Resort and has both indoor and outdoor seating. It's open seven days a week, serves hot food until 7 p.m. and is a popular stop for both West Highland Way walkers and Munro-baggers.

The Route

S Leave the lay-by and walk briefly east along the verge of the A82. Cross the road into the rough car park and follow the track towards Lagangarbh, the little white climbing hut that sits beneath Buachaille Etive Mòr. I stayed here back in 2019 after making arrangements through the Scottish Mountaineering Club, who have maintained the National Trust for Scotland-owned property since 1946. Cross the bridge over the River Coupall and pass to the left of the climbing hut on a footpath which winds towards the foot of the mountain.

2 After 200m, at NN221557, fork right, avoiding the left fork which leads to a series of climbing routes. There are hundreds of climbing routes on Buachaille Etive Mòr alone. On a clear day the route up the wide gully of Coire na Tulaich stands out up ahead as the path leading up to it is well maintained with gravel and stone pitching in places.

3 Just over 500m later, at NN219552, cross the Allt Coire na Tulaich and continue ascending on the stone-pitched path with the occasional short section of slabby rock and boulder scree to pass over. This doesn't amount to a scramble, although you might benefit from putting your hands down on the rocky sections, especially if you don't have walking poles. As you ascend, pause to look behind you as the most easterly Mamores Munros (page 185) and the western edge of Blackwater Reservoir both come into view. As more height is gained, the path veers away from the gully up to the right as it begins to zigzag up towards the ridge.

4 At 800m elevation the stone pitching merges with boulder scree and the route-finding can be more challenging. Towards the very top, at NN216542, the route hugs the underside of a steep crag wall back to the centre of the head of the corrie. From here, scramble up a rocky spur before picking up the path to the bealach.

Buachaille Etive Mòr (Stob Dearg) and Lagangarbh climbing hut from the bridge on the approach.

5 There is a brilliant view to Stob na Doire as you reach the top of the pass. Turn left to continue up red-tinged scree, heading east and then north-east to the summit of Stob Dearg. If you don't keep an eye on the map, you might overshoot the true summit, at NN222542, as there's a very attractive cairn and stone shelter on the final knoll a bit further along.
6 Retrace your steps to the bealach (**5**) and walk north-west up a stony path. The path swings left after reaching point 902m on the OS map. Follow the ridge south-west up Stob na Doire. It's rocky but not technical. The full length of Buachaille Etive Beag can be appreciated on your right along here, so it's a great opportunity to scout out the second half of the route.
7 From the summit of Stob na Doire, which is a Munro Top, there is an awesome view to the second Munro, Stob na Bròige. Walk south-west, taking care descending steep rocky terrain. When you reach the bealach, note the path on your right at NN200529 as this is your descent route for later. Begin the out-and-back to Stob na Bròige via another Munro Top – Stob Coire Altruim.
8 The ridge walking beyond Stob Coire Altruim is delightful as it undulates gently towards Stob na Bròige, which presents grand views down Glen Etive.
9 From Stob na Bròige, retrace your steps to the bealach at NN200529. Turn left on to the path into Coire Altruim – it is steep but largely stone-pitched so route-finding is straightforward. There are some natural rocky and slabby obstacles along the way so take care over these, they can be especially greasy in wet weather.
10 Cross the River Coupall at NN198541 and if you've decided to leave Buachaille Etive Beag for another day, turn right on the path that leads back to the A82. Otherwise, turn left and ascend gently for 400m to point NN195539. Turn right here to ascend grassy slopes beside a stream. The route can be wet and soggy but it's not technical. Towards the top the path becomes indistinct; simply aim for the lowest point of the bealach if you lose the trail.
11 Reach the top of the pass at Màm Buidhe and enjoy the view north-west to Aonach Eagach (pages 177 and 181) before walking north-east. Follow the path up Stob Coire Raineach, which keeps slightly to the right of centre up the broad ridge. The terrain is rocky with scree in places but it's non-technical. The summit of Stob Coire Raineach is marked by a cairn. Glancing to the south-east, it's hard to believe that you were even on the hills on the other side of the Lairig Gartain earlier that same day.
12 The profile of Stob Dubh from here is conical and inviting. Retrace your steps to Màm Buidhe (**11**) then walk west and then south-west over stony non-technical terrain. The view from the knoll en route at 902m is doubly outstanding, both back to Stob Coire Raineach and ahead to Stob Dubh. Continue over

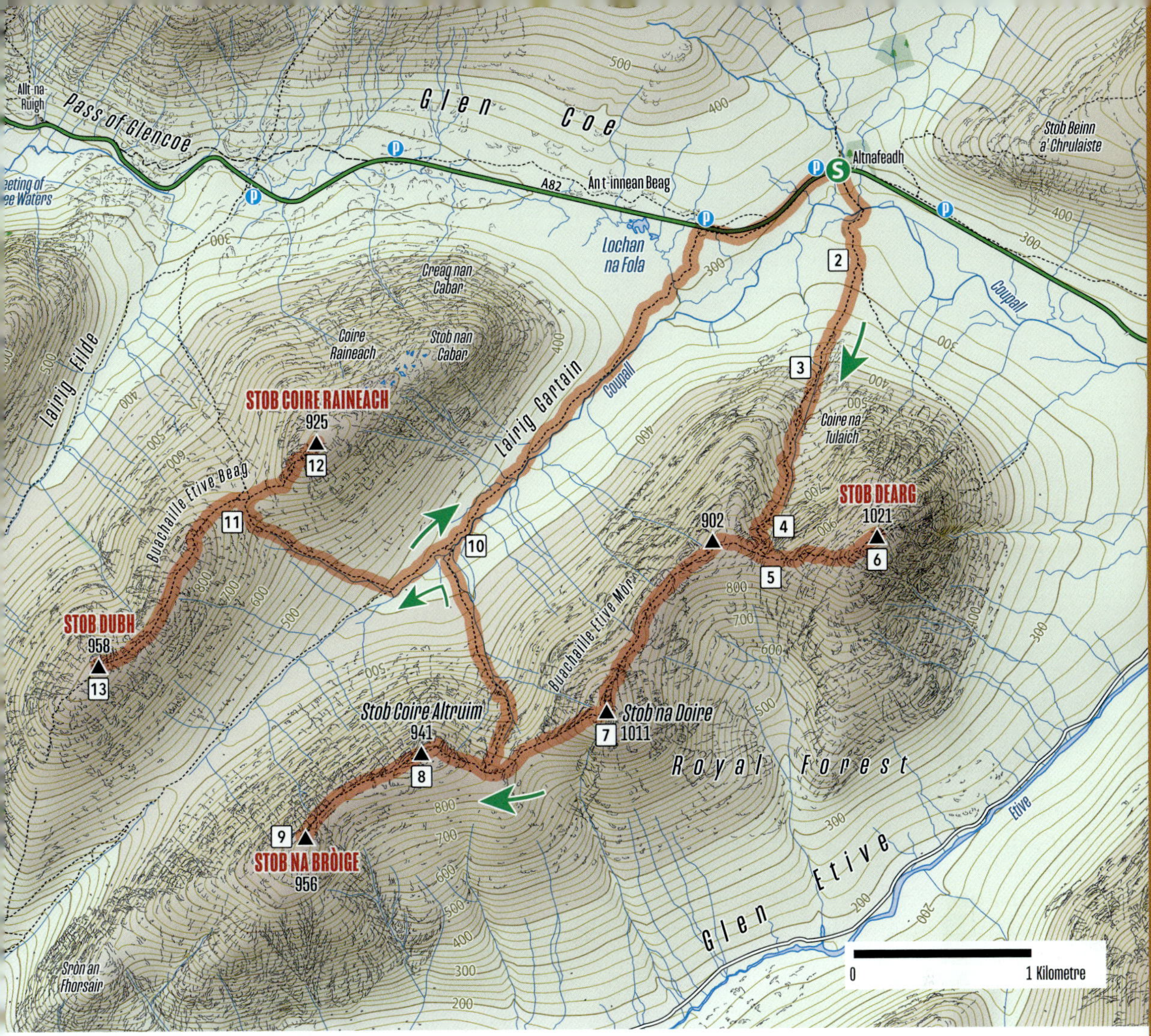

undulating ground along the ridge before the push to the final Munro summit of the day. From the top of Stob Dubh, enjoy the view to all the summits you've walked today. Additionally, I love the view to Bidean nam Bian (page 163) from this Munro.

13 Retrace your steps first to the bealach and then down to the Lairig Gartain at NN195539 before turning left and walking out of the valley. Meet the A82 and follow the grassy path below the main road back to the start.

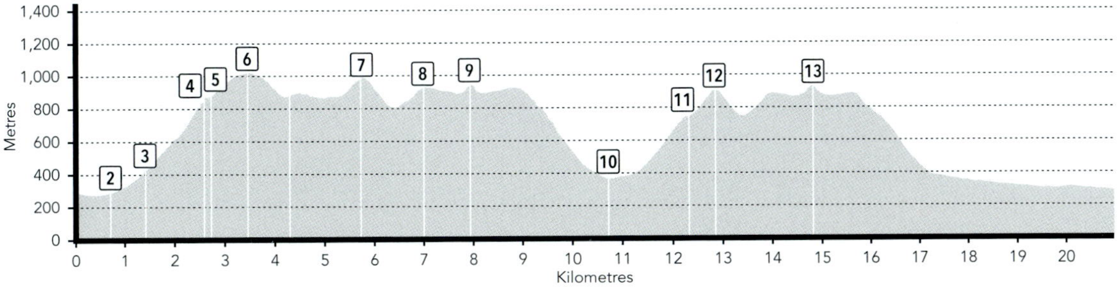

MEALL A' BHÙIRIDH 1,108m
CREISE 1,100m

35 North Black Mount Munros

HUB **Glencoe – 20-minute drive**

OVERALL 1 2 (3) 4 5 TERRAIN 1 2 3 (4) 5
DISTANCE 1 (2) 3 4 5 NAVIGATION 1 2 (3) 4 5

DISTANCE **10.7km** ASCENT **1,110m**
TIMES **6:35hrs** (Walker) **5:45hrs** (Trekker)
4:20hrs (Fastpacker) **2:45hrs** (Trail runner)

START LOCATION **Glencoe Mountain Resort car park, off the A82**
START LATITUDE/LONGITUDE **56.6323, -4.8281**
START GRID REFERENCE **NN266525**
START W3W **///eager.ourselves.sketch**

	PRONUNCIATION	NAME MEANING
Meall a' Bhùiridh	**'meowl a vuree'**	**hill of the bellowing**
Creise	**'kresher'**	**narrow peak**

Introduction

Neighbours to the famed Buachaille Etive Mòr, but lacking the same irresistible charm, Meall a' Bhùiridh and Creise are (perhaps) the overlooked peaks of the Glen Coe area. For some hillwalkers, the ugly pylons and metallic furniture of the ski centre spoil the scenery; for others, the chairlift is a welcome leg-up towards the summit. Either way, this out-and-back route's best section is undoubtedly the airy ridge between the two Munros. From the bealach, the views of Cam Ghleann gorge scything a deep gash through the landscape are quite remarkable.

Public Transport

Scottish Citylink buses linking Glasgow with Fort William stop at the bottom of the access road for the Glencoe Mountain Resort at the junction with the A82. The start point is a kilometre south of the bus stop.

Parking

Honesty box parking in the large car park at Glencoe Mountain Resort where there are hundreds of spaces. There are toilet facilities in the ski centre cafe.

Escape Route

None. Retrace your steps after the first Munro and return another day for the second.

Refreshments

White Corries Cafe sits within the Glencoe Mountain Resort and has both indoor and outdoor seating. It's open seven days a week, serves hot food until 7 p.m. and is a popular stop for both West Highland Way walkers and Munro-baggers.

Creise from the approach to Meall a' Bhùiridh.

View down Cam Ghleann from the ridge between Meall a' Bhùiridh and Creise.

The Route

S Leave the ski centre car park on the track to the south-south-west which passes between the ski centre buildings. After 100 metres at an open area where two tracks meet, continue south-south-west passing between two ski centre buildings and then turn immediately right on to a footpath. Cross a stream and then, at NN264522, the walking footpath crosses the downhill cycle path. Cross this with caution and follow the grassy path on the other side. Additional signage assists safe passage for both the downhill bikers and the walkers.

2 Walk under the chairlift on a stony path surrounded by heather. Glance behind you frequently to watch the vast expanse of Rannoch Moor spread out in the valley below. To the north-north-west Beinn a' Chrùlaiste looks impressive in the early morning light. When the terrain starts to level out, observe the footpath signage to avoid walking on the cycle path again by veering right at NN261517. Join a rickety makeshift boardwalk over wet terrain and ignore turn-offs to the ski buildings on the left.

3 At the path junction at NN260516, turn right and pass over a series of footbridges including one very bendy metal footbridge reminiscent of a runaway train funfair ride. After crossing this, fork left to walk north-west and then south-west following a grassy track over wet ground passing to the south of a small group of peaty lochans. Be cautious here because mountain bikers also use this part of the route, although the ground is flatter and boggier here which greatly decreases the risk.

4 After passing the lochans, walk west to point 724m on the OS map, where the path curves around to the left to climb south-south-west up Meall a' Bhùiridh. On the ascent of Meall a' Bhùiridh, look right across Glas Choire to the craggy north-east face of Creise. The lower part of the climb from here is grassy, while the higher part is over loose boulder scree. Keep heading south-south-west if you lose the path in low visibility. Remnants of the winter ski season are evident here and there along your ascent.

5 The summit of Meall a' Bhùiridh is marked with a cairn. From here, walk west-south-west down the ridge to the bealach at 932m. The ground here is loose at first, then, as the ridge narrows, stick to the crest and descend carefully over rockier terrain.

6 The hugely impressive deep gorge of Cam Ghleann can be seen down to your right from the bealach. Begin climbing up the other side from here over increasingly rocky terrain, where a small number of simple scramble moves are required to gain the long spine of interconnecting land between Clach Leathad and Creise.

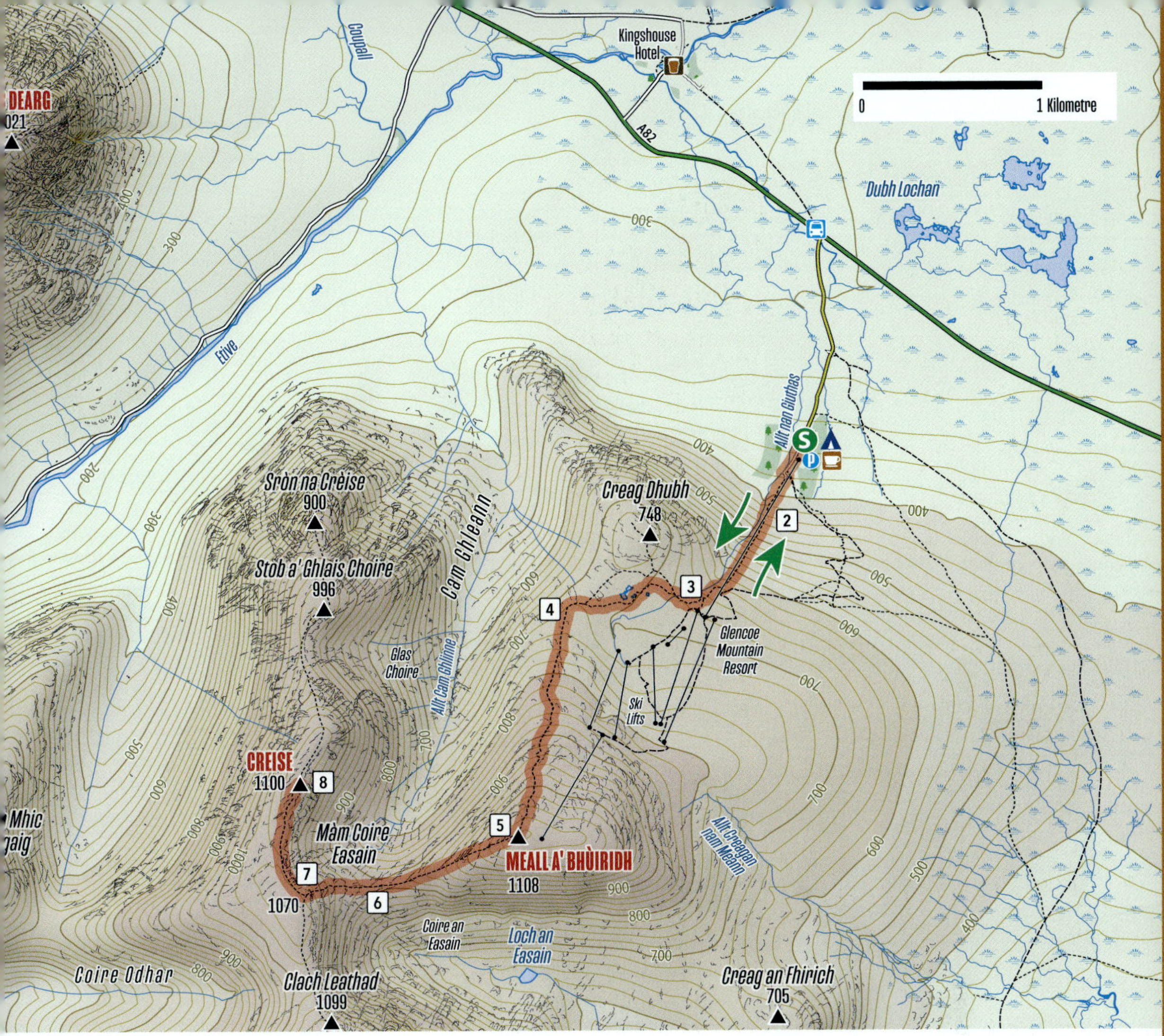

7 Follow the faint stony path north-west at first and slightly downhill. The path then turns north and then north-north-east along the edge of Màm Coire Easain and up to the summit of Creise which feels like a gentle final approach to the second Munro of the day. To the north-west, Buachaille Etive Mòr (page 167) demands your adulation.

8 Retrace your steps back to the start, unavoidably reascending Meall a' Bhùiridh along the way.

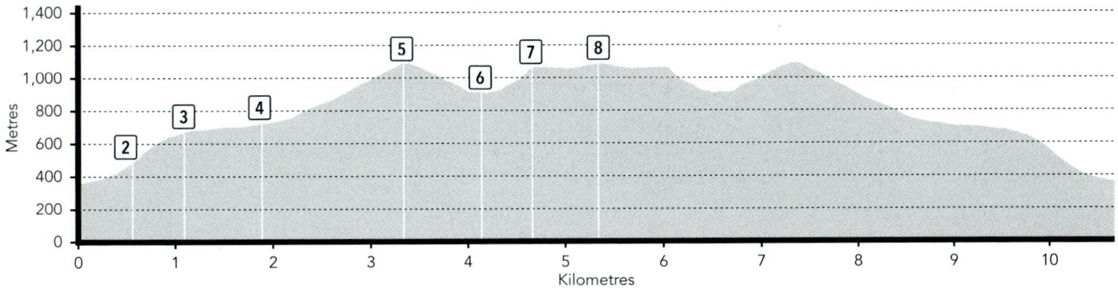

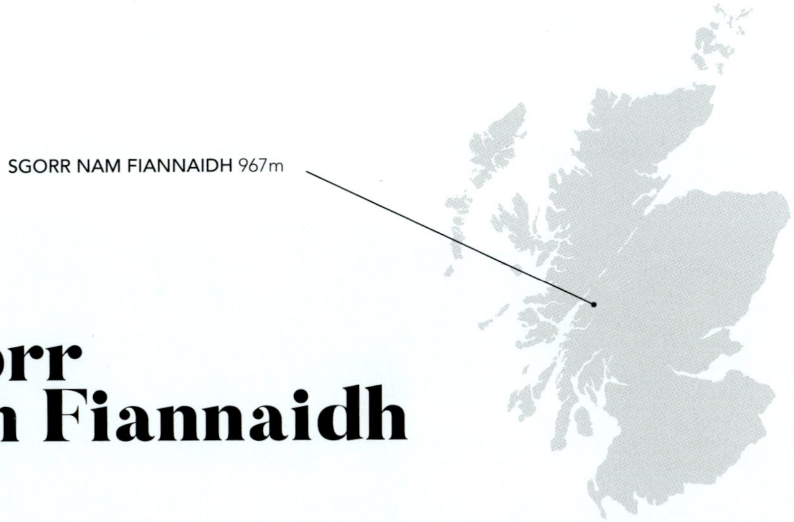

36 Sgorr nam Fiannaidh

HUB **Glencoe – 1km walk**

OVERALL 1 **2** 3 4 5 TERRAIN 1 2 **3** 4 5
DISTANCE **1** 2 3 4 5 NAVIGATION 1 2 **3** 4 5

DISTANCE **8.3km** ASCENT **975m**
TIMES **5:40hrs** (Walker) **4:55hrs** (Trekker)
3:40hrs (Fastpacker) **2:20hrs** (Trail runner)

START LOCATION **Loverose Way car park, Glencoe**
START LATITUDE/LONGITUDE **56.6821, -5.0918**
START GRID REFERENCE **NN107587**
START W3W **///lease.harnessed.workflow**

	PRONUNCIATION	NAME MEANING
Sgorr nam Fiannaidh	**'skor nam fee-anee'**	**peak of the Fingalian warriors**

Introduction

Aonach Eagach is a blade-like crest of frightening proportions: a snaking, serrated spine of jagged pinnacles, knobbly turrets and fangs of rock, with perpendicular drops on either side of the famed 'notched ridge'. Many don't realise that you don't have to take on the ridge if, for you, exposure makes you weak at the knees and takes most of the fun out of the adventure. When I scrambled Aonach Eagach I got to the other end and was mostly relieved to be back on wider ground. The feral goats that live high on the ridge seemed far more comfortable up there than I was. I've been back since, but only to each of the Munros which sit at either end of the ridge. You get all of the awesome views but without the sweaty palms. Here, Sgorr nam Fiannaidh is climbed from Glencoe village and Meall Dearg is described on page 181.

Public Transport

The nearest bus stop served by Citylink buses from Glasgow is in Glencoe village on the A82, one kilometre from the start point.

Parking

Free parking for around 30 cars in the car park on Loverose Way on the outskirts of Glencoe village. The nearest public toilets are a short distance away in the village.

Escape Route

None.

Refreshments

Grab a meal after your walk in the Glencoe Inn and Gathering. The fish supper is a classic.

Ascending Sgorr nam Fiannaidh.

Admiring the view across Aonach Eagach to Meall Dearg from the summit of Sgorr nam Fiannaidh.

The Route

S Exit the car park to the south-east on a gravel path through woodland which runs parallel to the road. Walk for just under 500m before turning left through a pair of gates. A fingerpost indicates that this is the way to the Pap of Glencoe – your route up Sgorr nam Fiannaidh shares the same trails for the first half of the walk. Follow the gravel track steeply uphill. In the summer months the bracken is thick on both sides of the wide trail here.

2 After 250m turn right and use the bridge to cross the stream. The trail leads south-east from here, on a traverse path with ground dropping from left to right. The path crosses a couple of streams before rising steeply north-east up the hillside. The path is clear here thanks to it being a popular route up Sgorr na Cìche (Pap of Glencoe).

3 The terrain eventually becomes rockier as you walk up white rocky slabs, which, while steep, present no technical difficulty. The terrain then becomes loose underfoot as you gain further height. Pause to appreciate the view behind you down Loch Leven past the Ballachulish Bridge to Loch Linnhe.

4 At NN123586, take the right-hand fork in the path to walk east-south-east, away from the Pap of Glencoe and uphill towards Sgorr nam Fiannaidh. The path becomes very stony here and after 700m elevation the trail ascends in tight zigzags up the mountainside. The steepness can feel quite unrelenting by this point but you're nearly there.

5 The terrain flattens out for the final 800m walk to the summit, with Bidean nam Bian (page 163) looking magnificent across Glen Coe to your right.

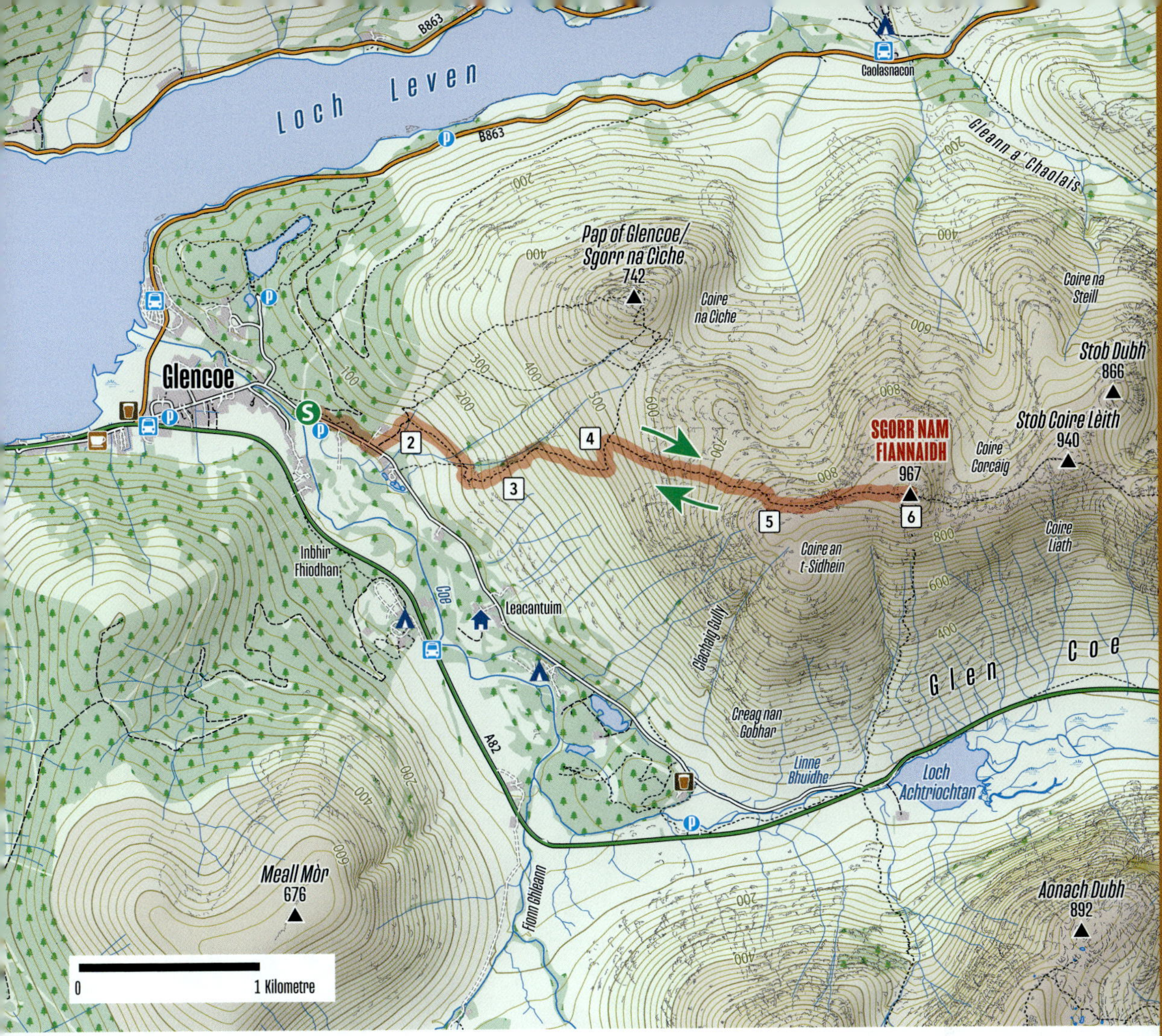

6 What remains of a trig pillar inside a storm shelter marks the summit of Sgorr nam Fiannaidh. The view down Aonach Eagach is slightly less dramatic from Sgorr nam Fiannaidh than it feels from the opposite end of the ridge on Meall Dearg (page 181), but you certainly gain an earnest understanding of how precipitous the drop to Glen Coe is. Return using the route of ascent, adding in a visit to the Pap of Glencoe if you're not ready for the day to be over.

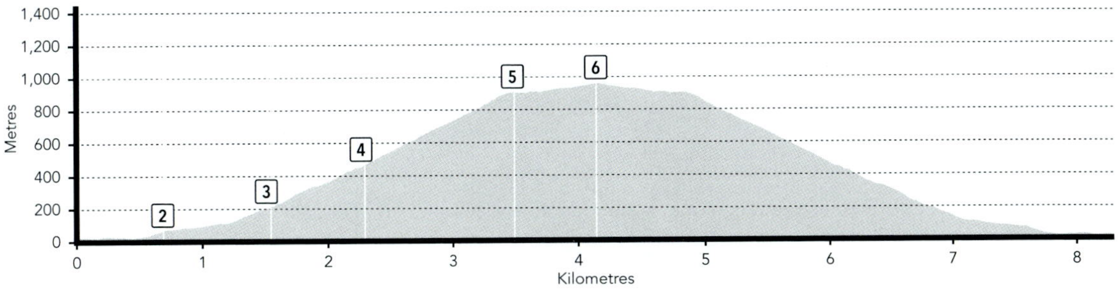

37 Meall Dearg

HUB **Glencoe – 5-minute drive**

OVERALL 1 2 ③ 4 5 TERRAIN 1 2 ③ 4 5
DISTANCE ① 2 3 4 5 NAVIGATION 1 2 3 ④ 5

DISTANCE **8.7km** ASCENT **920m**
TIMES **5:25hrs** (Walker) **4:45hrs** (Trekker)
3:30hrs (Fastpacker) **2:15hrs** (Trail runner)

START LOCATION **Lay-by on the B863 between Glencoe village and Kinlochleven**
START LATITUDE/LONGITUDE **56.7015, -5.0360**
START GRID REFERENCE **NN142607**
START W3W **///speak.switch.horseshoe**

	PRONUNCIATION	NAME MEANING
Meall Dearg	**'meowl jerreck'**	**red hill**

Introduction
Even in the summer months, the scramble across the ridge of Aonach Eagach is about as gnarly as it gets without being classified as rock climbing, and sadly there have been several fatalities here. Confident scramblers will relish the grade 2 challenge, but in truth you don't need to scramble the arête to enjoy Aonach Eagach's wonders. This hillwalker 'cheat' route and the route up Sgorr nam Fiannaidh on page 177 bag the ridge's twin Munros without encountering any exposure or technicality. Meall Dearg is ticked off via the 'back way' from the north, where my palms remain sweat-free but the views pack no less of a punch.

On Meall Dearg.

Public Transport
Local bus services operate between Fort William and Kinlochleven. Buses will stop outside Caolasnacon Caravan & Camping Park, a five-minute walk from the start point.

Parking
Free parking for three or four cars in a small lay-by on the B863 halfway between Glencoe village and Kinlochleven just west of the bridge over the Allt Gleann a' Chaolais. If full, there is another small lay-by east of the bridge. There are public toilets just over six kilometres away in both Glencoe village and Kinlochleven.

Escape Route
None.

Refreshments
Grab a meal after your walk in the Glencoe Inn and Gathering. The fish supper is a classic.

Aonach Eagach, Stob Coire Lèith and Sgorr nam Fiannaidh from the approach to Meall Dearg.

The Route

S Cross the road and walk east over the bridge across the Allt Gleann a' Chaolais. Straight afterwards, take the right-hand turn off the road on to a stony footpath, heading uphill through trees in a south-easterly direction. The path is wet in places and weaves up through heather and bracken with the Allt Gleann a' Chaolais on your right.

2 There are a number of splits in the path on the early part of the ascent: keep taking the right-hand forks which run closest to the stream. Once you've gained a bit of height, check out the impressive waterfalls high in the Coire na Steill. The route is stony underfoot for a while before becoming grassier as you progress further into the Gleann a' Chaolais, stretching away from Loch Leven in the valley below.

3 Cross a fair number of small burns followed by one notably larger one which is the Allt a' Chàthaidh Riabhaich. After this crossing, ascend east-south-east following a faint path. The Allt a' Chàthaidh Riabhaich is always reassuringly on your left.

4 At NN164595, the path peters out – walk south-east across the bumpy terrain, aiming to the right (south) of the lowest point marked by point 535m on the OS map until you reach a line of old iron fence posts. These can be followed south-south-west as you seek the higher reaches of Meall Dearg. Down to your left is the Feith nan Lab which runs through a wide empty basin, a great place to spot deer.

5 The iron fence posts are a great handrail during poor weather as the gradient steepens. The shoulder is rocky but only in patches, so it never becomes too bouldery underfoot. A faint path comes and goes as you weave towards the summit from the north-east ridge of Meall Dearg. Behind you, the Mamores look magnificent, especially the pointy profile of Stob Bàn (page 201), but Aonach Eagach has the most allure, the multiple jagged pinnacles reaching skyward to the right of Meall Dearg. The summit of the Munro is marked by a cairn.

6 For some, the grade 2 scramble of Aonach Eagach is too intimidating a prospect and rather than adding excitement to the walk, it can sap the joy and replace it with fear. This route provides a wonderful

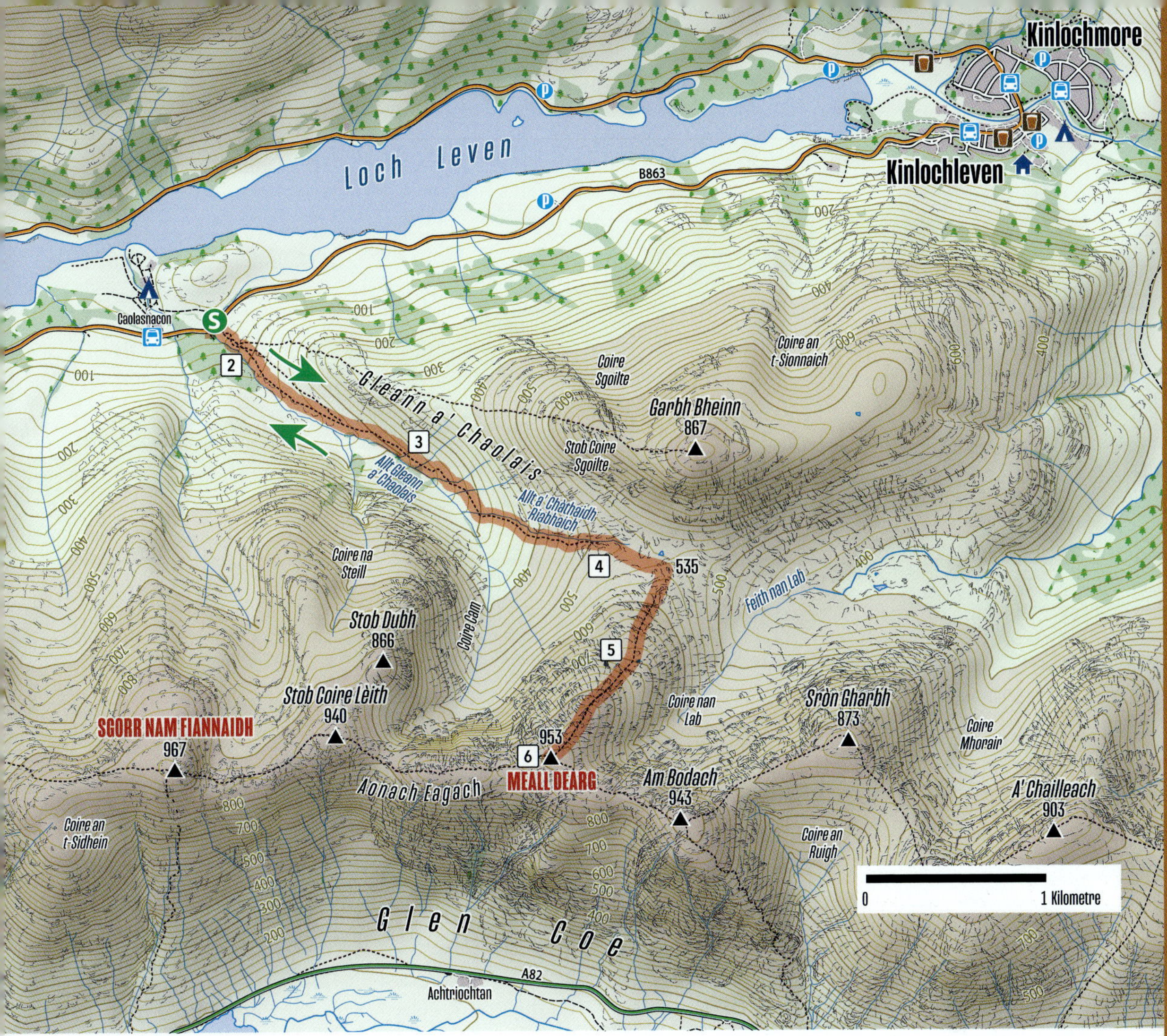

opportunity to appreciate the impressive view west along the ridge to Stob Coire Lèith and Sgorr nam Fiannaidh (page 177) without the trepidation of embarking on the ridge traverse itself. The view across Glen Coe to the higher Munros of Stob Coire Sgreamhach and Bidean nam Bian (page 163) and east to Am Bodach via The Chancellor on any other day would be breathtaking, but the view to Sgorr nam Fiannaidh reigns supreme. Retrace your steps back to the start point.

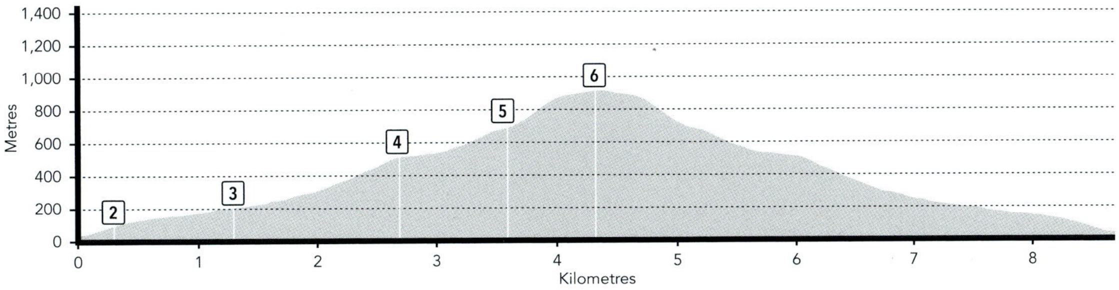

38 East Mamores

HUB **Glencoe – 15-minute drive**

OVERALL 1 2 3 **4** 5 TERRAIN 1 2 3 **4** 5
DISTANCE 1 2 **3** 4 5 NAVIGATION 1 2 3 **4** 5

DISTANCE **19.7km** ASCENT **1,980m**
TIMES **11:55hrs** (Walker) **10:20hrs** (Trekker)
7:45hrs (Fastpacker) **4:55hrs** (Trail runner)

START LOCATION **Grey Mares Waterfall car park, off Wades Road, Kinlochleven**
START LATITUDE/LONGITUDE **56.7166, -4.9626**
START GRID REFERENCE **NN187622**
START W3W **///peachy.lovely.lunge**

	PRONUNCIATION	NAME MEANING
Sgùrr Eilde Mòr	**'sgoor eylda more'**	**big peak of the hind**
Binnein Beag	**'bin-yen bake'**	**small peak**
Binnein Mòr	**'bin-yen more'**	**big peak**
Na Gruagaichean	**'na grooageechan'**	**the maidens**

Introduction

It was one of those sun-drenched, dreamy days in the mountains where everything turned to gold. The miles came easy, my legs felt strong, the endorphins flowed and my spirits soared. I felt so happy. Every turn seemed to serve up another spectacular landscape: jagged horizons, glassy lochans and pyramidal peaks. My most recent walk of these four Munros, the easternmost of the Mamores range, was one of my most memorable mountain days – a day to savour. Fingers crossed you'll have a similar experience. This challenging route starts in Kinlochleven, ticks off the loose screes of Sgùrr Eilde Mòr at the beginning and then completes the loop over three other Munros, including the small-but-mighty triangular outlier of Binnein Beag.

Na Gruagaichean from Binnein Mòr with Aonach Eagach beyond.

Public Transport

Local bus services operate between Fort William and Kinlochleven and stop within a five-minute walk of the start point.

Parking

Parking for around 20 cars in the Grey Mares Waterfall car park in Kinlochleven. The car park is free, but donations are encouraged. The nearest public toilets are a five-minute walk away, beside the main bridge over the River Leven.

Escape Route

This is a challenging route with a lot of ascent, however the same escape route can be used at multiple points in the walk. You can make your way back to the crossroads by the lochan between Sgùrr Eilde Mòr and Sgòr Eilde Beag at NN221656 and retrace your steps from there.

Above Sgùrr Eilde Mòr and Sgòr Eilde Beag from the summit of Binnein Beag. **Opposite** On the pathless ascent of Binnein Mòr with Binnein Beag behind.

Refreshments

The Tailrace Inn in Kinlochleven does a great steak and Guinness pie.

The Route

S Leave the car park via the north-western corner on a gravel path. At the first fork in the path turn left following a fingerpost which reads *Grey Mares Waterfall*. At the next fork in the path, turn right up the steps which lead out of the woodland briefly, noting the path on the left which is your route of descent later on.

2 Keep left, crossing a stream at NN189624, and continue walking gradually uphill through woodland in a north-easterly direction, passing through a gap in the deer fence. The route opens out on to the hillside with grass and heather either side of the path as you walk up a spur. The gravel path is good underfoot and quite gradual so great progress can be made.

3 At NN208634 there is a split in the path – take the left fork which leads to a track. Cross the track and walk north momentarily before the path veers north-east. The good gravel path continues for 1km as Loch Eilde Mòr appears down to your right. Cross the Allt Coire nan Laogh and the path turns sharply left and steeply uphill to the north-north-west with the stream down to your left.

4 The path turns north-east and then east-north-east. From here you can admire the Glen Coe skyline to the south featuring the iconic Buachaille Etive Mòr all the way around to the prominent Pap of Glencoe to the south-west. At NN222647 there's a cairn marking a left turn up a series of switchbacks and along the eastern flanks of Sgòr Eilde Beag. Take this and at the next split, at NN222649, walk straight ahead to the north. The view of Sgùrr Eilde Mòr over the lochan is marvellous.

5 At the path crossroads at NN221656, turn right to head east to begin the ascent of Sgùrr Eilde Mòr. The path becomes notably rougher on the ascent – the route becomes a more eroded line of scree, and the colours underfoot meld from grey to red as you get higher. Next comes a section of sustained scree walking which is not for the faint-hearted. Walking poles are very helpful on this part of the route. When you top out on to the ridge, turn left to walk along the crest to the summit which is marked by a cairn.

6 There are fantastic views north-north-east to the Grey Corries (page 213). The remaining three Munros of this route can also be appreciated from here. Binnein Beag has a wonderfully pointy profile. While Binnein Mòr, the highest Munro in the Mamores, has such bewildering bulk. Eyeing the ridge to the left leads to the twin peaks of Na Gruagaichean. Retrace your steps to the crossroads just above the lochan at NN221656 (**5**) and turn right to head downhill, north-north-west initially, following a path towards Binnein Beag. After 2km, you reach the lochan at the foot of Binnein Beag.

7 Turn right, crossing the stream outlet at the south-eastern end of the lochan. Walk the length of the lochan and ascend northwards to the summit of Binnein Beag. There are a couple of eroded lines to follow which pass around and over the bouldery terrain. The ascent is a piece of cake compared to the challenge of the screes on Sgùrr Eilde Mòr. At the summit of Binnein Beag there is a shelter and summit cairn. The view to Ben Nevis and the Càrn Mòr Dearg Arête is spectacular (page 195).

8 Retrace your steps to the stream crossing at the south-eastern end of the lochan (**7**). This time, fork right and follow the path west. Cross another stream and then 50m later, at NN217671, begin walking south-west over pathless ground, aiming for the leftmost of two lochans below the aptly named coire – Achlais (armpit) a' Bhinnein Mhòir.

9 Pass to the left of the lochans and ascend south-west between two ribs of craggy ground, which avoids the most complicated terrain. Once above the crags, walk west and then take a diagonal line on to the crest of Binnein Mòr's north ridge. Follow the airy ridge to gain the Munro summit of Binnein Mòr. A small amount of uncomplicated scrambling is required towards the top, but not in exposed positions. The highest point of the Munro is long, blocky and dramatic.

10 Leave the highest point of the Mamores to the south-south-west over peaty terrain which feels blissfully forgiving in comparison to the harder pathless walking of the ascent. After a short rise to Binnein Mòr's south top (1,062m) descend south-west to the bealach between this and Na Gruagaichean. Ascend the short distance to the bouldery top of your fourth Munro of the day. I love the view over the north-west summit of Na Gruagaichean to An Gearanach (page 205) and beyond to Ben Nevis (page 195).

Passing the lochan en route to Sgùrr Eilde Mòr.

11 Descend along the crest of the south ridge of Na Gruagaichean to Leachd na h-Aire. This is bouldery in the higher reaches but grassy beyond this. Descend west-south-west. The trail is clear until it hits a boulder field, after which it becomes indistinct. Walk south-west, shedding 300m of ascent before a path can be picked up once more from around 500m elevation. The path reaches a large track.

12 Turn left and walk along the track for 50m before turning right on to a footpath, passing through a slanting gate in the deer fence. After 650m at a path junction, turn left into the woodland. Cross multiple streams and pass through another gate. At the next path junction fork left downhill. After 350m, cross a bridge over the stream. The path soon rejoins the path of your initial ascent. Retrace your steps from here to the start.

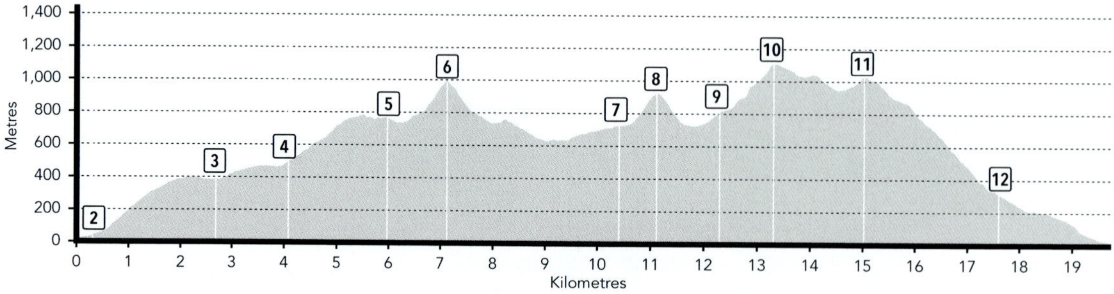

Water of Nevis
BINNEIN BEAG
943
8
7
9
Achlais a Bhinnein Mhòir
BINNEIN MÒR
1130
10
AN GEARANACH
982
An Garbhanach
975
STOB COIRE A' CHÀIRN
981
Coire an Easain
1062
Coire a' Chàirn
Coire na Ba
NA GRUAGAICHEAN
1056
11
AM BODACH
Coire nan Laogh
956
Sgòr Eilde Beag
Coire a' Bhinnein
SGÙRR EILDE MÒR
1010
5
6
Coire an Lochain
Allt Riabhach
Leachd an Uillt Riabhaich
Leachd na h-Aire
4
Sgùrr an Fhuarain
Allt Coire na Bà
Loch Eilde Mòr
Meall Beag
561
12
3
An Cumhann
Meall an Doire Dharaich
469
Meall na Duibhe
573
B863
2
Kinlochmore
Kinlochleven
Leven
0
1 Kilometre

HIGHLAND HUB 7

FORT WILLIAM

The seventh of the Highland hubs is Fort William, the largest town in Lochaber, known as the Outdoor Capital of the UK.

Direct trains link Fort William and Glasgow in under four hours and direct bus journeys are actually quicker, reaching the town in just over three hours.

With a steady stream of West Highland Way walkers arriving in the town every day, plus those staying there for hiking, climbing, cycling and water-based adventures, there's never a dull moment in Fort William.

There is a plethora of choice when it comes to finding somewhere to eat and drink plus multiple supermarkets to suit all budgets, alongside several outdoor shops to meet all your needs. I was once saved by one of these stores when the strap of one of my gaiters snapped and I managed to repair it here.

When it comes to Munros, Fort William is home to not only the highest, Ben Nevis, but some of the classic circuits like the Ring of Steall, and a personal favourite of mine, the Grey Corries.

WHERE TO STAY

£	££	£££
Glen Nevis Caravan and Camping Park	Fort William Backpackers	The Garrison

TRAVEL TIMES

39 Ben Nevis via the Càrn Mòr Dearg Arête	**10 minutes**
40 West Mamores	**15 minutes**
41 Ring of Steall	**20 minutes**
42 The Aonachs	**20 minutes**
43 The Grey Corries	**30 minutes**
44 West Loch Treig Munros	**35 minutes**
45 East Loch Treig Munros	**35 minutes**
46 Ardverikie Munros	**35 minutes**

Rest day ideas

- Visit the West Highland Museum to hear tales of the Jacobites and the social history of the area alongside hundreds of exhibits. This is a great rainy-day activity.
- Do some people watching on the High Street. I love watching walkers finish the West Highland Way, their foot pain dissipating as they approach the finish point with a great big smile.
- Catch a film at the independent Highland Cinema in Fort William – a great way to relax after a day in the hills.

Opposite View from Stob Choire Claurigh towards Stob Coire an Laoigh and Sgùrr Choinnich Mòr (route 43).
Overleaf Ben Nevis beyond Aonach Eagach from Stob Coire nan Lochan (route 33).

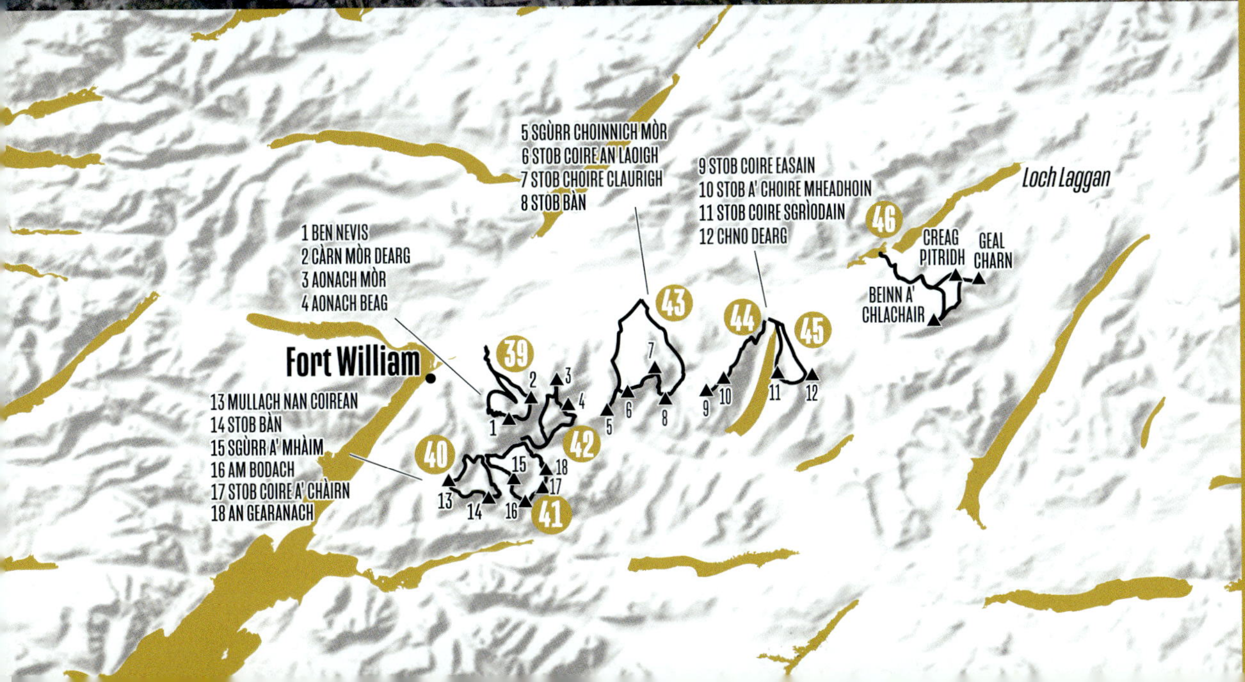

5 SGÙRR CHOINNICH MÒR
6 STOB COIRE AN LAOIGH
7 STOB CHOIRE CLAURIGH
8 STOB BÀN
9 STOB COIRE EASAIN
10 STOB A' CHOIRE MHEADHOIN
11 STOB COIRE SGRÌODAIN
12 CHNO DEARG
Loch Laggan
46
CREAG PITRIDH
GEAL CHARN
BEINN A' CHLACHAIR
1 BEN NEVIS
2 CÀRN MÒR DEARG
3 AONACH MÒR
4 AONACH BEAG
43
44
45
Fort William
39
13 MULLACH NAN COIREAN
14 STOB BÀN
15 SGÙRR A' MHÀIM
16 AM BODACH
17 STOB COIRE A' CHÀIRN
18 AN GEARANACH
40
42
41

39 Ben Nevis via the Càrn Mòr Dearg Arête

HUB **Fort William – 10-minute drive**

OVERALL 1 2 (3) 4 5 TERRAIN 1 2 3 (4) 5
DISTANCE 1 2 3 (4) 5 NAVIGATION 1 (2) 3 4 5

DISTANCE **21km** ASCENT **1,640m**
TIMES **10:45hrs** (Walker) **9:20hrs** (Trekker)
7:00hrs (Fastpacker) **4:35hrs** (Trail runner)

START LOCATION **North Face car park, near Torlundy**
START LATITUDE/LONGITUDE **56.8421, -5.0434**
START GRID REFERENCE **NN144764**
START W3W **///flickers.consented.infuses**

	PRONUNCIATION	NAME MEANING
Càrn Mòr Dearg	**'carn more jerreck'**	**big red hill**
Ben Nevis	**'ben nevis'**	**venomous mountain**

Introduction

Ben Nevis is Britain's highest peak, but its most popular route – the Mountain Track – can resemble a treadmill-like slog. This route via Càrn Mòr Dearg is far superior. It incorporates a rip-roaring ridge walk with incredible rock scenery and an adrenaline-inducing dose of adventure. The arête between the two Munros – a sweeping, curving, sharp spine of rock – is a grade-1 scramble, and undoubtedly the best way to reach the Ben's iconic summit.

View from the Ben Nevis plateau down Loch Linnhe towards Mull.

Public Transport

Scottish Citylink and local bus services pass through Torlundy which is one kilometre from the start point and links the area to major towns and cities including Glasgow, Inverness and Fort William. It's a 15-minute walk to the start from here.

Parking

Chargeable parking for around 35 cars in the North Face car park situated just off the A82 south of Torlundy. The car park is managed by Forestry and Land Scotland. The nearest toilet facilities are in nearby Caol, at the Nevis Range, or in Fort William. The access road to the car park is rough so be on the lookout for potholes.

Escape Route

No escape routes; retrace your steps after the first Munro and return another day for the second.

Refreshments

The Larder Cafe at the Highland Soap Company Visitor Centre is a great place to stop. You can grab some delicious food while also picking up some organic handmade soap souvenirs in quirky scents like Scottish bog myrtle, whisky and honey, and Hebridean seaweed.

Above The Càrn Mòr Dearg Arête and Ben Nevis from Càrn Mòr Dearg. **Opposite** Ben Nevis from Stob Choire Claurigh (see route 43).

The Route

S Leave the car park to the south-east. Cross the bridge over the Allt na Caillich and pass around the vehicle barrier. Walk along the track and after 100m turn right; a fingerpost reads *North Face Trail*. At the next junction continue straight on to the south and then fork left at NN145760 to ascend on a good path through the woodland.

2 At NN145756, a path merges in from the right as you continue straight ahead southwards. The path reaches a perfectly placed bench with a fantastic view over the woodland towards Fort William. Continue straight ahead, ignoring turn-offs for just over 500m until you reach NN147750 where you fork right following the fingerpost to *Allt a' Mhuilinn & CIC Hut*. Pass over a large stile.

3 Just over 500m after the stile, at NN151744, turn left off the track on to a grassy, boggy trail up the open hillside. The ground becomes firmer and rockier as you gain height. Lochan Meall an t-Suidhe comes into view on the other side of the valley to your right; this will become a key feature later on. Behind you is an expansive view down Loch Eil.

4 Cross a stream at around 840m in elevation and continue ascending. The ground becomes eroded and loose as you climb above 1,000m. Soon after this, you reach the ridge and can enjoy a marvellous view across Lochaber. Ascend over scree to Càrn Dearg Meadhonach.

5 From this elevated position, follow the trail down to the south and then ascend to gain the summit of the first Munro, Càrn Mòr Dearg. The view across to both Aonach Mòr and Aonach Beag (page 209) can be enjoyed from here but the sight down the arête to Ben Nevis is certainly the highlight.

6 If you use walking poles, put them away now. The ridge is blocky with an abundance of hand and footholds along the way. It's exposed but there is no requirement to make any complex scrambling moves in exposed positions. In fact, the scrambling across the long, undulating ridge is quite straightforward and never drops below 1,000m. There is a bypass path to the left for the most part.

7 Once you've reached the large cairn at the foot of the final climb, the ridge walk is over and it's time to ascend over bouldery terrain to the top of Ben

Ascending Càrn Mòr Dearg with the north face of Ben Nevis in the distance.

Nevis. Ascend over the boulders before a scree path becomes prominent on the right-hand side of the slope. Top out on the plateau and head to the trig pillar which marks the highest point in the United Kingdom. On a sunny weekend, don't be surprised if there are over a hundred visitors on the summit of Ben Nevis at any one time. Be sure to explore the full plateau, if only to distance yourself from the crowds. I particularly enjoy the view from the southern edge of the plateau to the Mamores (pages 185, 201 and 205).

8 Head off the summit plateau to the west following the well-used Mountain Track as it zigzags down towards Lochan Meall an t-Suidhe.

9 Fork right at NN147724. Most other hillwalkers will turn left here so you're likely to find some solitude on this part of the route. Many head north from the lochan down pathless wet slopes and cross the Allt a' Mhuilinn to rejoin the approach path, however my preference is to follow the path which loops around the northern crags of Ben Nevis towards the CIC Hut. Sticking to this path adds around 3km to the route but it avoids all the wet, pathless walking and it's a really pleasant trail.

10 Cross the stream just below the hut and walk downhill on the other side of the valley. Meet the turn-off for Càrn Mòr Dearg and retrace your steps back to the start.

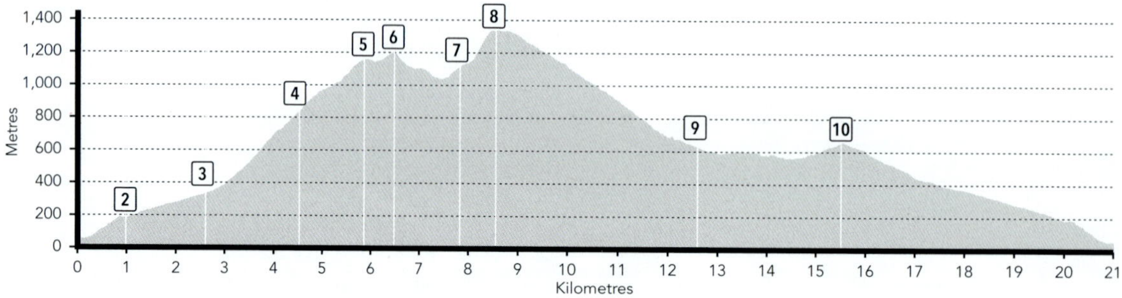

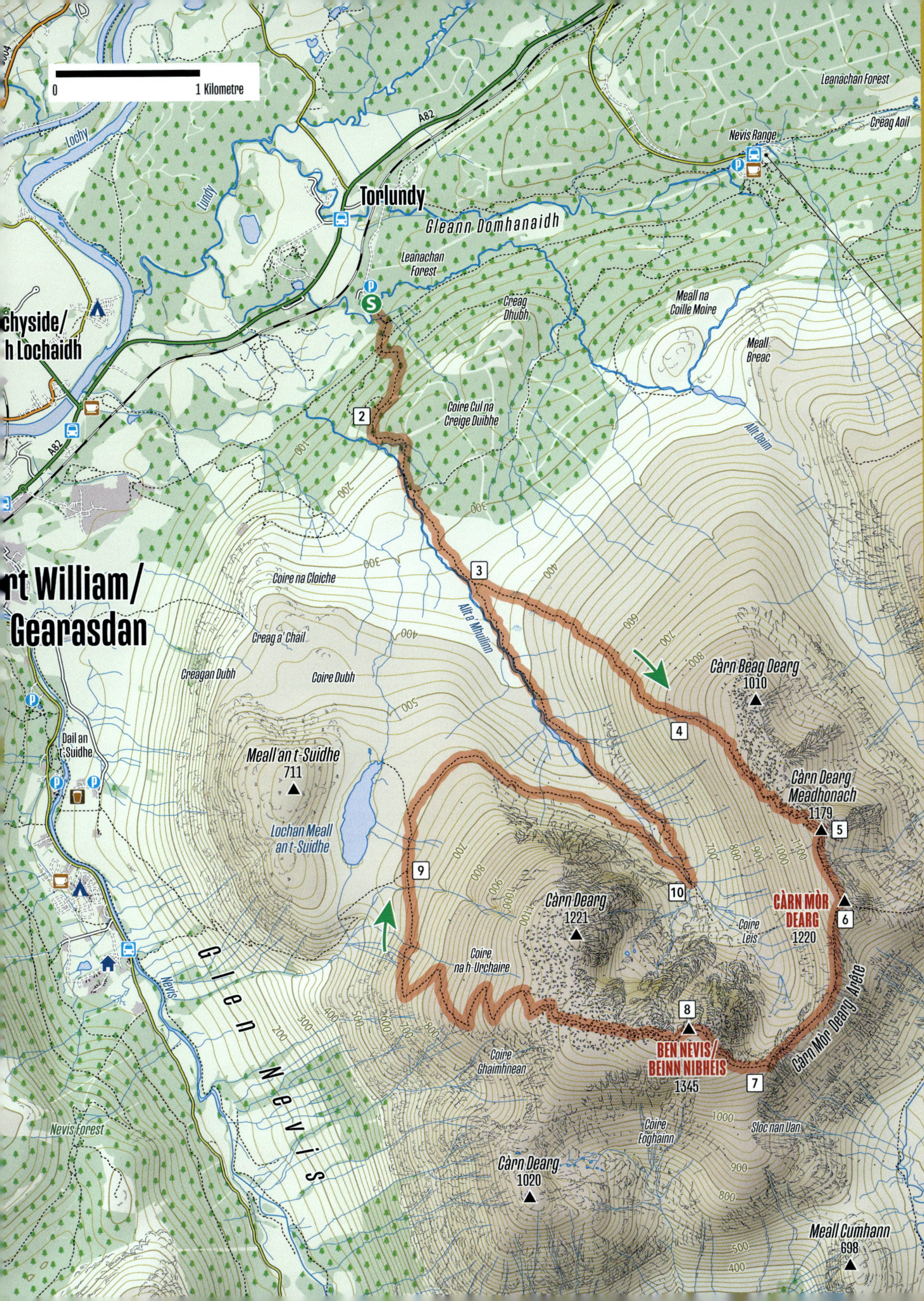

0
1 Kilometre
Leanachan Forest
Creag Aoil
Nevis Range
A82
Lochy
Torlundy
Lundy
Gleann Domhanaidh
Leanachan Forest
Creag Dhubh
Meall na Coille Moire
Meall Breac
chyside/
h Lochaidh
Coire Cul na Creige Duibhe
Allt Daim
rt William/
Gearasdan
Coire na Cloiche
Allt a' Mhuilinn
Creag a' Chail
Coire Dubh
Creagan Dubh
Càrn Beag Dearg
1010
Dail an t-Suidhe
Meall an t-Suidhe
711
Càrn Dearg Meadhonach
1179
Lochan Meall an t-Suidhe
CÀRN MÒR DEARG
1220
Càrn Dearg
1221
Coire Leis
Coire na h-Urchaire
Glen Nevis
Nevis
BEN NEVIS/ BEINN NIBHEIS
1345
Càrn Mòr Dearg Arête
Coire Ghaimhnean
Coire Foghainn
Sloc nan Uan
Nevis Forest
Càrn Dearg
1020
Meall Cumhann
698

40 West Mamores

HUB **Fort William – 15-minute drive**

OVERALL 1 2 **3** 4 5 TERRAIN 1 2 **3** 4 5
DISTANCE 1 **2** 3 4 5 NAVIGATION 1 2 **3** 4 5

DISTANCE **13.2km** ASCENT **1,175m**
TIMES **7:20hrs** (Walker) **6:20hrs** (Trekker)
4:45hrs (Fastpacker) **3:05hrs** (Trail runner)

START LOCATION **Lower Falls car park, Glen Nevis**
START LATITUDE/LONGITUDE **56.7696, -5.0370**
START GRID REFERENCE **NN144683**
START W3W **///dangerously.blackbird.spray**

	PRONUNCIATION	NAME MEANING
Stob Bàn	**'stob ban'**	**white peak**
Mullach nan Coirean	**'mull-uch nan kor-yen'**	**summit of the corries**

Introduction

Is west best, when it comes to the Mamores? These two Munros – the westernmost peaks of the range – serve up a superlative circuit, with fine views and a delightful ridge. There is a lovely dichotomy between the two mountains too. Stob Bàn is defined by its white quartzite screes, whereas Mullach nan Coirean is a rich red granite – the contrast is stark and intriguing. Between the two is an undemanding ridge traverse. It's the perfect place to meditate on the majesty of the Mamores.

Public Transport

In the summer months, a local bus service shuttles visitors from Fort William along Glen Nevis as far as the Lower Falls car park. The service does not operate during the winter months.

Parking

Chargeable parking for around 50 cars at the Lower Falls car park in Glen Nevis. Basic toilet facilities are provided.

Escape Route

Retrace your steps after the first Munro. Safe and comfortable escape routes are in short supply along the remainder of the route.

Refreshments

On your way out of the valley, stop at Glen Nevis Restaurant & Bar for a hot meal. The restaurant partners with a local butcher and fishmonger to deliver responsibly sourced meat and fish dishes.

Descending Mullach nan Coirean with Stob Bàn in the distance.

Looking back to Stob Bàn from the approach to Mullach nan Coirean.

The Route

S Leave the car park and walk north along the road briefly. Pass over a bridge and turn right before you reach a second bridge. Hop over a stile and pass through a gate on to a gravel path which leads south-east above the Allt Coire a' Mhusgain. Follow this path as it becomes muddy and then rocky, ascending uphill through scrubland.

2 Pass through a gate at NN150671 and continue up the left-hand side of the coire. After 1km the path begins zigzagging up the side of Sgùrr a' Mhàim (page 205). At the top of the switchbacks, traverse south-east, high above the Allt Coire a' Mhusgain opposite some dramatic waterfalls and crags.

3 Cross numerous small streams along this part of the route including the Allt Coire a' Mhusgain at the top of the valley at NN158655. Walk south-west from here on your approach to the bealach between Sgùrr an Iubhair and Stob Bàn at NN156653.

4 Turn right at the bealach at the large cairn and walk uphill to the right of a very small lochan which is surrounded by an array of shimmering quartzite. The long, narrow spine of Aonach Eagach (pages 177 and 181) can be seen to the south. Follow the eroded path as it leads around the edge of the Coire a' Mhusgain over pure white rock. The route here is steep and loose in places with some big steps up required, and while it doesn't constitute a scramble, you might want to put your hand down on occasion.

5 Towards the top of the ridge, the path leads west and narrows with drops on either side, but it is just wide enough to not feel too exposed. A cairn beside a line of slabby rock marks the path turning north where it's a straightforward short walk up to the summit of Stob Bàn.

6 Walk off the summit to the north following a path of loose white quartzite stones. At NN147655 beside a cairn, ignore the rockfall on your left which looks suspiciously path-like in low visibility and continue straight ahead before branching left just before the 912m north top. From here, head west-north-west down to the bealach (marked as 846m on OS maps) around the rim of Coire na Sleubhaich.

7 At the bealach walk west-south-west. When the route hits some red granite boulders, keep right here for the easiest line along the edge. The path then leads through increasingly reddening rock to a large cairn at NN139657. From here, walk south-west. The path drops and skirts around the head of Coire Dearg and you can't fail to notice some impressive gullies and rock needles on your right.

8 The path mildly undulates and then curves to the left of the crest as you're treated to the view down to the south-west where the Allt na Lairige Mòire runs beside the prominent West Highland Way path. Mullach nan Coirean is in view up ahead, framed by red-tinged crags. The ridge here undulates a final time before the eroded path of red granite leads north-west to the top of Mullach nan Coirean which is marked by a huge cairn. Sgùrr a' Mhàim (page 205) and Stob Bàn's sheer, pale grey faces command attention.

9 Descend to the north along a path which is stony at first. Soon the path bears north-east down the shoulder which is delectably narrow and airy but without too much exposure. At NN131671, there is a stile in the deer fence; ignore this and walk north-north-east. A second stile at NN134676 can also be ignored as the path weaves north-north-west.

10 The path becomes grassier and wetter as it approaches a stile at NN132681. Cross this stile and

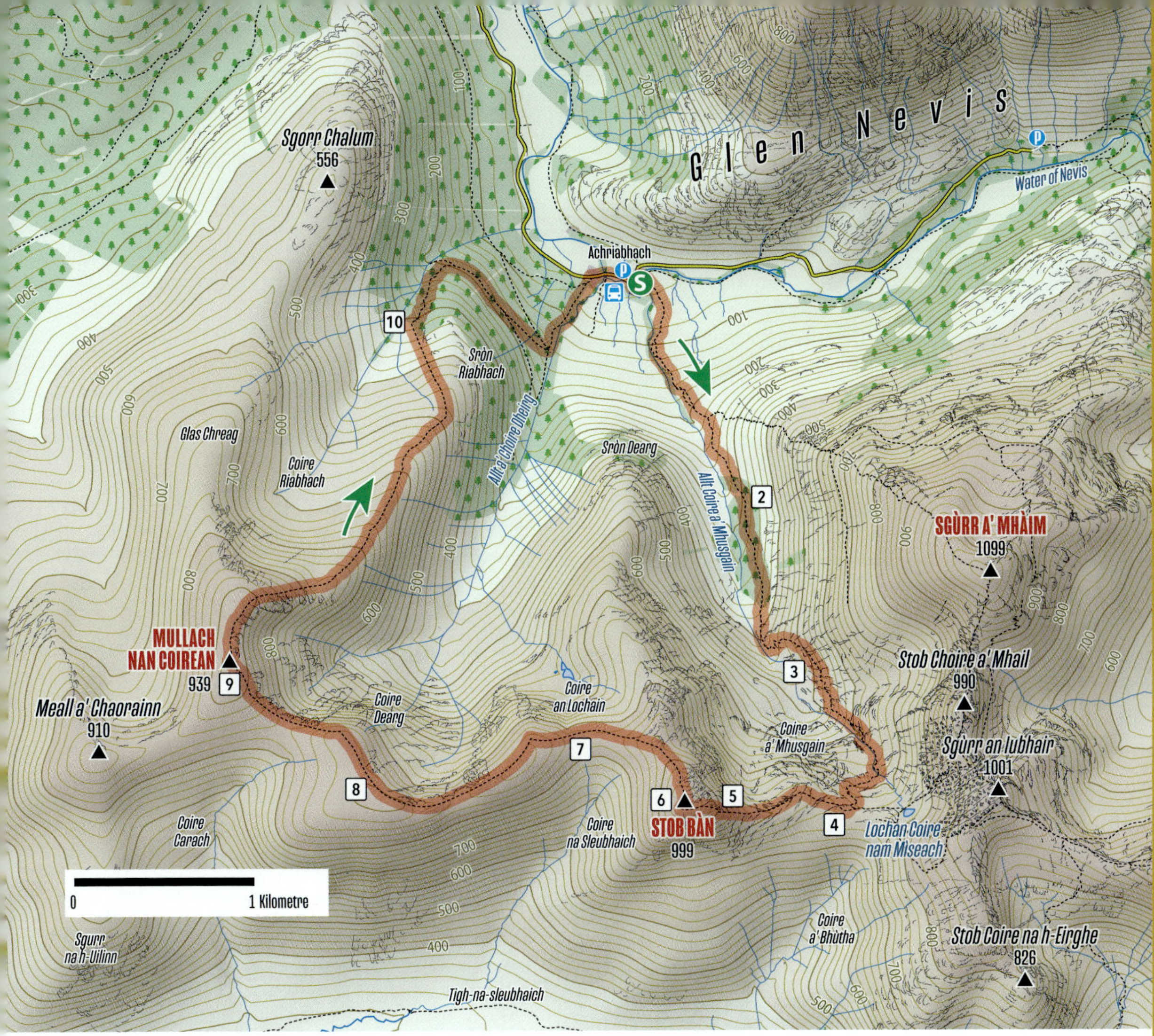

enter the woodland which leads to a stone-pitched path through heather and scrubland. At the bottom of this path, turn right on to a track. Walk on the track for 380m before taking the left-hand fork at a major junction in the track. Continue for 350m before rounding a hairpin in the track. Immediately after the hairpin bend, take a path on your right which leads north-east down to the road. Turn right at Achriabhach and walk for 300m back to the start.

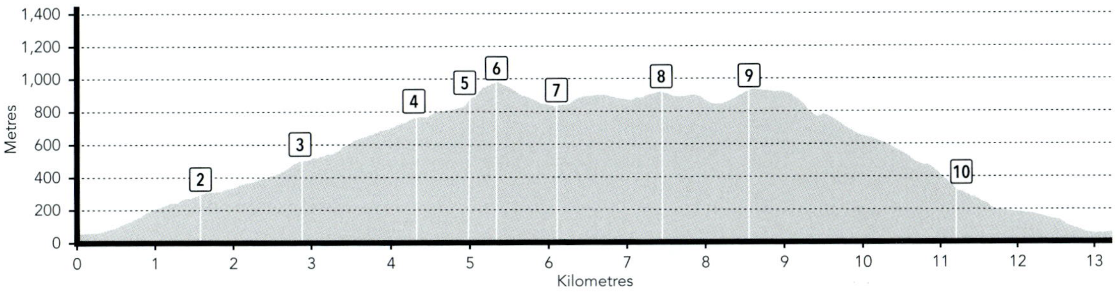

41 Ring of Steall

HUB **Fort William – 20-minute drive**

OVERALL 1 2 **(3)** 4 5 TERRAIN 1 2 3 **(4)** 5
DISTANCE 1 2 **(3)** 4 5 NAVIGATION 1 2 **(3)** 4 5

DISTANCE **16.7km** ASCENT **1,845m**
TIMES **10:45hrs** (Walker) **9:20hrs** (Trekker)
7:00hrs (Fastpacker) **4:25hrs** (Trail runner)

START LOCATION **Steall Waterfall car park, Glen Nevis**
START LATITUDE/LONGITUDE **56.7772, -5.0013**
START GRID REFERENCE **NN167691**
START W3W **///toasted.trailers.promoting**

	PRONUNCIATION	NAME MEANING
An Gearanach	**'an gerrenach'**	**the complainer**
Stob Coire a' Chàirn	**'stob corriya a carn'**	**peak of the corrie of the cairn**
Am Bodach	**'am bodach'**	**the old man**
Sgùrr a' Mhàim	**'sgoor a vame'**	**peak of the round hill**

Introduction
Compile a list of the best hikes in Scotland and the Ring of Steall will always make the cut. It's a Highlands classic and one of Great Britain's best horseshoe routes, with brutal ascents, epic ridges and captivating scenery combining to create a slice of Mamores magic. From the wobbly wire bridge river crossing near the start, to the sinuous ascent of the Devil's Ridge to Sgùrr a' Mhàim's summit, the Ring of Steall is a non-stop rollercoaster of hillwalking heaven.

Public Transport
In the summer months, a local bus service shuttles visitors from Fort William to the Lower Falls car park which is passed towards the end of the route. Start your walk from the bus stop and walk the latter part of this route first.

Parking
Free parking for around 40 cars in the Steall Waterfall car park which is situated at the very end of the Glen Nevis Road. This is a singletrack road with a few potholes so drive with care. The nearest toilet facilities are at Lower Falls car park.

Escape Route
Retrace your steps after the first Munro. Safe and comfortable escape routes are in short supply along the remainder of the route.

Refreshments
On your way out of the valley, stop at Glen Nevis Restaurant & Bar for a hot meal. The restaurant partners with a local butcher and fishmonger to deliver responsibly sourced meat and fish dishes.

Looking along the ridge towards Sgùrr an lubhair. © Jon Barton

The Route

S Leave the car park to the east, following the woodland path into the Steall Gorge. The path is rough and rocky but is popular and well walked. The John Muir Trust estimates that 60,000 visitors use this footpath every year. The path opens out into a meadow with the waterfall – An Steall – catching your eye up ahead.

2 Fork right on the path when the wire bridge comes into view. Cross the bridge with your feet on the central wire and hands outstretched either side. Using this bridge over the Water of Nevis is not for the faint-hearted – it's both higher and longer than it looks. The alternative is to paddle across the Water of Nevis further upstream, but this might be impossible after a lot of rain. Follow the path on the other side which leads to the left of a private climbing hut and across boggy ground to the foot of the waterfall.

3 Find a place to cross and then progress to a stretch of exceptionally boggy ground. Many a Ring of Steall walker completes the route with wet feet due to this section. It's a relief when you can progress beyond the bog on to a path, which ascends to the east at first. Cross the Allt Coire Chadha Chaoruinn and continue up the stony zigzag path. Cross the same stream again higher up as the path ascends under the craggy slopes.

4 At around 600m elevation, the smaller zigzags pause as the path traverses west before bending south-south-east over momentarily flatter ground with an epic view to Ben Nevis (page 195) behind you. The view to Sgùrr a' Mhàim can also be enjoyed to your right. It's the highest Munro of the day and from here it certainly looks it. All of a sudden you top out on to the summit of An Gearanach. The view north-east from here to the Grey Corries (page 213) is sublime.

5 Begin the ridge walk to An Garbhanach. As you approach the summit of An Garbhanach there are steep drops if you stick to the crest, so take the path to the left which provides easier and less exposed scrambling and skirts around the problems. Confident scramblers can take on the exposed scramble along the crest.

6 Descend from An Garbhanach to the bealach down an eroded path featuring loose scree. From the bealach, follow the path which zigzags up Stob Coire a' Chàirn, avoiding the need for scrambling. The view to Na Gruagaichean (page 185) from the summit is exceptional.

7 Leave the summit to the west initially, following a grassy trail which is interspersed with stony ground. The path soon turns south-west as you walk along the ridge over a minor top. The walk from the bealach up Am Bodach looks fearsome from this angle, but the path is clear and there are only a couple of simple scramble moves required towards the top. From the summit, you get the sense that you're slap bang in the middle of the Mamores, with wonderful views across the whole range.

8 Descend to the west while enjoying a view down Loch Leven as the ground drops away from right to left. Reach the bealach, which is long and flat. It offers some brief relief from the relentless cycle of steep ascending and descending required along the route. From the bealach, ascend north-west up Sgùrr an Iubhair on stony terrain. Sgùrr an Iubhair was once a Munro for a brief time back in the 1980s but it was relegated to a Munro Top in the 1990s.

9 Descend north-west to the bealach before rising northwards to start the Devil's Ridge. The ridge is deliciously airy with drops on either side. The crux of the ridge is just before the lowest point where you need to descend a ridged slabby section. There are shelves down the rock which are just wide enough for a foot. The best line is diagonally from right to left. Continue uphill from here over stony ground. The remainder of the ascent is straightforward uphill walking to the summit of Sgùrr a' Mhàim which is marked by a large cairn. This Munro offers the best view to Ben Nevis (page 195).

10 Descend to the north-west on a path which leads to some prime scree-surfing down loose shattered quartzite. When the scree ends, follow the eroded trail down grassy slopes as the gradient drops ceaselessly into Glen Nevis. Pass through a gate and continue descending until the ground

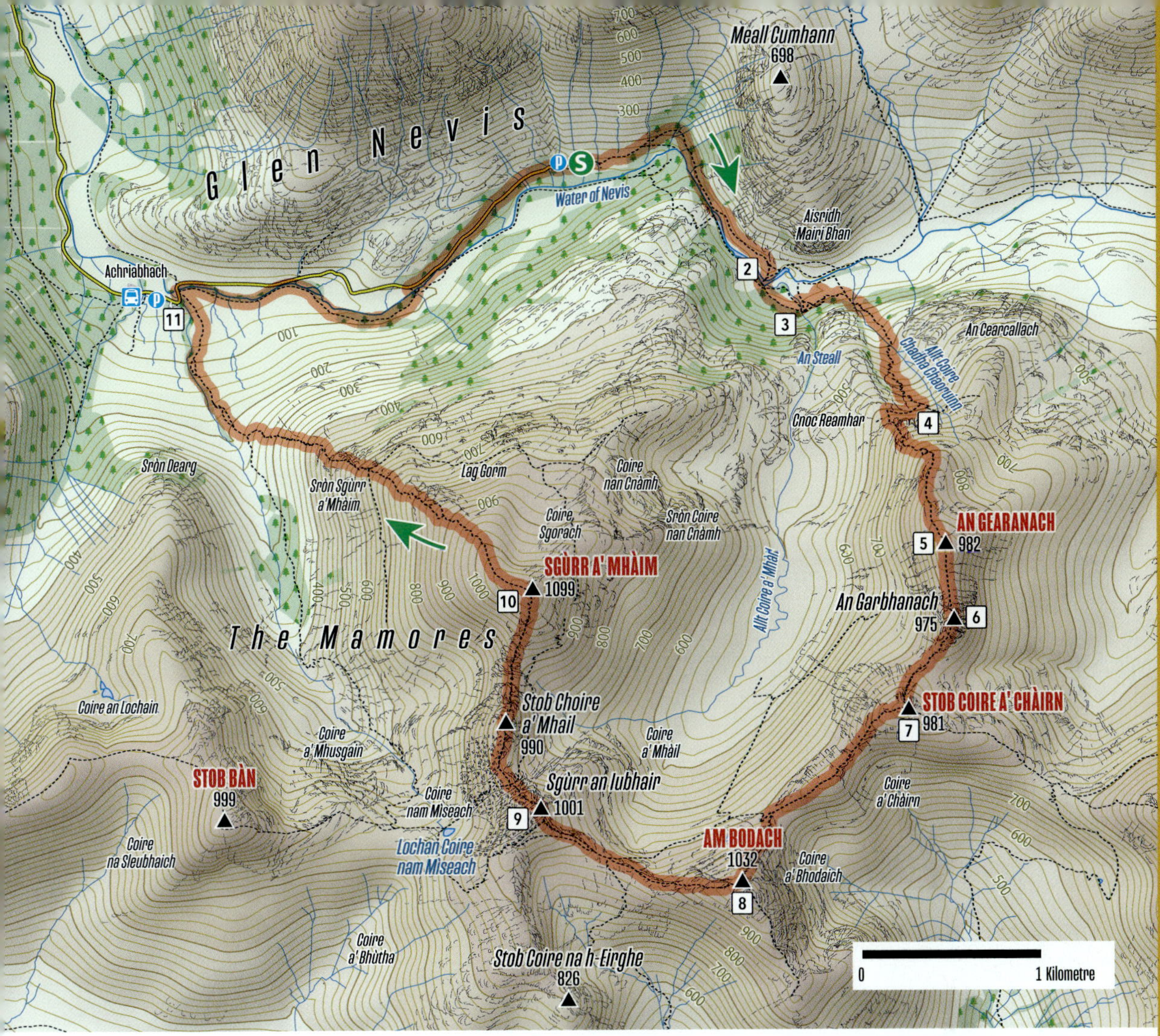

becomes wetter as you get closer to the Allt Coire a' Mhusgain.

11 At NN145683, pass through a gate and climb over a stile to the roadside next to Lower Falls car park. Make a hard right here to follow a path which runs along the south side of the river. A fingerpost reads *Riverside Path to Paddy's Bridge 1.6km* – the pleasant path runs alongside the Water of Nevis. Cross Paddy's Bridge and turn right on to the road, which can be followed for 1.3km back to the start.

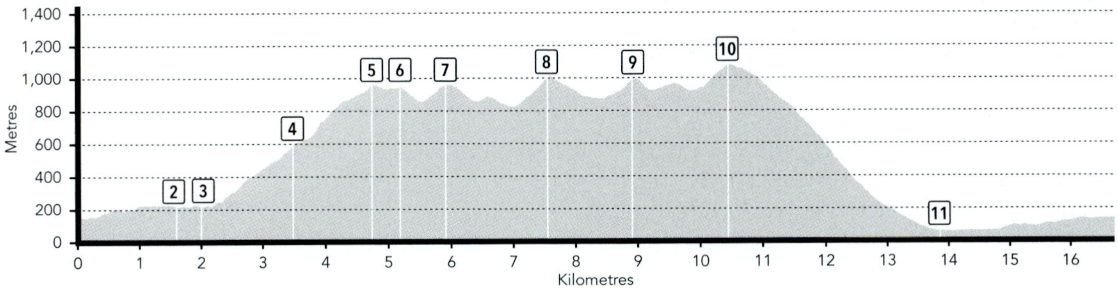

42 The Aonachs

HUB **Fort William – 20-minute drive**

OVERALL 1 2 3 (4) 5 TERRAIN 1 2 3 (4) 5
DISTANCE 1 2 (3) 4 5 NAVIGATION 1 2 3 (4) 5

DISTANCE **17.4km** ASCENT **1,425m**
TIMES **9:10hrs** (Walker) **8:00hrs** (Trekker)
6:00hrs (Fastpacker) **3:55hrs** (Trail runner)

START LOCATION **Steall Waterfall car park, Glen Nevis**
START LATITUDE/LONGITUDE **56.7772, -5.0013**
START GRID REFERENCE **NN167691**
START W3W **///toasted.trailers.promoting**

	PRONUNCIATION	NAME MEANING
Aonach Beag	**'oonach bake'**	**little ridge**
Aonach Mòr	**'oonach more'**	**big ridge**

Introduction

Destined to be forever overshadowed by nearby Ben Nevis (page 195), the twin peaks of Aonach Mòr and Aonach Beag deserve more acclaim – they are brutally built leviathans by every measure, except perhaps in direct comparison to their loftier neighbour. Confusingly, Aonach Beag (Little Ridge) is 13 metres higher than Aonach Mòr (Big Ridge), but both deliver unambiguously epic hillwalking terrain. The views across Glen Nevis from the Allt Coire nan Laogh are particularly impressive.

Public Transport

In the summer months, a local bus service shuttles visitors from Fort William to the Lower Falls car park. From the bus stop, take the riverside path to Paddy's Bridge and then follow the road to the start of the walk. This adds 5.6 kilometres to the route, bringing the total route length to 23 kilometres.

Parking

Free parking for around 40 cars in the Steall Waterfall car park which is situated at the very end of the Glen Nevis Road. This is a singletrack road with a few potholes so drive with care. The nearest toilet facilities are at Lower Falls car park.

Escape Route

Descend from Aonach Beag to the bealach and leave the out-and-back of Aonach Mòr for another day.

Refreshments

On your way out of the valley, stop at Glen Nevis Restaurant & Bar for a hot meal. The restaurant partners with a local butcher and fishmonger to deliver responsibly sourced meat and fish dishes.

View from Sgùrr Choinnich Mòr to Aonach Beag during a temperature inversion (see route 43).

The Route

S Leave the car park to the east, following the woodland path into the Steall Gorge. The path is rough and rocky but is popular and well walked. The John Muir Trust estimates that 60,000 visitors use this footpath every year. The path opens out into a meadow with the waterfall – An Steall – catching your eye up ahead.

2 Fork left when the path splits for the An Steall wire bridge and remain on the gravel path as it heads deeper into Glen Nevis. The path leads to an open area where you can walk along the edge of the Water of Nevis to observe the spectacular waterfall of An Steall from all angles.

3 Continue along the footpath on the northern bank of the Water of Nevis until you reach a footbridge over the Allt Coire Giubhsachan. Note for later that your route of descent rejoins the bottom path here, but for now cross the bridge and turn immediately left in front of the ruins of Steall, the remains of a settlement which dates back to the 1800s and was still in use just before World War II. Walk alongside the Allt Coire Giubhsachan for a short distance until the path meets the Allt Coire nan Laogh and follows it north-east. To your right, Binnein Beag (page 185) stands apart from its neighbours in the Mamores range, while an imposing Sgùrr a' Mhàim (page 205) looms over Glen Nevis behind you.

4 When a smaller stream meets the Allt Coire nan Laogh, cross it and ensure you keep to the path which follows the right-hand bank of the Allt Coire nan Laogh; it's easy to follow the wrong stream here as there is a path which follows the minor stream too. The path beside the Allt Coire nan Laogh is an eroded line interspersed with rocky sections which present no technical difficulty. As you gain higher ground, Stob Coire Bhealaich becomes more prominent up ahead.

5 At around 570m in elevation the path steepens as it begins to veer away from the Allt Coire nan Laogh to ascend a spur. Beyond 650m in elevation, if you lose the path in long grass, keep walking in a north-easterly direction. The ground runs away from right to left into the gorge which holds the Allt Coire nan Laogh.

6 When the path peters out for good around NN196701, walk east over pathless ground and trust that the quartzite-scattered slopes of Sgùrr a' Bhuic will soon come into view. The terrain becomes rockier as you ascend the upper slopes of Sgùrr a' Bhuic over shattered quartzite interspersed with grassy rakes.

7 Enjoy the view to the Grey Corries (page 213) from the summit of Sgùrr a' Bhuic, then descend north-north-east to the bealach keeping the dramatic drop to the Coire a' Bhuic on your right. Pick your way up the boulder-strewn slopes of Stob Coire Bhealaich. When the boulders relent, a grassy path can be picked up on the cliff edge to the right. Ascend beside the dramatic cliffs before skirting from east to west under point 1,048m on OS maps.

8 Pass over the top of Stob Coire Bhealaich and continue along the top of the cliffs as the path dips slightly and then curves to the right to begin the ascent of Aonach Beag. The route to the summit is straightforward thanks to a clear path along the edge of the cliffs. Reach the summit and look west from the edge of the summit plateau to Ben Nevis, Càrn Mòr Dearg and the sweeping arête which links them (page 195).

9 Descend Aonach Beag, keeping left of the cliffs to pick up a loose stony path down to the bealach between the two Munros. Cross the bealach and begin ascending Aonach Mòr. Note the fork in the path at NN193726 as this marks your route of descent. For now, continue north over gradual grassy slopes following the clear path to the cairned summit of Aonach Mòr.

10 Take a well-earned break before retracing your steps to NN193726. Fork right here to walk south-south-west on grassy ground which descends gradually towards a cairn at Seang Aonach Mòr at NN191722. This marks the start of a steeper descent to the bealach between this and Càrn Mòr Dearg. Take care descending the eroded steep loose channel of dusty red rock. The path is primarily on the left-hand side of the gully.

11 At NN186722 in the bealach, turn left and begin walking south towards Glen Nevis following a clear path. When the path becomes wet and indistinct at NN186716, cross the stream and pick up a grassy path on the other side.

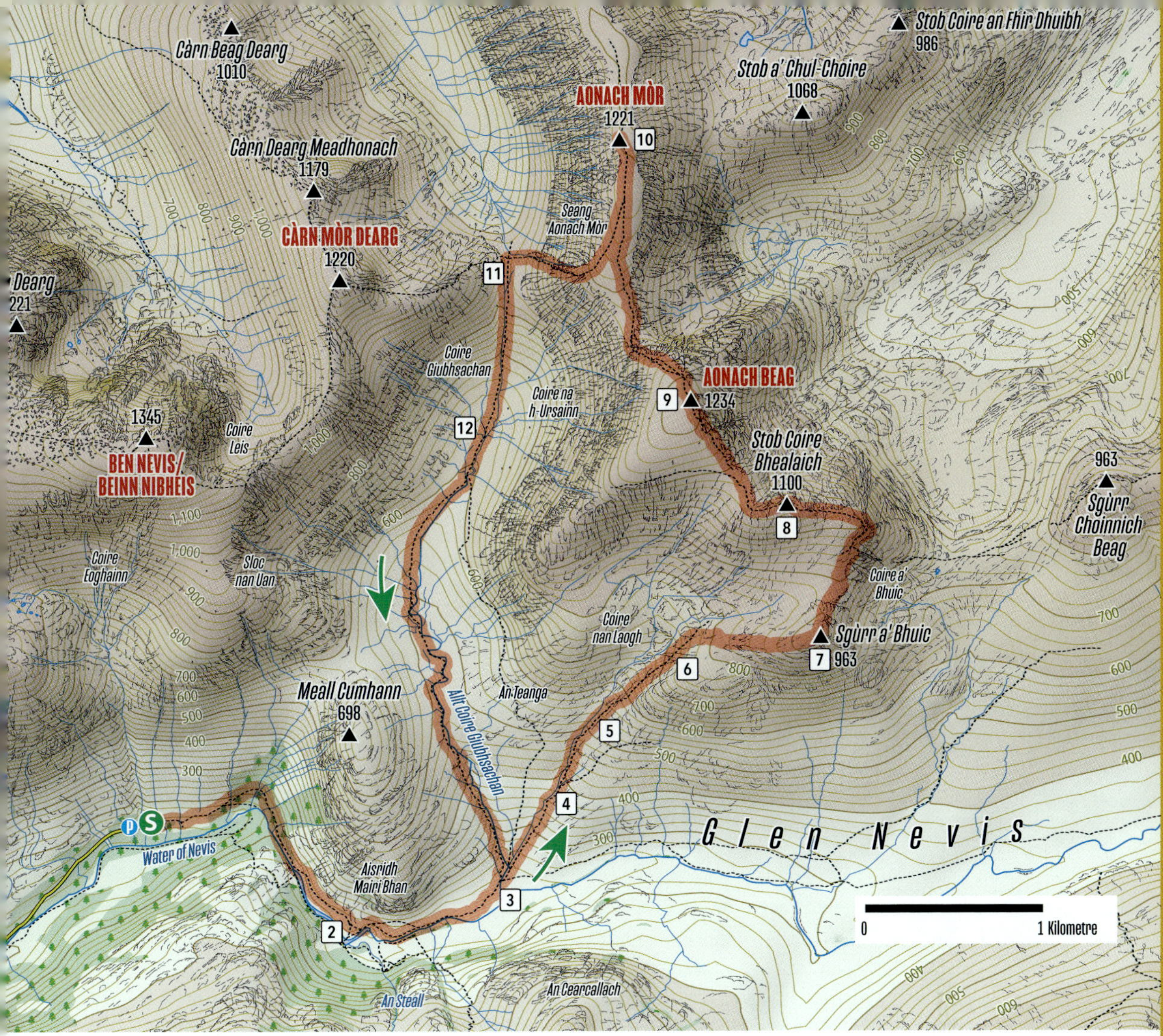

12 Remain on the western side of the Allt Coire Giubhsachan, enjoying marvellous views to Meall Cumhann and An Gearanach (page 205). The ground is very boggy, especially on the flatter parts of the route, and it is rocky here and there, but not technical. Once back at the bridge (**3**), turn right into Glen Nevis, retracing your steps through the glaciated valley back to the start.

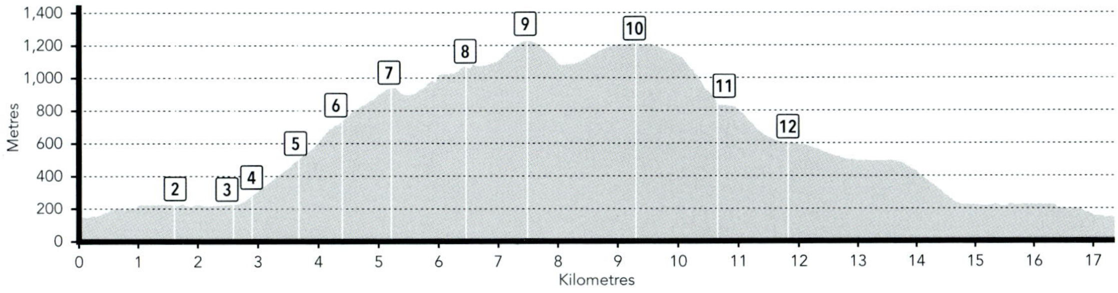

STOB BÀN 977m
STOB CHOIRE CLAURIGH 1,177m
STOB COIRE AN LAOIGH 1,116m
SGÙRR CHOINNICH MÒR 1,094m

43 The Grey Corries

HUB **Fort William – 30-minute drive**

OVERALL 1 2 3 **(4)** 5 TERRAIN 1 2 3 **(4)** 5
DISTANCE 1 2 3 4 **(5)** NAVIGATION 1 **(2)** 3 4 5

DISTANCE **27.2km** ASCENT **1,915m**
TIMES **13:05hrs** (Walker) **11:20hrs** (Trekker)
8:30hrs (Fastpacker) **5:35hrs** (Trail runner)

START LOCATION **Parking area south of Corriechoille**
START LATITUDE/LONGITUDE **56.8727, -4.8682**
START GRID REFERENCE **NN252793**
START W3W **///escapades.educated.radically**

	PRONUNCIATION	NAME MEANING
Stob Bàn	**'stob ban'**	**white peak**
Stob Choire Claurigh	**'stob horriya cloree'**	**peak of the corrie of clamouring**
Stob Coire an Laoigh	**'stob corriya an loyee'**	**peak of the corrie of the calf**
Sgùrr Choinnich Mòr	**'sgoor connich more'**	**big peak of the moss**

Introduction

The rocks are distinctively grey, but your mood never will be when hiking in the Grey Corries. This quartzite-rich, scree-scarred mountain range to the east of Ben Nevis (page 195) is ruggedly unique. A snaking labyrinth of arêtes links these four Munros, exploring a mesmerising, silver-hued world of rock – with a long ridge section between Stob Choire Claurigh and Stob Coire an Laoigh never once dropping below 1,000 metres in elevation.

Public Transport

There is a railway station and a bus stop in Spean Bridge with Scottish Citylink and local bus services linking to major towns and cities including Glasgow, Inverness and Fort William, but using these will add around 10 kilometres to the overall route length and for most people this would require a multi-day outing. In this case, the bothy in Lairig Leacach can be used (many bothies in Scotland are maintained by the Mountain Bothies Association – always follow the Bothy Code, see page xvi). My recommended itinerary involves a 13-kilometre walk to the bothy on day one then a 24-kilometre Munro walk, including the walk back to Spean Bridge, on day two.

Parking

Free parking for six cars in a parking area beside a wide track junction next to the Leanachan Forest, south of Corriechoille. The drive to the parking area is via a gated track which is extremely rough and around a kilometre in length. If you have lowered suspension, don't attempt to drive on this track.

Walkers descend Stob Choire Claurigh bound for Stob Coire an Laoigh.

Above A walking party passes the Lairig Leacach bothy en route to Stob Bàn. **Opposite** Beginning the descent of Sgùrr Choinnich Mòr in front of the broad flanks of Stob Coire an Laoigh with Stob Choire Claurigh in the distance.

If full, there are a few more spaces further up the track (NN255788, ///hopefully.trucked.screening). The nearest toilet facilities are in Spean Bridge.

Escape Route

There are a few possible escape routes.

- Simply return via the route of ascent from Stob Bàn.
- From the bealach between Stob Bàn and Stob Choire Claurigh (NN264729), follow the stream back down to the bothy.
- Descend to the north from Stob Choire Claurigh via Stob Coire Gaibhre back into the forest.
- Make a decision on Stob Coire Easain to cut the out-and-back to Sgùrr Choinnich Mòr from the route and return for it another day.

Refreshments

'Dine on the line' at the Old Station Restaurant at Spean Bridge railway station – the building dates back to the Victorian era.

The Route

S Leave the parking area on the track to the south-south-east. After 700m pass through a gate to enter the plantation and keep an eye out for the Wee Minister on your right. The original stone statue, reported to be of Reverend John McIntosh, was said to bring good luck to climbers and walkers. It has since been replaced by a wooden replica.

2 The track exits the plantation leading into the Lairig Leacach. Walk through the pretty glen, crossing the Allt Leachdach via a substantial bridge. Walk for a further 4.3km to reach Lairig Leacach bothy. Some walkers may prefer to split this route by staying at the bothy, which comes at the expense of carrying a heavier pack for the remainder of the route.

3 Ford the Allt a' Chùil Choirean next to the bothy and continue walking along the track. At NN281734, turn right off the track and on to a path which leads south-west through heather. The ground is soft and somewhat boggy at first but firms up as you gain height.

4 Cross the high point on the ridge, marked as 769m on OS maps, and drop slightly to look up at the striking profile of Stob Bàn. Follow the path as it crosses a small re-entrant and continues uphill. The ground steepens and the path passes through a series of small crags, avoiding any technicality.

5 Top out on Stob Bàn and enjoy the view across to all of the upcoming Munros. Stob Bàn is the smallest of the group but it has a great triangular profile. I particularly like the view across to the Mamores range (pages 185 and 201). Begin your descent by briefly walking west-south-west before picking up the start of a steep path heading north then north-west down loose quartzite scree. The drop to the bealach is short but very steep.

6 Cross the bealach, passing around the small lochan, and begin the climb up Stob Choire Claurigh following a grassy trail heading north-east. As you approach the higher reaches of the mountain, quartzite blocks prevail.

7 It's surprising how small Stob Bàn looks from the summit of Stob Choire Claurigh. Head south-west from the summit with one of my favourite Munro views stretching out right in front of you.

The Grey Corries, with Stob Bàn on the left, from the descent of Stob a' Choire Mheadhoin (see route 44).

The incredible ridge walk to Stob Coire an Laoigh doesn't drop below 1,000m and is backed by mountains increasing in height beyond the Grey Corries – the Aonachs (page 209) and Ben Nevis (page 195). The terrain along the ridge is a mixture of grass and boulder scree. First, pass Stob a' Choire Lèith before walking on to Stob Coire Cath na Sìne and along the ridge to Caisteal.

8 Keep glancing behind you to behold the snake-like ridge unfurl back to Stob Choire Claurigh. It's just as pleasing a view in the reverse direction. Dramatic cliffs plunge down on your right which can be observed safe in the knowledge that the route doesn't approach them. The ridge narrows slightly and then broadens for the final push up Stob Coire an Laoigh.

9 From the third Munro summit of the day, descend boulder scree to the west-north-west down to the bealach at the foot of Stob Coire Easain. Ascend to the summit from where it is possible to cut the walk short. If you're doing this, skip the next part of the directions and begin the descent into the valley.

10 Hardy Munro-baggers will want to continue on to Sgùrr Choinnich Mòr because it's only 1.7km away. That said, a spot of technical ground comes next so progress will likely be slowed. Leave Stob Coire Easain to the south-west on what starts off as a grassy line avoiding the rocks. A section of steeper, rocky ground soon presents itself and a handful of simple scramble moves over larger rocks are required to reach the grassy bealach. The ascent is short, in terms of vertical height gain, but it's steep and passes over a few false summits along the way. From the summit of Sgùrr Choinnich Mòr, the pointy tops of Sgùrr Choinnich Beag and Sgùrr a' Bhuic add a lot of character to the scene.

11 Retrace your steps to the top of Stob Coire Easain (**10**) and descend north-north-east over stony ground initially. The route, which passes over Beinn na Socaich, soon becomes grassier and wetter the further down you descend. Ignore any faint paths which lead off to the right too early and instead aim for NN235764 where you can pass through a gate in the deer fence which protects the plantation. The trail leads northwards through dense undergrowth to reach a forestry track.

12 Turn right on to the track, signposted *Spean Bridge/Aonach Mòr*. Cross a bridge over a stream and follow the wide track northwards for 700m to another junction. Turn right here, signposted *The Lairig/Corriechoille*. At the next junction, fork right which is identically signposted. Fork left at the final track junction for the last kilometre of walking. Exit the plantation using a gate and arrive back at the start point.

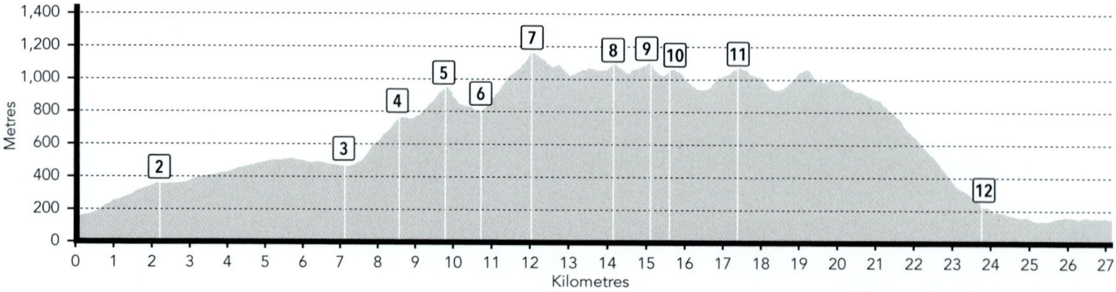

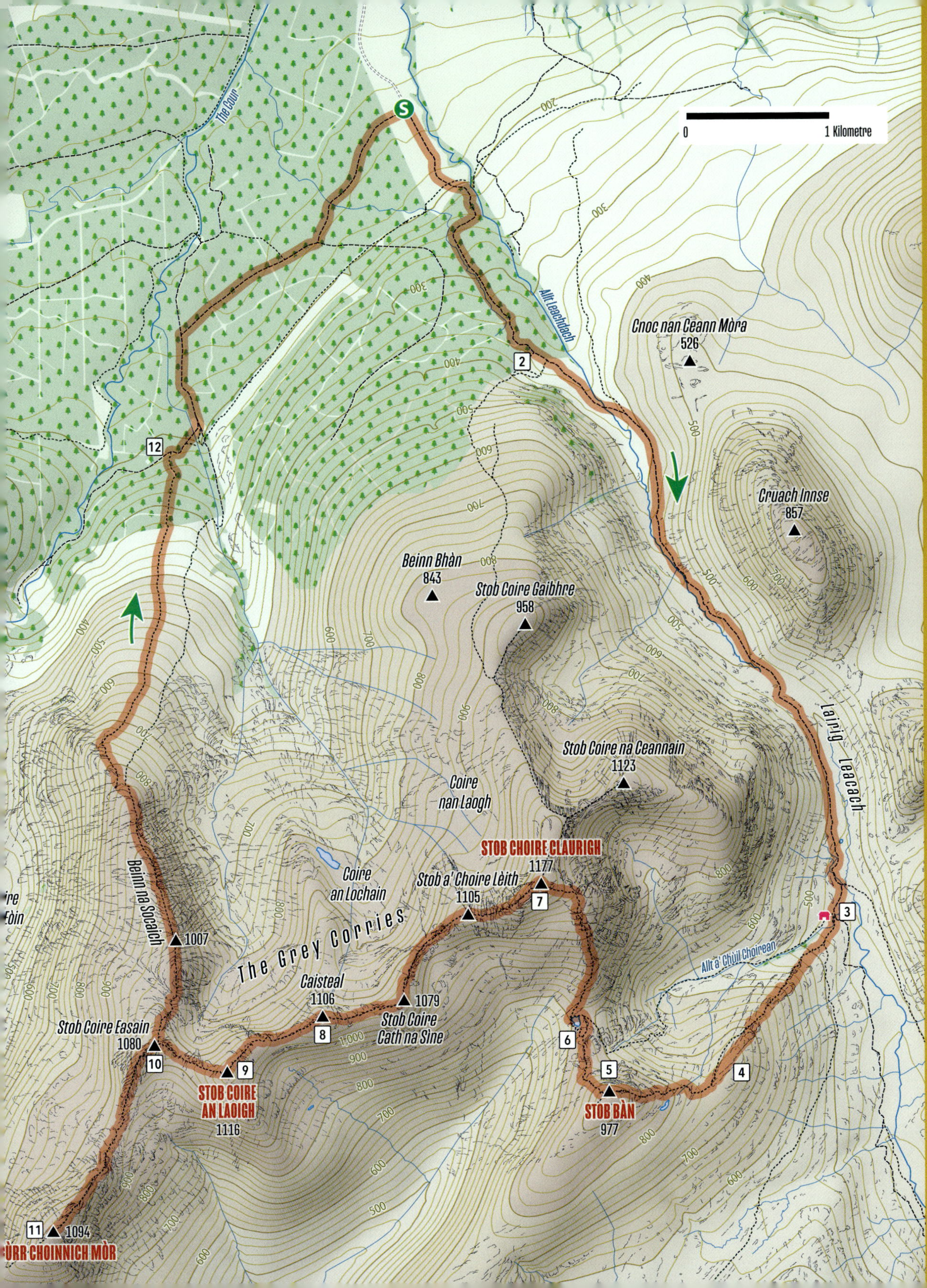

0
1 Kilometre
S
The Cour
Allt Leachdach
Cnoc nan Ceann Mòra
526
Cruach Innse
857
Beinn Bhàn
843
Stob Coire Gaibhre
958
Stob Coire na Ceannain
1123
Lairig Leacach
Coire nan Laogh
STOB CHOIRE CLAURIGH
1177
Stob a' Choire Lèith
1105
Coire an Lochain
The Grey Corries
Beinn na Socaich
1007
Caisteal
1106
1079
Stob Coire Cath na Sìne
Stob Coire Easain
1080
STOB COIRE AN LAOIGH
1116
STOB BÀN
977
Allt a' Chùil Choirean
1094
ÙRR CHOINNICH MÒR

44 West Loch Treig Munros

HUB **Fort William – 35-minute drive**

OVERALL 1 2 **(3)** 4 5   TERRAIN 1 2 **(3)** 4 5
DISTANCE 1 2 **(3)** 4 5   NAVIGATION 1 **(2)** 3 4 5

DISTANCE **17.2km** ASCENT **1,270m**
TIMES **8:35hrs** (Walker) **7:25hrs** (Trekker)
5:30hrs (Fastpacker) **3:35hrs** (Trail runner)

START LOCATION **Walkers' car park, Fersit**
START LATITUDE/LONGITUDE **56.8656, -4.7084**
START GRID REFERENCE **NN349781**
START W3W **///stripped.tablets.zoos**

	PRONUNCIATION	NAME MEANING
Stob a' Choire Mheadhoin	**'stob a horriya veeyan'**	**peak of the middle corrie**
Stob Coire Easain	**'stob corriya issen'**	**peak of the corrie of the little waterfall**

Introduction

These two fine Munros are an inseparable pair – the best of comrades with a never-ending partnership and an uncannily synchronised presentation. Known collectively as the 'Easains', or sometimes 'This Yin and That Yin', everyone thinks of this duo as a pair, and they are almost always climbed together. Located on a long, lofty ridge on the north-western shores of Loch Treig, the optimum way to tackle these twin summits is as an out-and-back from Fersit. The highlight is the stunning view across the distant Grey Corries – a sight to behold.

Public Transport

The nearest public transport is at the railway station in Tulloch which is on the West Highland Line. Starting the walk from here adds just over 11 kilometres of road walking to the total walk length.

Parking

Free parking for around 20 cars in the gravel walkers' car park in Fersit. The nearest toilet facilities are in Spean Bridge.

MTB-ability

The first and last two kilometres of the route can be cycled.

Escape Route

While it is possible to descend from the bealach between the two Munros into Coire Easain Beag by dropping steeply north-west and then walking back either just above or through Coire Làire, this is less pleasant in my view.

On the approach to Stob a' Choire Mheadhoin.

Refreshments

En route back to Fort William, stop at the Glen Roy Cafe in Roybridge. The cafe looks distinctly unpretentious from the outside, while inside the warmest of welcomes and some great cakes await you.

The Route

S Leave the car park to the south and walk along a track. Pass through a gate and continue for 700m before forking right at a track junction. After a further 200m, just before the dam, fork right again. Enjoy the view down Loch Treig from this slightly higher track which runs parallel to the one below and offers better views.

2 After 1km, the track drops slightly. Turn right to rejoin the bottom track and then turn right again to pick up a grassier track which leads north and then north-west up the hillside. The track becomes wetter and grassier as you gain elevation, before it bends west and then south-south-west, passing below a concrete pillar used to survey the local hydro scheme.

3 At NN336766 turn left, leaving the track, to walk south-west up Meall Cian Dearg. The route passes over a boggy flat section before becoming rockier. Hoist yourself up some big rocky steps and take care on the most eroded ground as the gradient steepens here. There is no graded scrambling required but you may benefit from putting your hand down to assist your ascent on occasion.

4 Top out on Meall Cian Dearg. The ground levels out here for a more gradual section of the ascent. Be sure to stop and enjoy the view behind you down Loch Laggan and over to Beinn a' Chlachair (page 227). Follow a clear stony trail, firstly to the south-south-west, then south-west and then south, over the firm ground of the ridge.

On the ascent of Meall Cian Dearg en route to Stob a' Choire Mheadhoin with Loch Treig below.

5 Pass point 978m on OS maps and begin the final part of the ascent of Stob a' Choire Mheadhoin, heading south-west. The ground steepens towards the very top; the summit of Stob a' Choire Mheadhoin is marked by a cairn. The nearby Munro Stob Coire Sgrìodain (page 223) can be appreciated from here but the view which draws the eye the most is the one south-west to the next Munro, Stob Coire Easain.

6 Descend rather gently at first over stony ground to the saddle between Stob a' Choire Mheadhoin and Stob Coire Easain. Glance left down to Loch Treig and right down Coire Easain Beag before you head uphill over increasingly reddening rock on the ascent of Stob Coire Easain. The terrain becomes rockier as you approach the summit, but no obstacle requires scrambling; there are simply a few big rocky steps along the route. The summit is marked by a cairn.

7 Enjoy the spectacular view west from here to the Grey Corries (page 213) and retrace your steps over Stob a' Choire Mheadhoin back to the start.

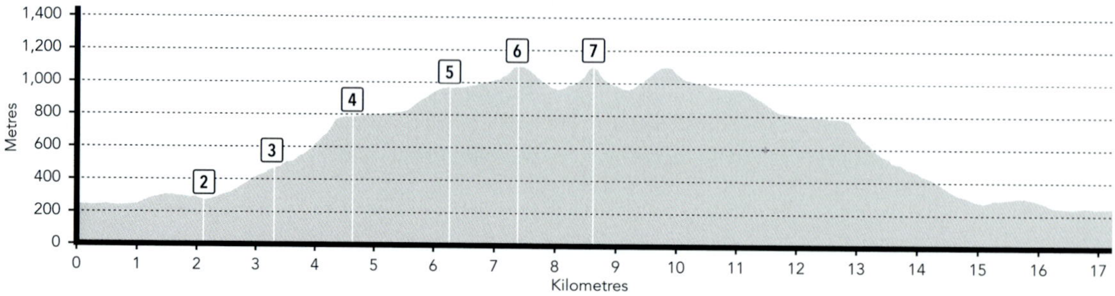

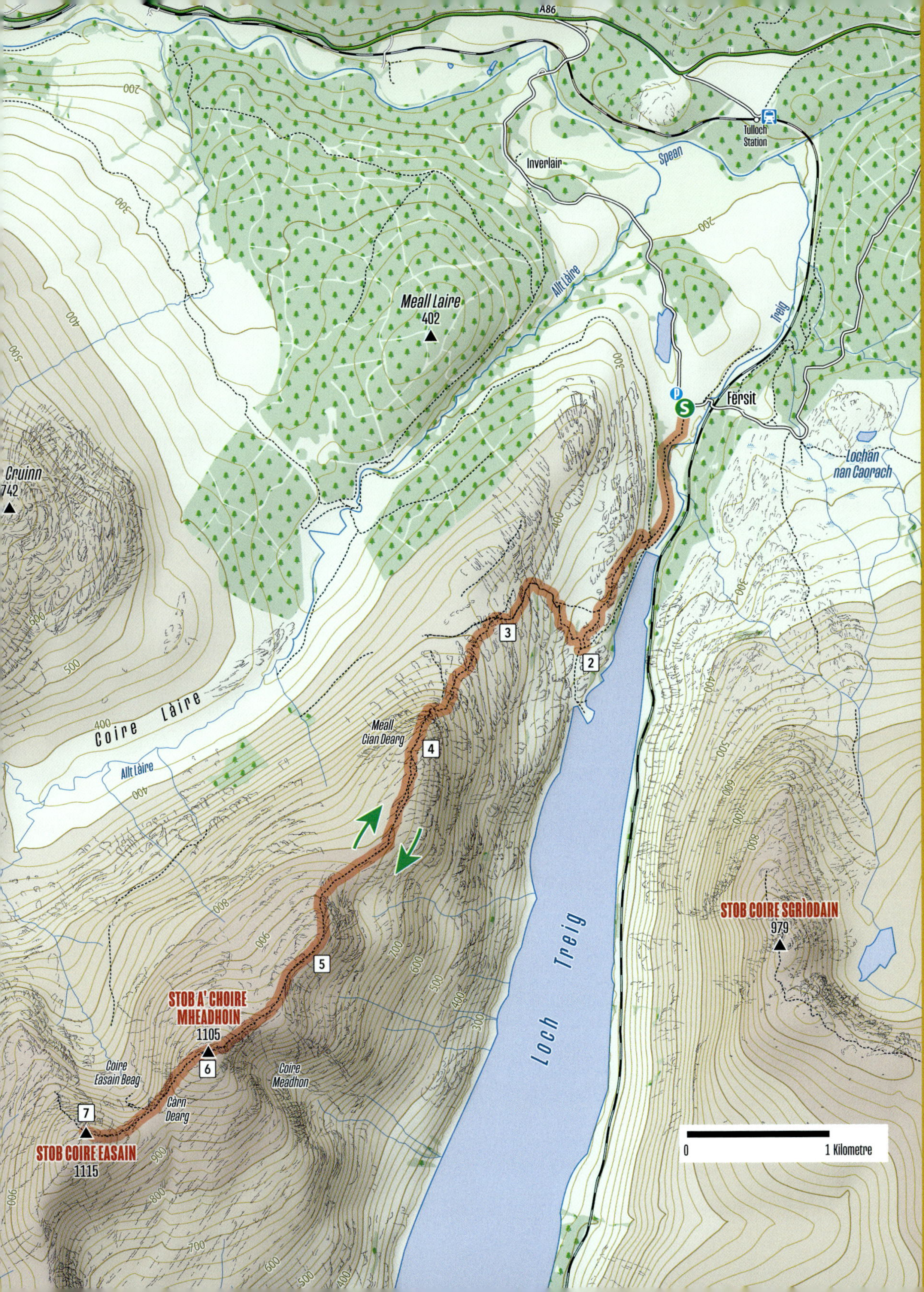
A86
Tulloch Station
Spean
Inverlair
Allt Laire
Meall Laire
402
Treig
Fersit
Lochan nan Caorach
Cruinn
742
Coire Làire
Allt Làire
Meall Cian Dearg
STOB COIRE SGRÌODAIN
979
Loch Treig
STOB A' CHOIRE MHEADHOIN
1105
Coire Easain Beag
Coire Meadhon
Càrn Dearg
STOB COIRE EASAIN
1115
0
1 Kilometre

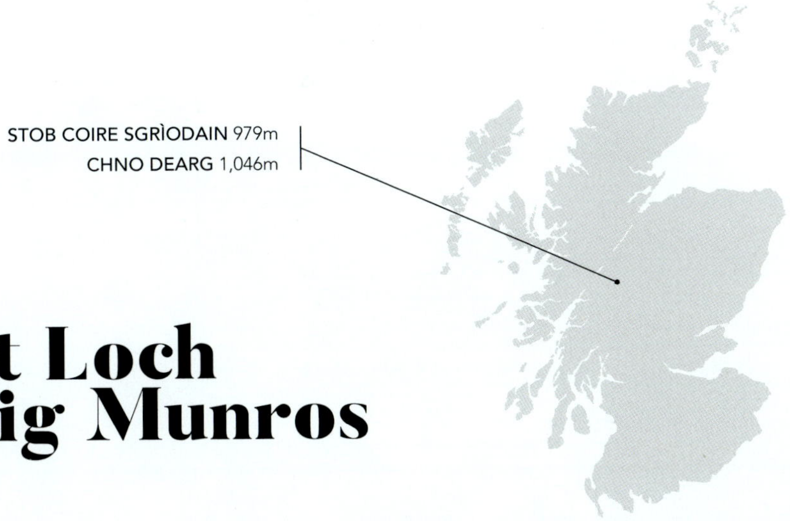

45 East Loch Treig Munros

HUB **Fort William – 35-minute drive**

OVERALL 1 2 **3** 4 5 TERRAIN 1 2 **3** 4 5
DISTANCE 1 **2** 3 4 5 NAVIGATION 1 2 3 **4** 5

DISTANCE **13.6km** ASCENT **995m**
TIMES **6:45hrs** (Walker) **5:50hrs** (Trekker)
4:20hrs (Fastpacker) **2:50hrs** (Trail runner)

START LOCATION **Walkers' car park, Fersit**
START LATITUDE/LONGITUDE **56.8656, -4.7084**
START GRID REFERENCE **NN349781**
START W3W **///stripped.tablets.zoos**

	PRONUNCIATION	NAME MEANING
Stob Coire Sgrìodain	**'stob corriya sgriden'**	**peak of the scree corrie**
Chno Dearg	**'knor jerreck'**	**red hill**

Introduction

This is the hillwalking equivalent of that old football cliché – it's a game of two halves. Well, this is a walk of two halves. One half is exciting and full of character; the other descends into a boggy slog devoid of major interest. The fun is front-loaded. The first half of this route is a delightful ascent of Stob Coire Sgrìodain, featuring a sweeping ridgeline, superb views and lots of rocky intrigue. But the second half, over the rounded lump of Chno Dearg, is anticlimactic in comparison.

Public Transport

The nearest public transport is at the railway station in Tulloch which is on the West Highland Line. Starting the walk from here adds just over 11 kilometres of road walking to the total walk length.

Parking

Free parking for around 20 cars in the gravel walkers' car park in Fersit. The nearest toilet facilities are in Spean Bridge.

Escape Route

No escape routes; retrace your steps after the first Munro and return another day for the second.

Refreshments

En route back to Fort William, stop at the Glen Roy Cafe in Roybridge. The cafe looks distinctly unpretentious from the outside, while inside the warmest of welcomes and some great cakes await you.

Enjoying the view across Loch Treig from Stob Coire Sgrìodain.

Stob Coire Easain and Stob a' Choire Mheadhoin (route 44) from Stob Coire Sgrìodain.

The Route

S Leave the car park on a track to the east to cross a bridge over the River Treig. Cross a second bridge, this time over the railway line, before passing through a gate in the track. Walk uphill on the tree-lined track ignoring all minor turn-offs as you pass a line of cottages and latterly a farm.

2 At NN357780, take a right-hand fork in the track which leads south past a corrugated outbuilding. The track transforms into a path which crosses grassy terrain which is boggy in places as you loosely follow the line of the Allt Chaorach Beag.

3 As the ascent steepens, the path forks at NN359763. Take the right fork here, crossing two small streams, and walk south-south-west through heathery terrain towards the foot of Sròn na Gàrbh-bheinne. The path leads up a wide grassy gully between the crags before ascending rockier terrain where some very mild scrambling is required to gain Sròn na Gàrbh-bheinne.

4 Continue south-south-west where the path is now dry and stony. The twin peaks of Stob Coire Easain and Stob a' Choire Mheadhoin (page 219) stand proudly to the south-west. The summit of Stob Coire Sgrìodain is marked by a cairn and your eye is drawn to the dramatic cliffs which appear to fall spectacularly into Loch Treig.

5 A broad ridge curves around the head of the valley above Lochan Coire an Lochain to Chno Dearg. Leave the summit of Stob Coire Sgrìodain heading south and then south-east walking over point 958m on OS maps. This is the first in a series of small knolls which make up the curve. The ground is stony, but the path undulates gently around to the bealach between Chno Dearg and Meall Garbh.

6 The path forks in the bealach. Take the right-hand fork which leads uphill over soft, mossy ground to Chno Dearg. The summit is marked by a large cairn and the view is similar to that of Stob Coire Sgrìodain, just without the impressive cliffs. Chno Dearg is marginally closer to the Creag Meagaidh range which can be spotted on the other side of Glen Spean.

7 Leave the summit to the north-north-west initially, descending over grassy terrain interspersed with rocks. After 300m, descend north-west over a mixture of grassy, heathery and mossy terrain. Hopping down from, or bypassing, one small crag is the only real obstacle on this long, wet and boggy plod. On the plus side, the gradient of the descent is fairly mild in comparison to most Munro walks. A series of faint trails can be picked up along the way, but these peter out regularly, and the descent is largely pathless. On a clear day, the car park and

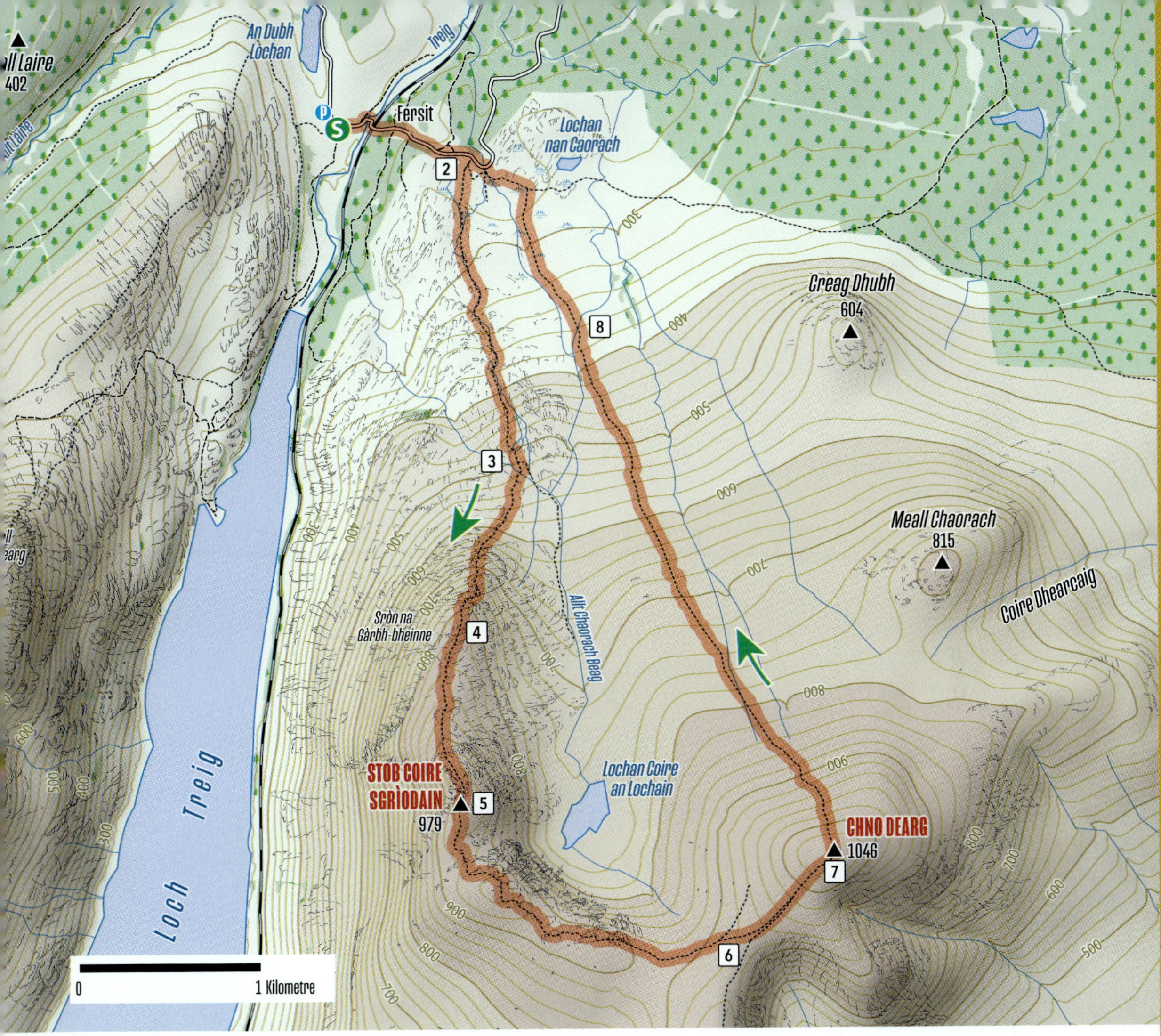

nearby An Dubh Lochan are in sight and can be used as a navigation aid on the higher slopes.

8 At NN363770 cross two streams, which are the rightmost branches of the Allt Chaorach Beag following a divide upstream. Pick up a clear path on the eastern bank of the leftmost branch of the stream which drops gently to a track. Turn left here and walk towards a farm, fording a couple of streams along the way. Turn left on the main track in front of the farm and retrace your steps to the start.

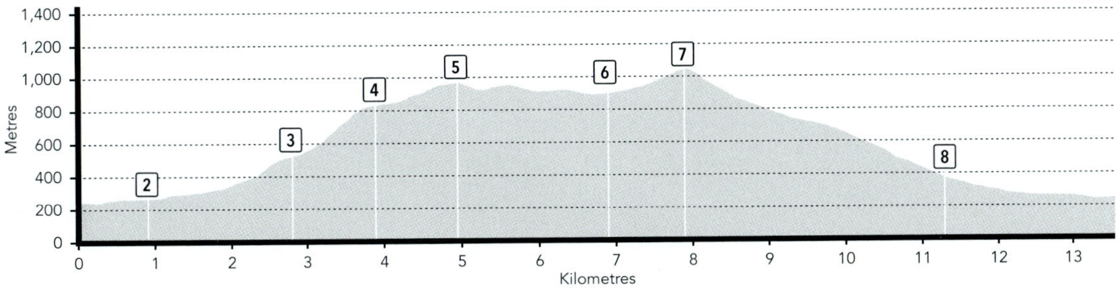

BEINN A' CHLACHAIR 1,087m
GEAL CHARN 1,049m
CREAG PITRIDH 924m

46 Ardverikie Munros

HUB **Fort William – 35-minute drive**

OVERALL 1 2 3 4 5 TERRAIN 1 2 3 4 5
DISTANCE 1 2 3 4 5 NAVIGATION 1 2 3 4 5

DISTANCE **26.3km** ASCENT **1,295m**
TIMES **10:30hrs** (Walker) **9:00hrs** (Trekker)
6:45hrs (Fastpacker) **4:35hrs** (Trail runner)

START LOCATION **Large lay-by off the A86 at the south-western end of Loch Laggan**
START LATITUDE/LONGITUDE **56.9121, -4.5758**
START GRID REFERENCE **NN432830**
START W3W **///unzipped.originate.supposing**

	PRONUNCIATION	NAME MEANING
Beinn a' Chlachair	**'beyn a chlacha'**	**stonemason's hill**
Geal Charn	**'ghee-yal harn'**	**white hill**
Creag Pitridh	**'craig peetree'**	**Petrie's crag**

Introduction

When I first climbed this trio of remote Munros – located in a wild no-man's-land between Loch Laggan and Loch Ericht – I stumbled across a humongous film crew shooting the James Bond blockbuster *No Time To Die*. I later found out that high-octane chase scenes were shot on the Ardverikie Estate, to the south of Loch Laggan – a touch of Hollywood glamour in the remote Scottish Highlands. There is also currently a planning application proposing the UK's largest pumped storage hydroelectric scheme in this very spot. If successful, this will dramatically change this sensitive wild place forever and it's worthy of note that this route will undoubtedly require diversions.

Beinn a' Chlachair from Geal Charn summit cairn.

Public Transport

There are no public transport services close to this walk.

Parking

Free parking for around 15 cars in the lay-by off the A86 at the south-western end of Loch Laggan. The nearest public toilets are in Spean Bridge.

MTB-ability

The first and last 5.4 kilometres of the route can be cycled.

Escape Route

Cut the route short from Bealach Leamhain to pick up the descent path and return for Geal Charn and Creag Pitridh another day.

Refreshments

En route back to Fort William, stop at the Glen Roy Cafe in Roybridge. The cafe looks distinctly unpretentious from the outside, while inside the warmest of welcomes and some great cakes await you.

Above Lochan na h-Earba, on the approach to Beinn a' Chlachair. **Opposite** Ascending the pathless section of Beinn a' Chlachair with Creag Pitridh behind.

The Route

S Leave the car park to the west following a path through the trees which runs parallel to the A86. Take the first left across the River Spean using the large bridge which leads to a wide track. Follow the track until it splits 250m after the bridge – fork left here, remaining on the wide track. Pass through a gate and take pleasure in the easy walking along the gorse-lined track. Down to your right you'll see the Abhainn Ghuibinn and up to your left is the craggy-topped Binnein Shuas. The wetlands either side of the track offer perfect conditions for cotton grass to grow in the spring.

2 At the next track junction, turn right. Beinn a' Chlachair is the mountain right in front of you to the south-east as you walk along the broad track. Creag Pitridh and its higher neighbour beyond, Geal Charn, soon come into view to the east. After 2km the track splits again at a small reservoir; follow the track to the left here and remain on it as Lochan na h-Earba comes into view.

3 Cross the bridge and pause to admire the sandy lochside and then fork right to follow the track south. At NN463810, at a small cairn, fork right on to a grassy path which soon runs parallel to the Allt Coire Pitridh. While it may feel like you're overshooting your target of Beinn a' Chlachair, it's beneficial to gain as much height on the path as possible as you'll be approaching on the left-hand side of the Coire Mòr a' Chlachair, keeping clear of the crags on all sides of the bowl.

4 At NN472804, note the cairn and the trail leading up to the left as this will be your descent route from Creag Pitridh later. For now, continue straight on along the track and cross the stream. Continue uphill along the track until you see a cairn at NN477802 which marks a trail off to the right. Follow this trail down to a stream crossing and beyond, over heather and bilberry-strewn terrain. Tread cautiously as there are lots of subterranean streams in this section and losing a boot in a hole is not an uncommon occurrence.

5 The trail becomes indistinct as it crosses boggy ground. Aim south to follow the left-hand side of a prominent stream running down from higher ground. As the stream heads underground at around 850m elevation, continue walking south to reach the rim of the coire, avoiding the temptation to swing west too early. As you approach the ridge, the terrain becomes rockier underfoot and the gradient eases. Once on the ridge, curve around to the west to reach the highest point of Beinn a' Chlachair over blocky terrain.

6 Looking south from the summit of Beinn a' Chlachair you can see the long ridge of (the other) Geal-Chàrn, Aonach Beag and Beinn Eibhinn (page 241) and just to the left beyond this set of hills lies their taller neighbour Ben Alder (page 259). Panning

Creag Pitridh (left) and Geal Charn (right) from the summit of Beinn a' Chlachair.

across to the west on a clear day you can see beyond the Loch Treig Munros (pages 219 and 223) to a selection of the highest mountains in the UK. Retrace your steps from the summit briefly before walking high along the ridge following a faint trail to the north-east, visiting the 977m north-east top of Beinn a' Chlachair and continuing on towards the top of the crags which guard the north-eastern aspect of the Munro.

7 The steepest part of the descent starts at NN490795. My preferred line is north-north-west from here. Scrambling can be completely avoided by keeping left. The slope is steep, but it is possible to descend over grass around the gaps in the crags without needing to use your hands. Once at the bottom of the crags, pick up a good path which forks at NN489799 at Bealach Leamhain; take the right fork here and start ascending. Follow the path up to the south-eastern edge of the col between Creag Pitridh and Geal Charn.

8 A small cairn marks the start of a right-hand turn up Geal Charn at NN491810. Follow the faint trail east up the hillside. From the 900m contour, a prominent eroded line uphill appears. Much of the last 80m of vertical ascent is boulder-strewn as the trail turns north-east to the summit but there are short sections of soft grassy terrain to break it up. Geal Charn, also known as Mullach Coire an lubhair, has a trig pillar and the summit itself is marked by a cairn.

9 Retrace your steps to NN491810 (**8**) and walk across the col over a knoll marked 837m on OS maps. Drop slightly before continuing up the southern flanks of Creag Pitridh following an eroded trail interspersed with slabby rock which presents no technical difficulty. In fact, it is by far the easiest climb of the day. The summit of this notably smaller Munro boasts impressive views down to Lochan na h-Earba, Loch Laggan and the Monadhliath mountain range beyond.

10 Follow an eroded trail off the summit to the west over rocky ground before the trail becomes grassier and begins to curve south-west. In a few places, the trail becomes indistinct as it passes over wet ground, but it keeps coming back if you keep heading south-west aiming for the point just below the stream confluence at NN472804. Once on the main path, retrace your steps back to the start.

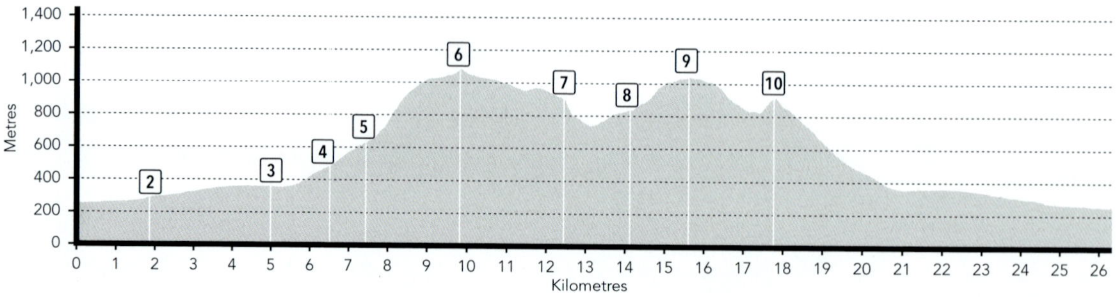

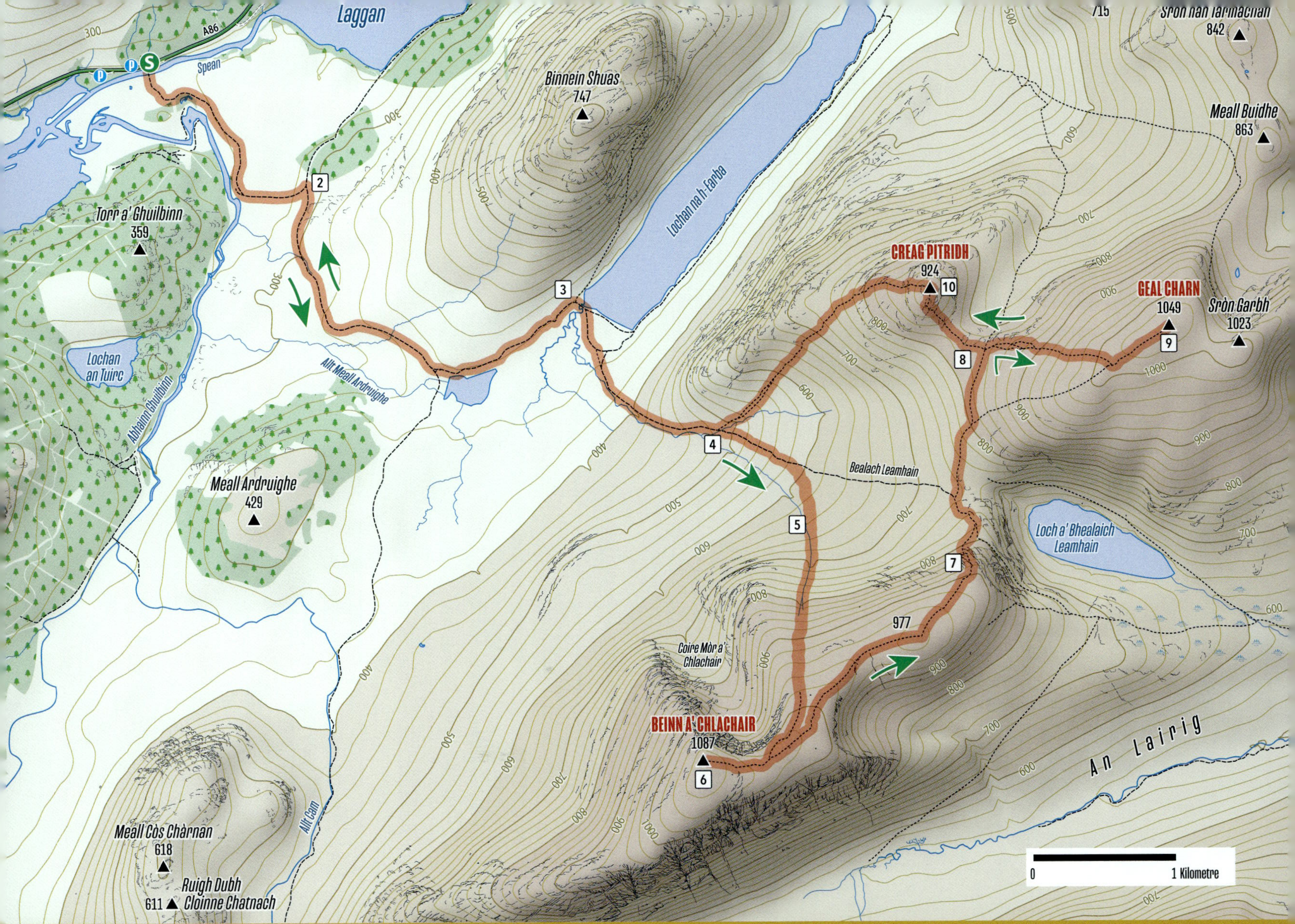
Laggan
A86
Spean
S
Binnein Shuas
747
Lochan na h-Earba
Torr a' Ghuilbinn
359
Lochan an Tuirc
Abhainn Ghuilbinn
Allt Meall Ardruighe
Meall Ardruighe
429
CREAG PITRIDH
924
GEAL CHARN
1049
Sròn Garbh
1023
Meall Buidhe
863
Sròn nan Iarmachan
842
Bealach Leamhain
Loch a' Bhealaich Leamhain
Coire Mòr a' Chlachair
BEINN A' CHLACHAIR
1087
977
An Lairig
Allt Cam
Meall Cos Chàrnan
618
Ruigh Dubh Cloinne Chatnach
611
0
1 Kilometre

HIGHLAND HUB 8

CORROUR

The eighth of the Highland hubs, and one of my personal favourites, is Corrour. This is the quirkiest hub as it is only accessible by train, on the West Highland Line, or by a long walk-in.

Unlike the other hubs, there are limited facilities in Corrour due to its position north of Rannoch Moor, a vast, characterful, wet 'n' wild moorland surrounded by mountains, but that adds to its appeal. Bring the basics with you and everything else will fall into place. The Station House acts as a cafe, restaurant, meeting point, gift shop and WiFi spot. If you're staying in Corrour for the three walks, make sure you resupply before you arrive in a big town like Pitlochry (if arriving via Rannoch) or Fort William (if arriving from any of the stations to the north or west).

I highly recommend staying in Loch Ossian Youth Hostel, a wonderful eco-hostel which was converted from a boathouse to a hostel in 1931. Countless Munro-baggers and nature lovers have come through its doors, staying just one or two nights at a time but leaving with new friends and brimming with memories. Now popular with long-distance cyclists too, the hostel utilises renewable energy and has ecologically sound water and waste disposal systems, respecting the sensitive environment. John Dickson, former chairman of Hostelling Scotland, said 'Loch Ossian is the epitome of youth hostelling. It embodies everything that we stand for: affordable, comfortable accommodation providing access to the most remote yet most beautiful areas in Scotland.' Loch Ossian Youth Hostel was named Hostel of the Year in *The Great Outdoors* Reader Awards in 2024.

Although there are no campsites at the hub, an alternative would be to stay at the Bunroy Park Campsite in Roybridge and get the train each day.

WHERE TO STAY

£	££	£££
–	Loch Ossian Youth Hostel	The Signal Box, Station House

TRAVEL TIMES

47 Beinn na Lap	**0 minutes**
48 Munros North-East of Loch Ossian	**0 minutes**
49 Loch Ossian Munros	**0 minutes**

Rest day ideas

- Go bird watching on the shore of Loch Ossian – see if you can spot black-throated diver, goldeneye and heron.
- Keep your joints oiled with a 14-kilometre loop of Loch Ossian on good tracks with minimal undulation past the memorial to Sir John Stirling Maxwell, a pioneer in planting trees in peat.
- Cosy up to watch Danny Boyle's cult classic film *Trainspotting*, which was partially filmed in the Corrour area.

Opposite Corrour railway station with a train on the platform and Leum Uilleim beyond.
Overleaf Corrour railway station with Leum Uilleim behind it from the ascent of Beinn na Lap (route 47).

Corrour
An Coire Odhar
ScotRail

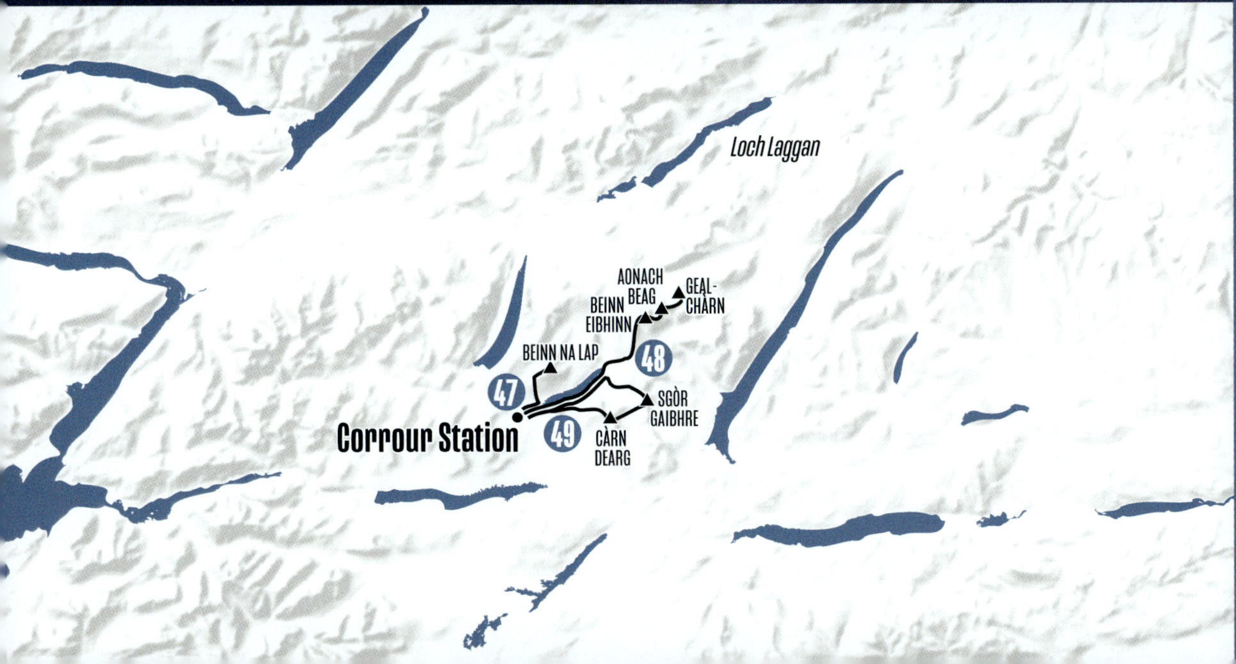
Loch Laggan
AONACH BEAG
GEAL-CHÀRN
BEINN EIBHINN
BEINN NA LAP
48
47
SGÒR GAIBHRE
Corrour Station
49
CÀRN DEARG

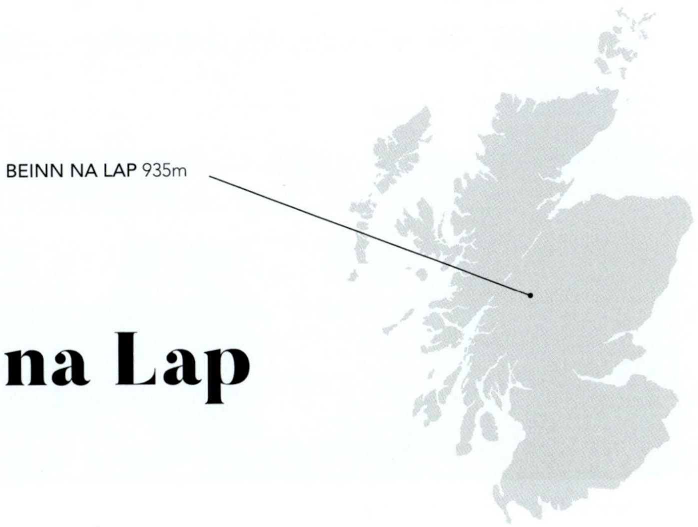

47 Beinn na Lap

HUB **Corrour – at start**

OVERALL 1 **2** 3 4 5 TERRAIN 1 **2** 3 4 5
DISTANCE **1** 2 3 4 5 NAVIGATION 1 **2** 3 4 5

DISTANCE **9.9km** ASCENT **585m**
TIMES **4:20hrs** (Walker) **3:45hrs** (Trekker)
2:50hrs (Fastpacker) **1:50hrs** (Trail runner)

START LOCATION **Corrour railway station**
START LATITUDE/LONGITUDE **56.7604, -4.6905**
START GRID REFERENCE **NN356664**
START W3W **///nooks.boat.flush**

	PRONUNCIATION	NAME MEANING
Beinn na Lap	**'beyn na lap'**	**dappled hill**

Introduction

Wildly remote, yet easily accessible – Beinn na Lap is a bit of a contradiction. Located in the isolated land between Loch Ossian and Loch Treig, this 935-metre mountain initially feels far-flung and cut-off; but, on closer inspection, it is one of the most uncomplicated Munros out there. The proximity of Corrour railway station makes the logistics infinitely easier and, to make a weekend of it, you can even stay overnight in the delightful Loch Ossian Youth Hostel – something I think everyone would benefit from doing at least once in their lifetime. Plus, Beinn na Lap is neither high nor technical, which means it's an excellent candidate for a final Munro celebration with friends and family.

Public Transport

The walk starts at Corrour railway station. Take the train from Rannoch (duration 12 minutes) for the shortest journey from the south; however, you can start your journey anywhere along the West Highland Line. There are toilets in the Station House for customers of the on-site cafe.

Parking

This route is not accessible by car – see page 232.

MTB-ability

The first and last two kilometres of the route are cyclable.

Escape Route

None.

Refreshments

Corrour Station House is the only establishment of its kind for miles. Whether you fancy a tangy slice of lemon drizzle and a brew, or venison casserole with a glass of Malbec, you can experience either one here at the self-dubbed 'UK's remotest restaurant'.

Beinn na Lap rising up to the right of Loch Ossian, seen from Ben Alder (see route 51).

View to Beinn na Lap and Loch Ghuilbinn from Beinn a' Chlachair (see route 46).

Walkers ascending Beinn nap Lap with Loch Ossian behind.

The Route

S Leave the train station to the north-east, passing around the northern edge of Loch Coir' a' Bhric Beag. Follow the wide track on an easy gradient towards Loch Ossian. If you're staying at Loch Ossian Youth Hostel and wish to stow your overnight gear, fork right after 1.3km (NN368669) and then turn left just after passing a small electrical switch room outbuilding to reach the hostel which is nestled within a copse of trees. If you're heading straight for the Munro however, take the left fork at that first junction and follow the wide track which leads around to a bridge over a stream.

2 At the fork in the track just after the bridge, turn left, and after 70m peel off the track to follow a wet and squelchy path heading north-north-west up the grassy hillside. The soft and muddy terrain in the lower part of the climb firms up and becomes slightly stony as a prominent eroded line scores the slope.

3 At 740m in elevation the trail curves north-east to gain the shoulder of Beinn na Lap. The walking is more gradual here as the trail weaves around rocks along the ridge. Pass to the right of a small lochan and arc northwards to the summit of Beinn na Lap which is marked by a shelter cairn. Views to Beinn na Lap's closest Munro neighbours extend northwards

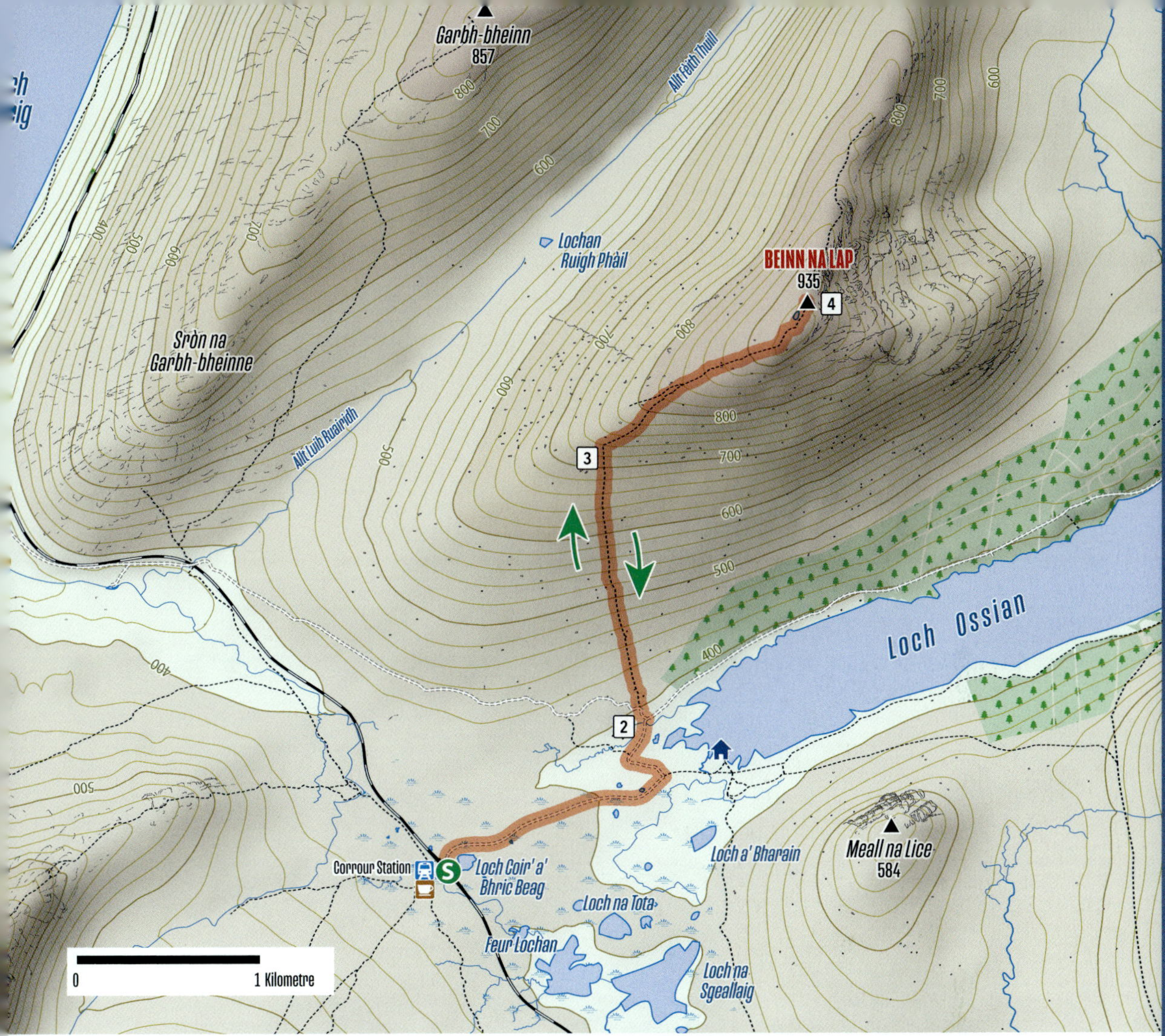

to Chno Dearg (page 223) and south-east across Loch Ossian to Càrn Dearg (page 247).

4 Retrace your steps back to the station. As you walk, you can appreciate the size of the station buildings, backed by the Corbett Leum Uilleim but juxtaposed against the ample plane of moorland that stretches up from Rannoch Moor and Blackwater Reservoir.

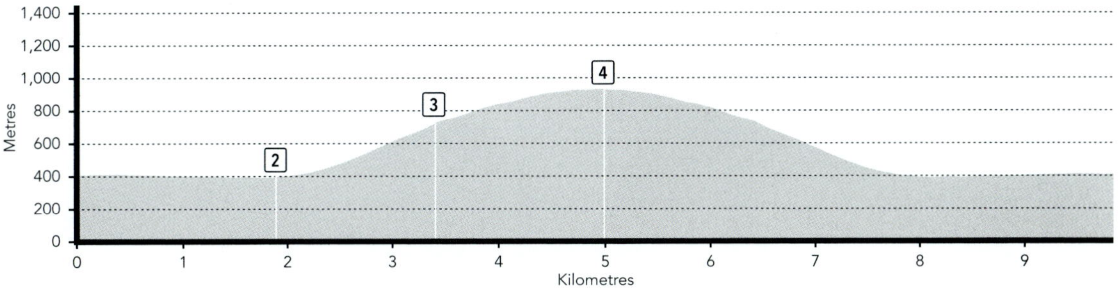

48 Munros North-East of Loch Ossian

HUB **Corrour – at start**

OVERALL 1 2 3 **4** 5 TERRAIN 1 2 **3** 4 5
DISTANCE 1 2 3 4 **5** NAVIGATION 1 2 3 **4** 5

DISTANCE **32.9km** ASCENT **1,375m**
TIMES **12:05hrs** (Walker) **10:20hrs** (Trekker)
7:45hrs (Fastpacker) **5:20hrs** (Trail runner)

START LOCATION **Corrour railway station**
START LATITUDE/LONGITUDE **56.7604, -4.6905**
START GRID REFERENCE **NN356664**
START W3W **///nooks.boat.flush**

	PRONUNCIATION	NAME MEANING
Beinn Eibhinn	**'beyn ey-vin'**	**delightful hill**
Aonach Beag	**'oonach bake'**	**little ridge**
Geal-Chàrn	**'ghee-yal harn'**	**white hill**

Introduction

Never underestimate Scotland's mighty mountains – and always approach them with humility and respect. That's what I learnt from one of my research walks of these three Munros from Corrour railway station. It was 5 June 2024, and out of nowhere a brutal snowstorm transformed the hills into a perilous wintry expedition. To avoid slipping from a remote frosty ridge, I made the decision to bail and come back another day – an important lesson in safety for all Munro-baggers. This trio of Munros can also be tackled from Dalwhinnie in the north-east, but I prefer the approach from Corrour, it's a real treasure. There's a long walk-in along Loch Ossian before climbing into the gloriously rugged hills.

Public Transport

The walk starts at Corrour railway station. Take the train from Rannoch (duration 12 minutes) for the shortest journey from the south; however, you can start your journey anywhere along the West Highland Line. There are toilets in the Station House for customers of the on-site cafe.

Parking

This route is not accessible by car – see page 232.

MTB-ability

The first and last nine kilometres of the route are cyclable.

Escape Route

None.

View to Aonach Beag and Geal-Chàrn from Beinn Eibhinn.

Above Ben Alder (left), Bealach Dubh (centre), with Geal-Chàrn (Loch Ossian) and Aonach Beag on the right from Geal-charn (Drumochter Pass), see route 53. **Opposite** Beinn Eibhinn from Loch Ossian Youth Hostel.

Refreshments

Corrour Station House is the only establishment of its kind for miles. Whether you fancy a tangy slice of lemon drizzle and a brew, or venison casserole with a glass of Malbec, you can experience either one here at the self-dubbed 'UK's remotest restaurant'.

The Route

S Leave the station to the north-east, passing around the northern edge of Loch Coir' a' Bhric Beag. Follow the wide track on an easy gradient towards Loch Ossian. After 1.3km, fork right at the track junction. If you're staying at Loch Ossian Youth Hostel and wish to stow your overnight gear, take your next left after passing a small electrical switch room outbuilding to reach the hostel which is nestled within a copse of trees. If you're heading straight for the Munros however, ignore the forks in the track and walk the full length of the south-eastern edge of Loch Ossian which is mostly wooded and ever so peaceful.

2 With the hostel now 5km behind you, pass Corrie Odhar cottage and fork left immediately afterwards. Walk north for 200m and cross the bridge over the River Ossian. After 60m, fork right, away from the shooting lodge and walk north-east, crossing another bridge. After a further 300m, pass through a gate and then after another 300m, cross a further bridge, passing to the right of Uisge Labhair Hydro-electric Station. Follow the track round to the east alongside the Uisge Labhair.

3 Cross the bridge over the Allt Feith a' Mheallain where bikes can be left if you've cycled this far. Continue straight ahead on a footpath for 450m then, at NN433705, begin your pathless ascent of Beinn Eibhinn via Creagan na Craoibhe and Meall Glas Choire. Peeling off the path here ensures you avoid the peat hags and the boggiest ground. Navigate over soft and wet moorland, passing to the right of a fenced enclosure before aiming directly for the crags of Creagan na Craoibhe. The gradient is mild at first as you ascend north-north-east over grassy and heathery ground around an abundance of sphagnum moss in a beautiful melange of colour.

4 Navigate around the series of boulders and crags which comprise Creagan na Craoibhe on soft ground which presents no technical difficulty as you aim for the top of Meall Glas Choire. Be sure not to stray too far east towards the coire.

5 From the top of Meall Glas Choire, descend in the same direction into a small dip and then continue your ascent north-north-east at first, before curving east-north-east towards point 1,019m on OS maps.

Descending Geal-Chàrn.

Pass to the south of a pair of small lochans and begin ascending gradually east, with the Munros of Sgòr Gaibhre and Càrn Dearg (page 247) visible to your right. Once you reach the edge of Coire a' Charra Mhòir, walk south-east along the rim to gain the first of two high points on Beinn Eibhinn. Continue for 200m to the south-east summit which, in 2012, was surveyed to be a smidgen higher than the first top and is therefore considered the true summit of the Munro.

6 Drop down the narrowing ridge to the bealach. Walk north-east following an eroded line over some rocky but non-technical terrain to the summit of Aonach Beag. You really get the sensation that after a long walk-in followed by a largely pathless grind up the first Munro, the summits start to come thick and fast.

7 Leave the summit of Aonach Beag to the east following a clear path over grassy terrain before dropping downhill over stony terrain which is loose in places but not technical. The ridge beyond the bealach is wide and grassy. Keeping to the right of the drop into Coire na Coichille, walk north-east to reach the cairned summit of Geal-Chàrn which sits on the western edge of a vast plateau at 1,132m.

8 While it is possible to pick your way down to the Uisge Labhair through Coire a' Chàrra Bhig, the going is relentlessly steep, so my preference is to retrace my steps over the first two Munros and descend via the route of ascent. The height regain is minimal and traversing this seldom-walked ridge twice in one day is a privilege as, due to the complexity involved in getting here, few will ever return.

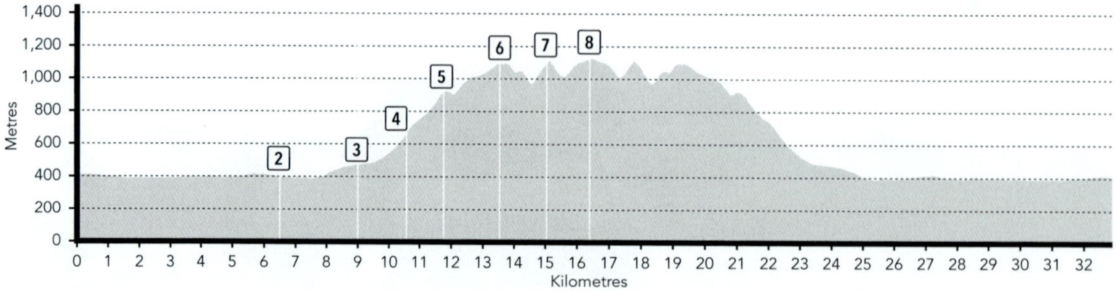

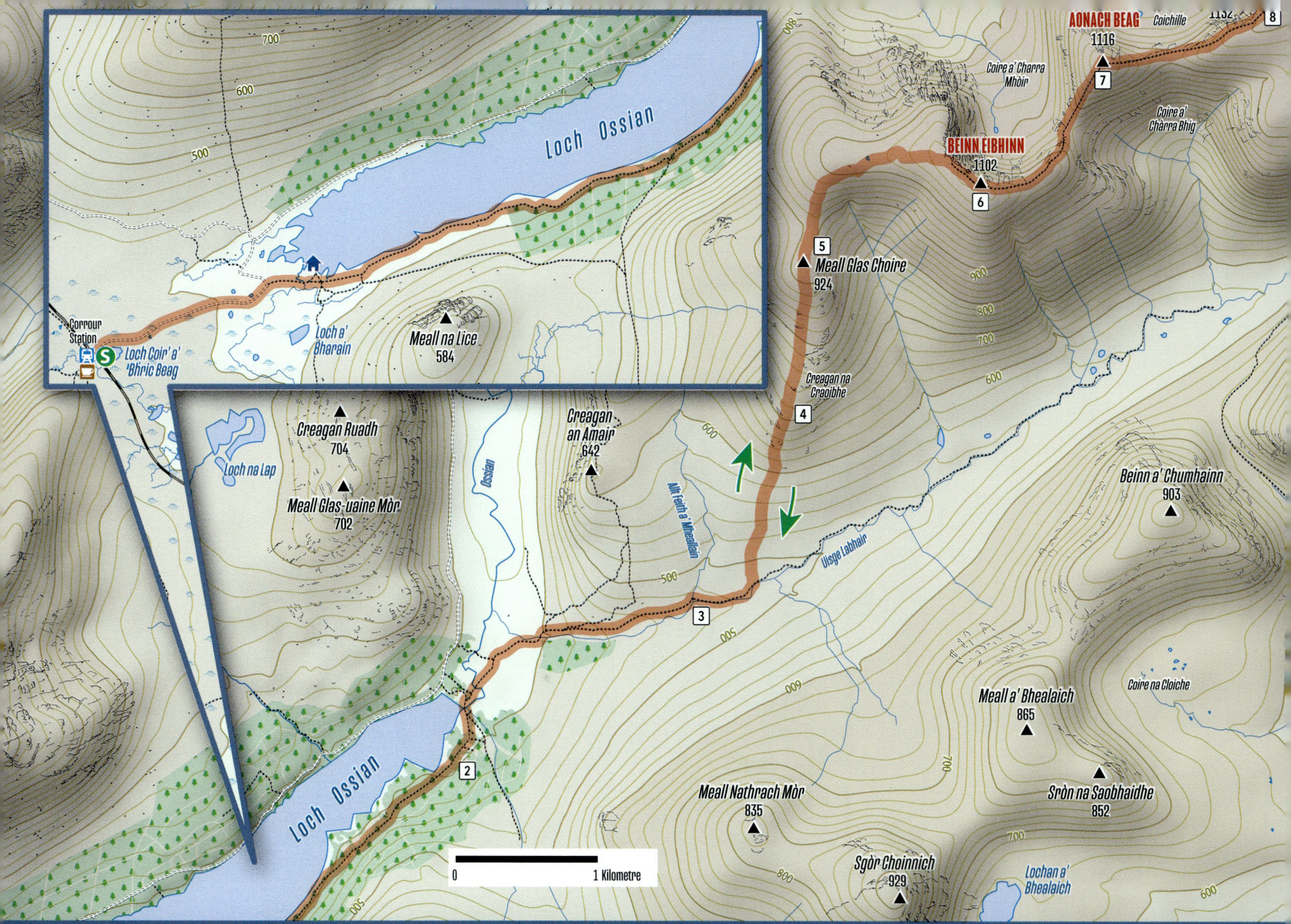

Loch Ossian
Corrour Station
Loch Coir' a' 'Bhric Beag
Loch a' Bharain
Meall na Lice
584
Creagan Ruadh
704
Loch na Lap
Meall Glas-uaine Mòr
702
Ossian
Creagan an Amair
642
Allt Feith a' Mheallain
Creagan na Craoibhe
Meall Glas Choire
924
BEINN EIBHINN
1102
AONACH BEAG
1116
Coichille
Coire a' Charra Mhòir
Coire a' Chàrra Bhig
Uisge Labhair
Beinn a' Chumhainn
903
Coire na Cloiche
Meall a' Bhealaich
865
Sròn na Saobhaidhe
852
Meall Nathrach Mòr
835
Sgòr Choinnich
929
Lochan a' Bhealaich
Loch Ossian
0
1 Kilometre
2
3
4
5
6
7
8

49 Loch Ossian Munros

HUB **Corrour – at start**

OVERALL 1 2 3 **(4)** 5 TERRAIN 1 2 **(3)** 4 5
DISTANCE 1 2 3 **(4)** 5 NAVIGATION 1 2 3 **(4)** 5

DISTANCE **21.3km** ASCENT **880m**
TIMES **7:50hrs** (Walker) **6:40hrs** (Trekker)
5:00hrs (Fastpacker) **3:25hrs** (Trail runner)

START LOCATION **Corrour railway station**
START LATITUDE/LONGITUDE **56.7604, -4.6905**
START GRID REFERENCE **NN356664**
START W3W **///nooks.boat.flush**

	PRONUNCIATION	NAME MEANING
Càrn Dearg	**'carn jerreck'**	**red hill**
Sgòr Gaibhre	**'skor guy-rer'**	**peak of the goats**

Introduction

They aren't the stars of the Loch Ossian area – and some hillwalkers think of them as unspectacular and uninspiring – but Càrn Dearg and Sgòr Gaibhre still create an enjoyable circular route. Starting at Corrour railway station, you'll climb the twin peaks to the east of the loch, before returning via the southern shoreline path. Celebrate ticking off two more summits with a flat white and slice of cake at Corrour Station House cafe, or perhaps an overnight stay at the achingly idyllic Loch Ossian Youth Hostel – a true get-away-from-it-all destination.

Public Transport

The walk starts at Corrour railway station. Take the train from Rannoch (duration 12 minutes) for the shortest journey from the south; however, you can start your journey anywhere along the West Highland Line. There are toilets in the Station House for customers of the on-site cafe.

Parking

This route is not accessible by car – see page 232.

Escape Route

No escape routes; retrace your steps after the first Munro and return another day for the second.

Refreshments

Corrour Station House is the only establishment of its kind for miles. Whether you fancy a tangy slice of lemon drizzle and a brew, or venison casserole with a glass of Malbec, you can experience either one here at the self-dubbed 'UK's remotest restaurant'.

Ascending Càrn Dearg with Loch Ossian behind.

The Route

S Leave the station to the north-east, passing around the northern edge of Loch Coir' a' Bhric Beag. Follow the wide track on an easy gradient towards Loch Ossian. After 1.3km, fork right at the track junction. If you're staying at Loch Ossian Youth Hostel and wish to stow your overnight gear, take your next left after passing a small electrical switch room outbuilding to reach the hostel which is nestled within a copse of trees. If you're heading straight for the Munros however, fork right 120m after the switch room on to another track and then turn left after 30m on to a footpath which exits back out on to the wider track further up. Turn left here and walk along the wide gravel track for 2km, gaining height gradually.

2 At NN393670, turn left beside Peter's Rock, a memorial to a young man who sadly perished in Loch Ossian during the winter of 1979. A faint path leads eastwards towards Meall na Leitire Duibhe across moorland opposite Beinn na Lap (page 237). The trail becomes indistinct as it crosses boggy ground, so continue east to gain the broad ridge which links Meall na Leitire Duibhe and Càrn Dearg. On a clear day, in the distance behind you, Binnein Beag (page 185) can be identified due to its prominent detached position from the rest of the Mamores Munros opposite Ben Nevis (page 195).

3 When you've gained the ridge, walk south-south-east along the left-hand side of the crest following a clear eroded line that leads through rockier but uncomplicated terrain. The path then curves south-east to gain the summit of Càrn Dearg which is marked by a tall cairn.

4 Look south towards Rannoch and the immense blanket-bog-covered watershed before leaving the summit to the north-east following a faint path through grassy terrain interspersed with rocky sections which present no technical difficulty. Head downhill then pass to the right of a small knoll before dropping to the bealach named Màm Bàn. It can be boggy in the wide hollow but as soon as you start your ascent of Sgòr Gaibhre, you'll find it's a straightforward grassy pull right to the summit.

5 As your feet crunch on the eroded ground that surrounds the summit cairn, your eye is drawn north-east to the Ben Alder massif (page 259). Leave the summit to the north-west down a heavily eroded section of the route to reach Bealach nan Sgòr. It's very boggy here and finding a traverse route around the south-western flanks of Sgòr Choinnich is no easy task, especially in poor visibility. Stick loosely to the 800m contour on a right-to-left slant over heather and grass with tussocks and bilberry – it is tough going but is better than dropping into the peat bog basin down in the valley.

6 From NN431686, below the summit of Meall Nathrach Mòr, begin to veer left to pick up an all-terrain vehicle track which descends westwards and passes just to the left of the crags on Meall Nathrach Beag.

7 Descend to the Allt a' Choire Chreagaich and cross it 200m downstream of the sluice. Turn right on to a track which enters the woodland. Follow the track left, passing a collection of buildings, and cross a yard before meeting another track. Turn left to follow this track for 5km back to the hostel or 6.7km back to the station.

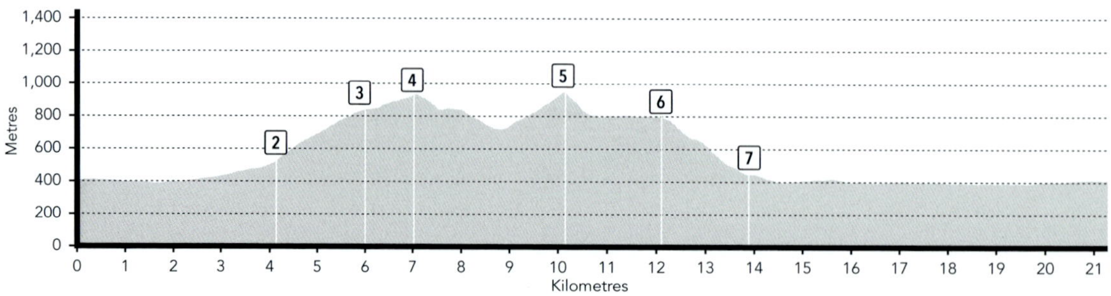

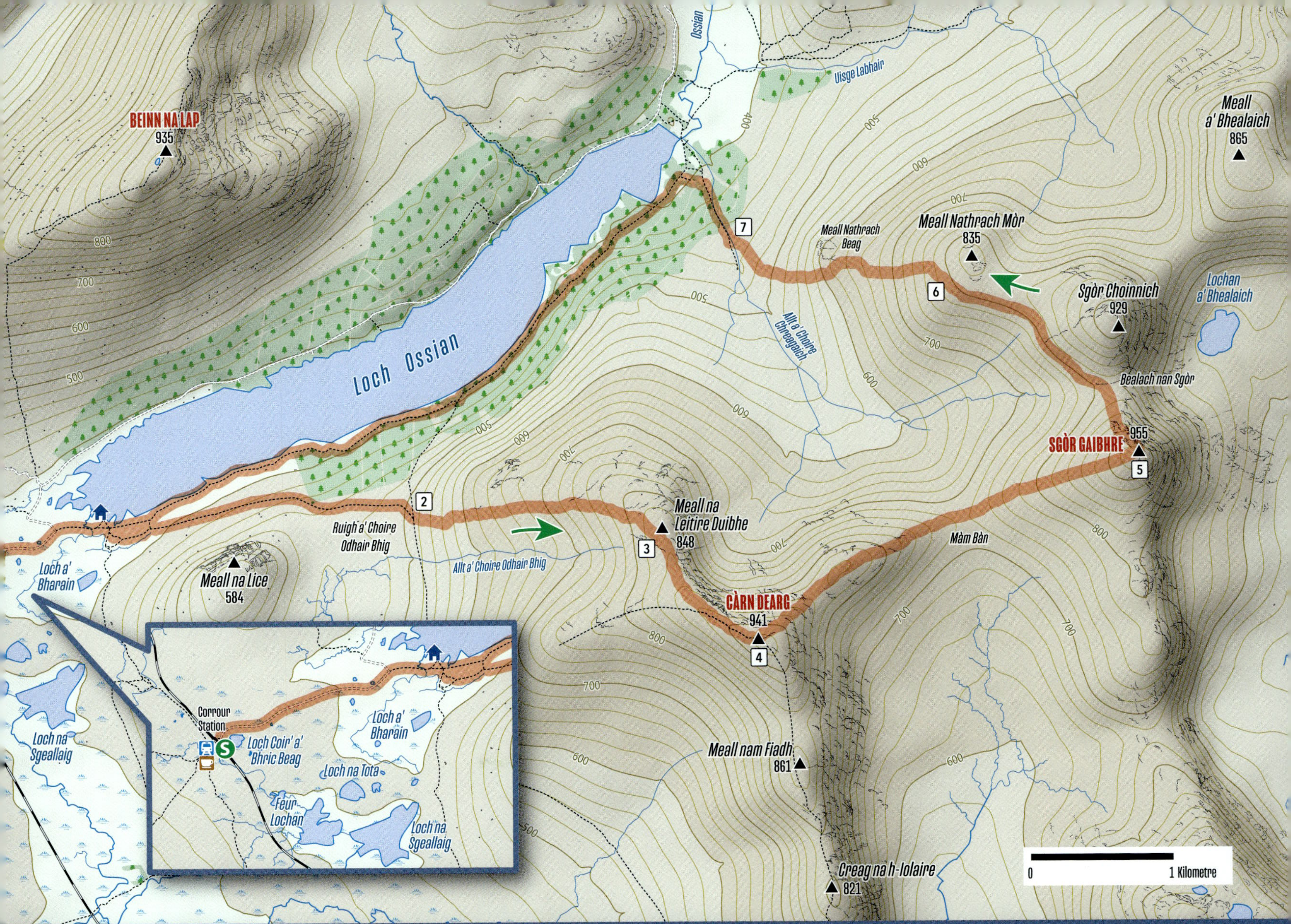

BEINN NA LAP
935
Loch Ossian
Ossian
Uisge Labhair
Meall a' Bhealaich
865
Meall Nathrach Mòr
835
Meall Nathrach Beag
Sgòr Choinnich
929
Lochan a' Bhealaich
Bealach nan Sgòr
SGÒR GAIBHRE
955
Allt a' Choire Chreagaich
Meall na Leitire Duibhe
848
Ruigh a' Choire Odhair Bhig
Allt a' Choire Odhair Bhig
Màm Bàn
CÀRN DEARG
941
Meall na Lice
584
Loch a' Bharain
Loch na Sgeallaig
Meall nam Fiadh
861
Creag na h-Iolaire
821
Corrour Station
Loch Coir' a' 'Bhric Beag
Loch a' Bharain
Loch na Tota
Feur Lochan
Loch na Sgeallaig
0
1 Kilometre
1
2
3
4
5
6
7

HIGHLAND HUB 9
PITLOCHRY

The ninth Highland hub is Pitlochry; the 'place of the Sentinel Stone'. It is a thriving town in the heart of Scotland and is larger than most other hubs so has an abundance of supermarkets and places to eat out.

Hostelling Scotland operates a hostel in Pitlochry in a lovely, elevated position above the town. As a solo traveller, I find hostels are great places to meet like-minded people. The ability to cook your own meals helps to keep a trip within budget too.

Like Braemar, Pitlochry has a lot of Victorian architectural influences since the royal family first began to visit the area in the mid 19th century. The town has a great buzz about it when you walk around the streets in summer. Like all good towns, there are walking trails right on the doorstep and a river; in this case the River Tummel runs to the south of town.

Pitlochry is well connected to the Central Belt thanks to its railway station and regular bus services. Direct trains and buses from Glasgow and Edinburgh reach the town in under two hours.

Many of the routes from this hub involve a long walk-in to the foot of the Munros and a mountain bike is often helpful. As a result, many of the walks are long, so it's nice to have Schiehallion in the group – a popular shorter walk with a sprinkling of excitement from scrambling over the boulders on its higher reaches.

WHERE TO STAY

£	££	£££
Blair Castle Caravan Park	Pitlochry Youth Hostel	The Claymore Guest House

TRAVEL TIMES

50 Meall Chuaich	**35 minutes**
51 Ben Alder Range	**40 minutes**
52 East Drumochter Munros	**35 minutes**
53 West Drumochter Munros	**30 minutes**
54 Schiehallion	**40 minutes**
55 Beinn Dearg	**15 minutes**
56 Càrn a' Chlamain	**15 minutes**
57 Beinn a' Ghlo	**20 minutes**

Rest day ideas

- Catch an eclectic mix of performers at the Pitlochry Festival Theatre. It started off life as a post-war tent theatre but is now hosting classic productions like *The Sound of Music* and *The Wind in the Willows.*
- Take a trip back in time at Blair Castle and Gardens. Wander around the 30 rooms that are open to the public along with the picturesque gardens.
- Take in the waterfalls at the Linn of Tummel at the confluence of the River Tummel and River Garry. Keep an eye out for red squirrels in the woodland.

Opposite Beinn Iutharn Mhòr and Carn an Righ from Carn nan Gabhar (route 57).
Overleaf Ben Alder from the descent of Beinn Bheòil (route 51).

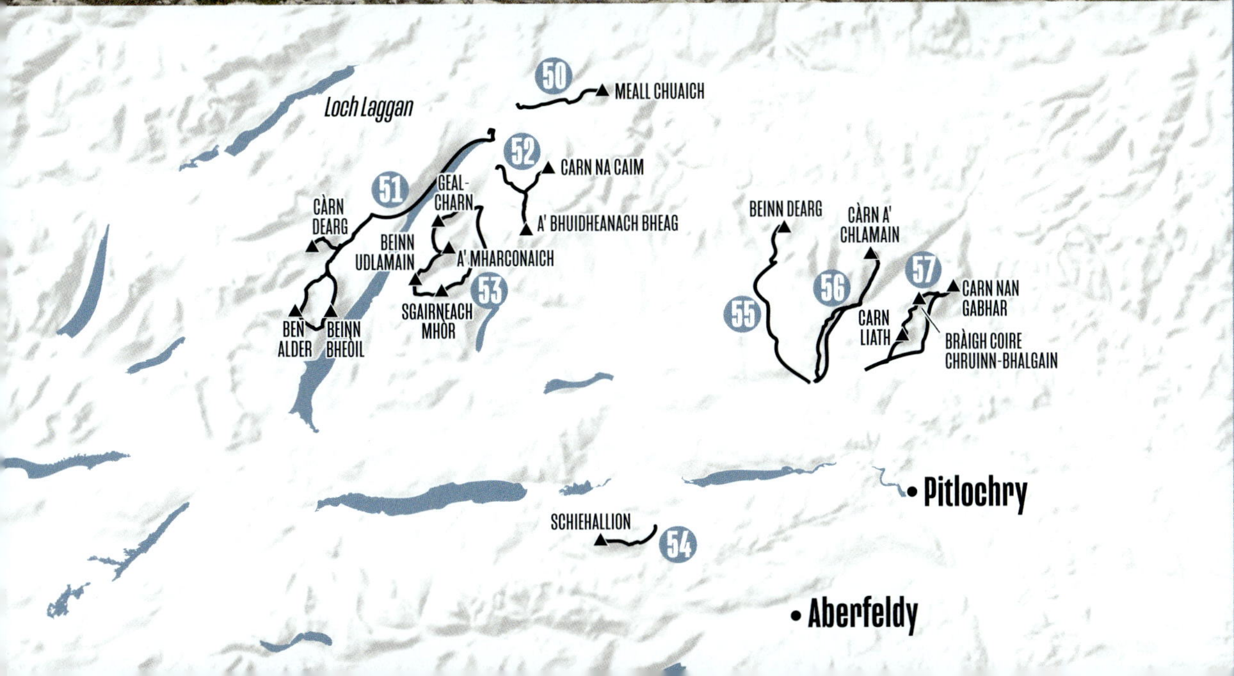
50
MEALL CHUAICH
Loch Laggan
52
CARN NA CAIM
51
GEAL-CHARN
CÀRN DEARG
A' BHUIDHEANACH BHEAG
BEINN UDLAMAIN
A' MHARCONAICH
53
SGAIRNEACH MHÒR
BEN ALDER
BEINN BHEÒIL
BEINN DEARG
CÀRN A' CHLAMAIN
56
57
CARN NAN GABHAR
55
CARN LIATH
BRÀIGH COIRE CHRUINN-BHALGAIN
Pitlochry
SCHIEHALLION
54
Aberfeldy

50 Meall Chuaich

HUB **Pitlochry – 35-minute drive**

OVERALL 1 **(2)** 3 4 5 TERRAIN 1 **(2)** 3 4 5
DISTANCE 1 **(2)** 3 4 5 NAVIGATION 1 **(2)** 3 4 5

DISTANCE **14.7km** ASCENT **600m**
TIMES **5:20hrs** (Walker) **4:35hrs** (Trekker)
3:25hrs (Fastpacker) **2:20hrs** (Trail runner)

START LOCATION **Lay-by (number 94), A9, north of Dalwhinnie**
START LATITUDE/LONGITUDE **56.9512, -4.2147**
START GRID REFERENCE **NN653866**
START W3W **///giraffes.report.holiday**

	PRONUNCIATION	NAME MEANING
Meall Chuaich	**'meowl koo-eek'**	**hill of the quaich (cup)**

Introduction

Munros like this are worth their weight in gold. Their bite-sized routes are perfect for a quick half-day jaunt into the hills, neatly filling those little gaps in your Munro-bagging itinerary. Got a few hours spare? Fancy splitting up your long drive with a short 'n' sweet blast up a solitary peak? Meall Chuaich is ideal for this sort of add-on mission. Located next to Dalwhinnie, and isolated to the west of the main Cairngorm plateau, some see Meall Chuaich as an inconveniently located, boring lump, while others relish the unique perspective it offers across the surrounding Highlands.

Meall Chuaich with the aqueduct in the foreground.

Public Transport

Start your walk from Dalwhinnie railway station, on the Highland Main Line. Pick up the track beside the aqueduct from Dalwhinnie, just south of the garage (NN637840, ///blazed.afford.importers). Using this approach would add around eight kilometres to the walk.

Parking

Along the A9 the lay-bys are conveniently numbered in ascending order from south to north. Look out for number 94, north of Dalwhinnie, where there is comfortable free parking for around eight cars. If the lay-by is full, try number 93 (on the other side of the road) or number 92 (900 metres to the south). There are no nearby toilet facilities.

MTB-ability

The first and last 4.5 kilometres of the route can be cycled on the hydro track.

Escape Route

None.

Refreshments

A five-minute drive south of the start point, head to The Apiary in Dalwhinnie, a smashing cafe where muddy boots and paws are welcome.

Approaching the large summit cairn of Meall Chuaich.

The Route

S Leave the lay-by and walk north-east along the wide grassy verge of the A9 for 150m. Turn right, pass through a gate and walk south-south-east gently uphill along a wide gravel track. After 300m, at a track junction, fork left and follow the track along the edge of the aqueduct for 1.6km. Pass through a gate and cross a bridge over to the southern side of the aqueduct.

2 The track leads past Cuaich Power Station and beyond it to a bridge over the Allt a' Choire Chàim. Be sure to look right towards Coire Cam and Meall Chuaich's nearest Munro neighbour – Carn na Caim – which is far more impressive when viewed from this angle than the side it's commonly walked from, along with A' Bhuidheanach Bheag (page 267). After crossing the bridge, fork left to continue walking east-north-east. Here on the other side of the dam, the calm waters of the aqueduct morph into the glorious natural chaos of the Allt Cuaich which curves left and right as the water travels out of Loch Cuaich. Cross another bridge, this time over the Allt a' Choire Chàis.

3 Fork left shortly after the bridge. Remain on the main hydro track for nearly a kilometre, ignoring a couple of minor turn-offs, and fork right at NN689871, 300m short of the edge of Loch Cuaich. Follow this track past a small wooden estate building and across a bridge over the Allt Coire Chuaich. After 50m, a small cairn marks the start of the ascent path of Meall Chuaich to the left. Take this path and navigate north-east over a short section of soggy ground before beginning your ascent in earnest following a clear eroded line east through a sea of heather.

4 When the path forks, walk left; a fingerpost which reads *Walkers' Path* has been erected to catch those who miss the turning. There are a couple more boggy patches to navigate along the ascent which curves to the north-east, but careful foot placement will ensure you won't need to reach for your snorkel.

5 Take the right-hand fork at NN705877; the ground becomes stonier as you approach the higher reaches of the mountain. In the final stages of the ascent there are a number of parallel lines you could take to the summit, but all trails lead east. Whichever you choose, the terrain presents no technical difficulty.

6 The ground plateaus towards the top and you can't miss the summit as it's marked by a huge cairn which means you'll always have shelter to eat your lunch at the summit whichever way the wind blows. The superb view from the top across to Ben Macdui and the high Cairngorms to the north-east is finally unmasked, as it was blocked by the slopes of Meall Chuaich throughout your ascent. This solitary Munro appears sandwiched between two distant giants – Ben Macdui and Ben Alder (page 259) to the south-west. Retrace your steps back to the start.

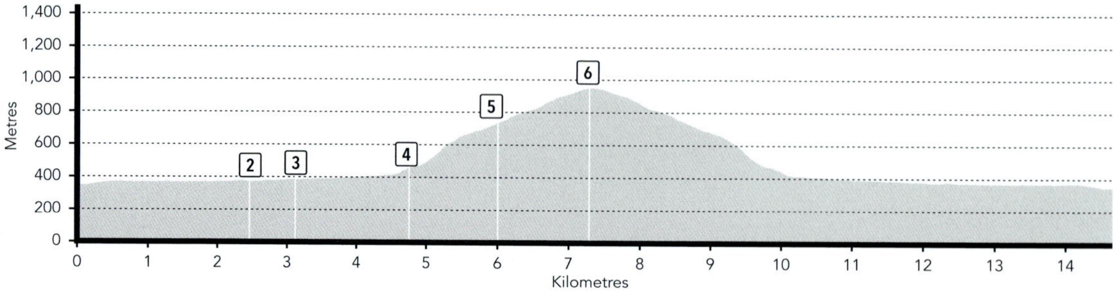

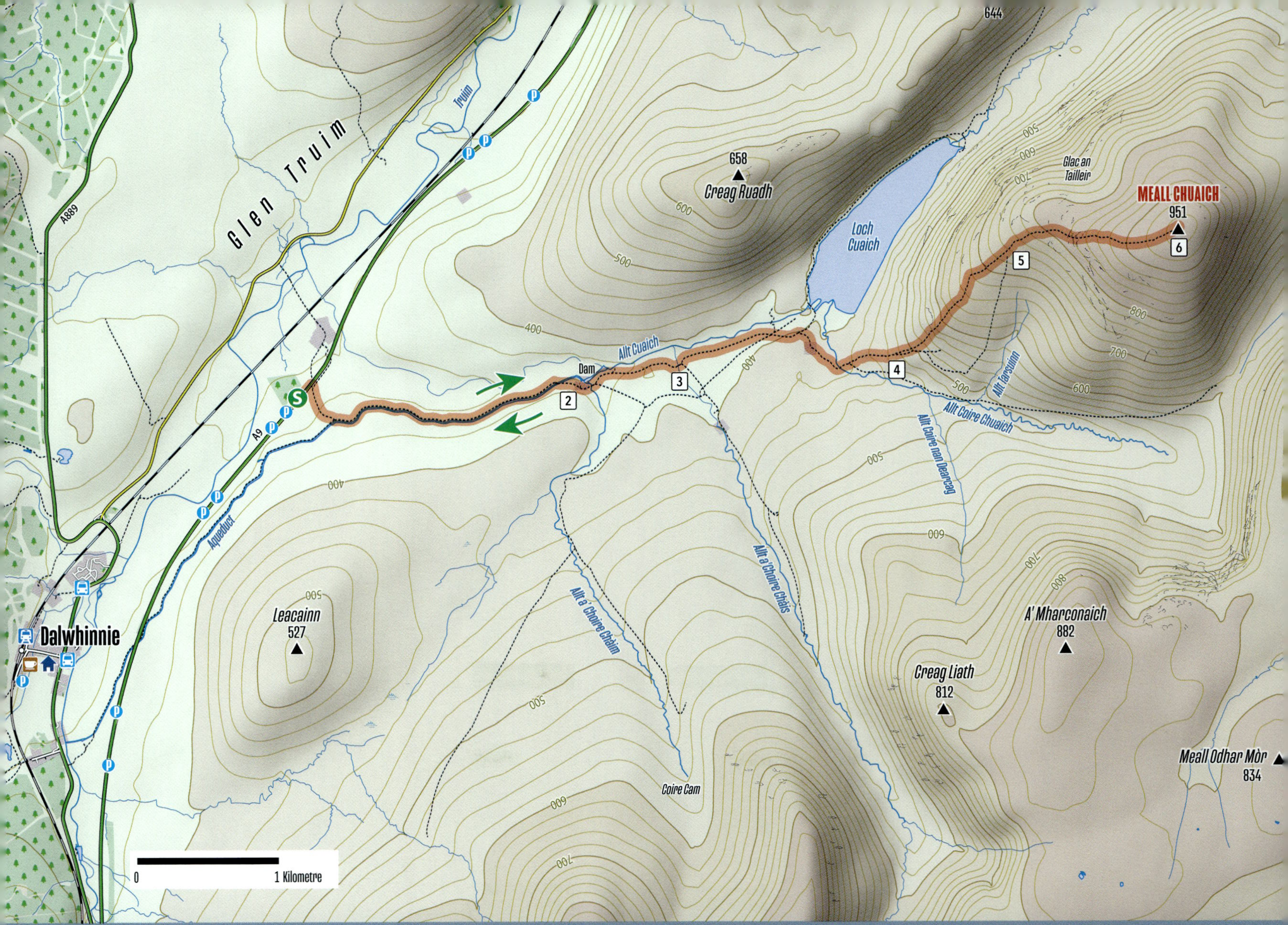
MEALL CHUAICH
951
Glac an Tailleir
Loch Cuaich
Creag Ruadh
658
644
Glen Truim
Truim
A889
A9
Aqueduct
Dam
Allt Cuaich
Allt Tarsuinn
Allt Coire Chuaich
Allt Coire nan Dearcag
Allt a' Choire Chàis
Allt a' Choire Chàim
Coire Cam
Leacainn
527
Creag Liath
812
A' Mharconaich
882
Meall Odhar Mòr
834
Dalwhinnie
0
1 Kilometre

51 Ben Alder Range

HUB **Pitlochry – 40-minute drive**

DAY 1 (DALWHINNIE/CÀRN DEARG/WILD CAMP AT CULRA)

OVERALL 1 2 3 **(4)** 5 TERRAIN 1 2 **(3)** 4 5
DISTANCE 1 2 3 **(4)** 5 NAVIGATION 1 2 3 **(4)** 5

DISTANCE **22km** ASCENT **810m**
TIMES **7:40hrs** (Walker) **6:35hrs** (Trekker)
4:55hrs (Fastpacker) **3:25hrs** (Trail runner)

DAY 2 (CULRA/BEN ALDER/BEINN BHEÒIL/DALWHINNIE)

OVERALL 1 2 3 4 **(5)** TERRAIN 1 2 3 4 **(5)** (Water)
DISTANCE 1 2 3 4 **(5)** NAVIGATION 1 2 3 **(4)** 5

DISTANCE **34.5km** ASCENT **1,165m**
TIMES **11:25hrs** (Walker) **9:45hrs** (Trekker)
7:15hrs (Fastpacker) **5:05hrs** (Trail runner)

START LOCATION **Hillwalkers' car park, near Dalwhinnie railway station**
START LATITUDE/LONGITUDE **56.9330, -4.2469**
START GRID REFERENCE **NN633846**
START W3W **///partied.pimples.anguished**

	PRONUNCIATION	NAME MEANING
Càrn Dearg	**'carn jerreck'**	**red hill**
Ben Alder	**'ben ollder'**	**hill of rock and water**
Beinn Bheòil	**'beyn veeyorl'**	**hill of the mouth**

Introduction

There's more than one way to skin a cat – and there's more than one way to tick off the Ben Alder Munros to the west of Loch Ericht. This is superlative hillwalking terrain, with majestic mountain scenery at every turn. The hills are hard-won, the ridges are edgy and exposed, and the burn crossings are unforgiving, yet the rarefied air leaves a palpable taste of adventure and escapism on the lips.

There are many ways you could tackle these Munros, with different route options, schedules and overnight stays. But I'd recommend a two-day expedition. Starting in Dalwhinnie, day one involves the long walk-in or cycle-in to Culra and a steep out-and-back hike to Càrn Dearg, before spending the night wild camping near Culra (there used to be a bothy at Culra, but it is now closed due to asbestos problems – and is consequently very dangerous). Day two focuses on bagging Ben Alder and Beinn Bheòil – one of Scotland's classic walks featuring the rip-roaring Long Leachas ridge and a high-level loop of Loch a' Bhealaich Bheithe – before the long yomp back to civilisation in Dalwhinnie.

Beinn Bheòil and Loch Ericht from Sròn Coire na h-Iolaire.

Public Transport

Start your walk from Dalwhinnie railway station, on the Highland Main Line.

Parking

Free parking for around eight cars in the hillwalkers' car park 250m south of Dalwhinnie railway station. There are no nearby toilet facilities. This car park used to be perfectly placed for walkers and cyclists to cross the railway line and pick up the track to Ben Alder Lodge. This historic right of way is now blocked by Network Rail so walkers must take a circuitous detour through Dalwhinnie to pick up the track further down. Local residents appreciate the continued use of the hillwalkers' car park to keep local roads free from overcrowding.

MTB-ability

The route is cyclable from Dalwhinnie as far as Culra (17.5 kilometres). If you don't want to wild camp overnight near Culra, using a bike to get back to Dalwhinnie at the end of day one and back out to Culra at the start of day two is a great option. See below for distances and timings; GPX files are available for download.

MTB two-day option

DAY 1 (DALWHINNIE/CÀRN DEARG/DALWHINNIE)

DISTANCE **35.1km** (bike) and **5.1km** (walk)
ASCENT **415m** (bike) and **580m** (walk)
TIMES (walk section only) **3:20hrs** (Walker) **2:55hrs** (Trekker) **2:10hrs** (Fastpacker) **1:25hrs** (Trail runner)

DAY 2 (DALWHINNIE/CULRA/BEN ALDER/BEINN BHEÒIL/ CULRA/DALWHINNIE)

DISTANCE **35.1km** (bike) and **17.7km** (walk)
ASCENT **415m** (bike) and **1,040m** (walk)
TIMES (walk section only) **7:40hrs** (Walker) **6:40hrs** (Trekker) **5:00hrs** (Fastpacker) **3:20hrs** (Trail runner)

Escape Route

No escape routes on day one. On day two, from Bealach Breabag between Ben Alder and Beinn Bheòil, descend north to Loch a' Bhealaich Bheithe and walk along the eastern side of the loch.

Refreshments

A short walk from the train station in Dalwhinnie you'll find The Apiary, a smashing cafe where muddy boots and paws are welcome. Hot food and cakes are served with a smile.

The Route

Day 1 (Dalwhinnie/Càrn Dearg/wild camp at Culra)

S From the car park, walk or cycle north back towards the station. At the T-junction, turn right where a fingerpost reads *Village Hall 400 yards.* At the next T-junction turn right and walk for 700m before turning right again on the road just beyond Dalwhinnie Service Station where a fingerpost reads *Loch Ericht ¼ mile.* Use the underpass to travel beneath the railway and turn right immediately afterwards.

2 After 300m the track merges with another track; follow this track south-west beside Loch Ericht for 8km to Ben Alder Lodge. Along the way, pass to the right of the first of many ornate estate buildings and enjoy the long undulating track beside Loch Ericht. Latterly, ignore the driveways which lead down to the main lodge with its striking turrets and crenellations. Instead, continue along the trail as it ascends gently through woodland to a gate and a bridge over the Allt Coire na Mèine. Keep right as the track continues uphill out of the woodland.

3 Ignore the right turn at NN559784 and continue straight ahead towards Loch Pattack. A left-hand turn at NN548786 is the most direct route to Culra for walkers; cyclists will prefer to continue along the track which runs along the shore of Loch Pattack. **If taking the walkers' path**, walk south-west alongside the Allt a' Chaoil-rèidhe until you reach the bridge at NN524764. Cross the bridge and turn left, following the track to Culra. **If selecting the cycle path**, continue on the main track to NN545789 and turn left to cycle along the southern edge of Loch Pattack, crossing two footbridges along the way. At the south-west tip of the loch, fork left to cycle south for 2.5km along the rough track to reach the old bothy building at Culra where bikes should be left.

4 My preference is to stow overnight gear at Culra and hike Càrn Dearg with a lighter pack. Pass to the left of the building following a faint line over grassy terrain, heading west-north-west up the lower slopes

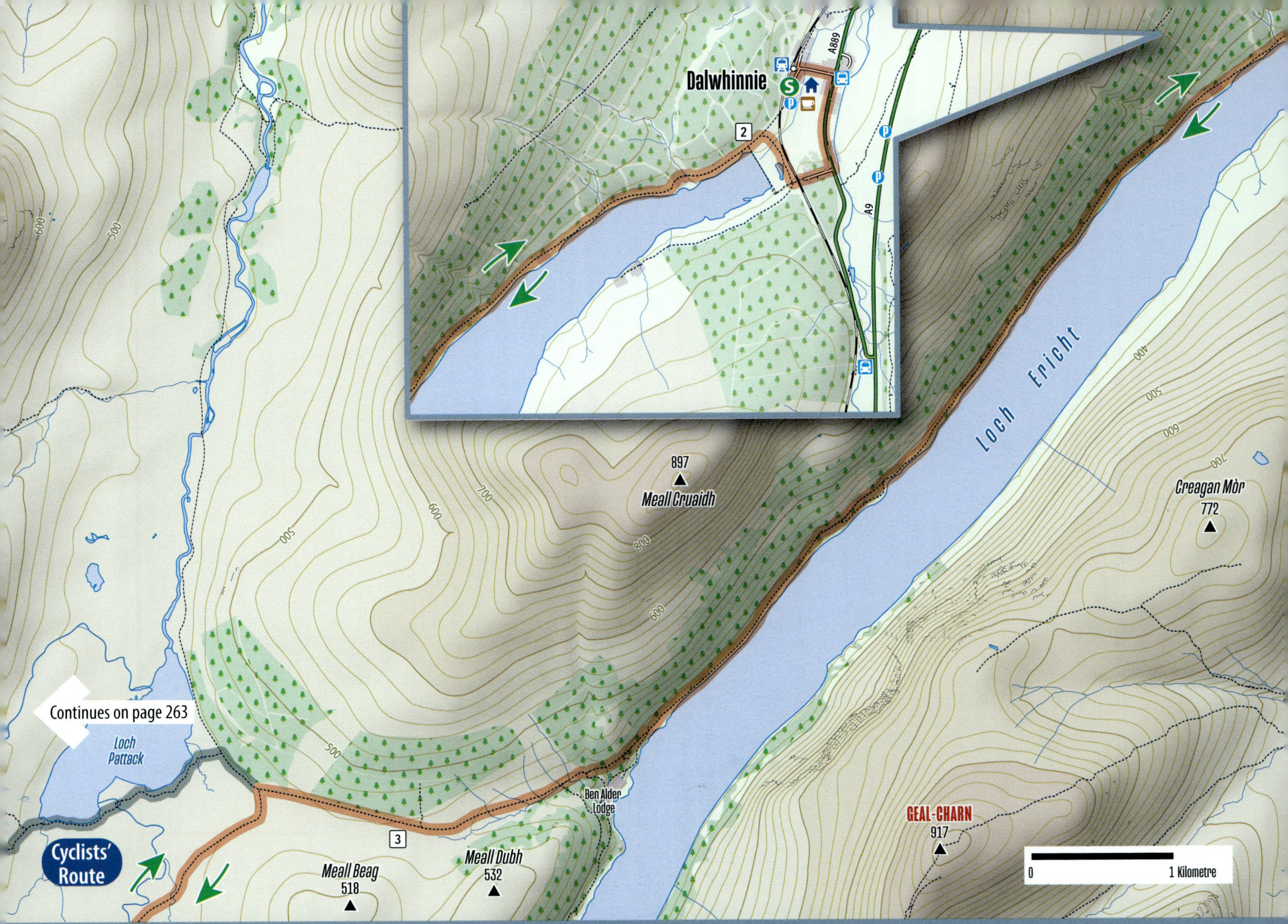

Continues on page 263

of the Munro. When the trail becomes indistinct, walk north-west up the slope which is a mix of grass, heather and bilberry. It's a tough, direct climb so walking in zigzags is recommended, with frequent breaks to marvel at the flatlands at the foot of the climb and how small the buildings near the river appear as you stretch up and away from them.

5 At 700m elevation the steep slopes begin to relent somewhat until you top out on the ridge at NN512769 to a view of Geal Charn (page 227) and Loch a' Bhealaich Leamhain to the north. Embark on the final portion of the ascent, heading south-west along a path which is stony, grassy and non-technical, running parallel to the long ridge of Beinn a' Chlachair.

6 The summit is marked by a cairn, beside which is the perfect place to sit and lap up the view across to Ben Alder and Beinn Bheòil with Loch a' Bhealaich Bheithe parting them. Retrace your steps to Culra (**4**) where you can camp nearby overnight and walk Ben Alder and Beinn Bheòil in the morning. Or close out the long day by making your way back to Dalwhinnie.

Day 2 (Culra/Ben Alder/Beinn Bheòil/Dalwhinnie)
Leave Culra using the track to the north-east until you reach the bridge over the Allt a' Chaoil-rèidhe.

7 Cross the bridge and pick up the path on the other side which leads south-east before it meets another path. Turn right here to walk south-west on a good path which follows the Allt a' Chaoil-rèidhe. Remain on the path until it reaches a small cairn at NN515744 which marks a smaller path on the right. Take this path heading south-west through moorland towards a stream.

8 The path ends when you reach Allt a' Bhealaich Bheithe at NN511740. Use your own judgement here to identify a suitable crossing point, exploring both upstream and downstream for options. If in doubt, use the escape route described on page 260 as your route of ascent, I once used this route myself when the burn was in spate, and I walked Ben Alder as an out-and-back. After crossing the stream, pick up a faint path which leads south-west initially across the heathery ground towards the Long Leachas ridge. The route then curves westwards and passes under the crags on the southern side of the knoll marked as point 727m on OS maps. The trail is indistinct over wet ground but then becomes more prominent as you join the ridge at NN504737.

9 Follow the path as it leads up the crest of the ridge. This part of the route is comprised mostly of steep walking interspersed with a couple of mild scrambling moves or big steps for those with long legs. Pass over point 848m and drop down briefly into a small depression. Some simple scrambling up a short rocky section follows. The route then levels out again before continuing to ascend steeply once more. The path then traverses to the right slightly to avoid some more difficult scrambling and leads to the foot of a short gully. Ascend this and emerge back on to the crest. The path leads up the narrowing ridge with some simple scrambling before passing to the right once more to avoid the small rock stacks on the highest part of the ridge. Prepare for the view down the serrated spine of the ridge towards distant Loch Pattack and Loch Ericht to blow your socks off when you top out.

10 Head south and then south-south-east across the stony plateau aiming for higher ground, keeping right of the cliffs of Coire na Lethchois. Walk over boulder scree to reach a cairn marking an epic view down over Garbh-choire Beag across to

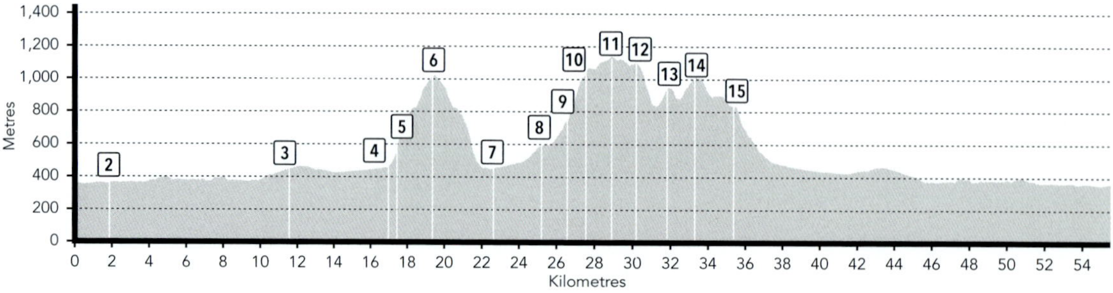

Loch Pattack
Cyclists' Route
An Lairig
Meall Beag
518
Continued from page 261
Allt a' Chaoil-rèidhe
Meall Mòr
528
CÀRN DEARG
1034
Culra
Allt a' Bhealaich Bheithe
Coire na Lethchois
Garbh-choire Beag
Loch a' Bhealaich Bheithe
BEN ALDER
1148
BEINN BHEÒIL
1019
Garbh Choire
Sròn Bealach Beithe
1104
Sròn Coire na h-Iolaire
956
Bealach Breabag
Loch Ericht
0
1 Kilometre
1
2
3
4
5
6
7
8
9
10
11
12
13
14
15

Above Mountain bike at the old bothy building at Culra. **Opposite** Climbing the Long Leachas ridge on Ben Alder.

Beinn Bheòil. From here, walk south-south-west over grassier terrain, with the precipitous cliffs on your left. Patches of boulder scree return along the sustained but gradually ascending pull to the summit of Ben Alder. Pass a row of ruined stone shelters which are evidence of mapping parties during the principal triangulation activities of the 19th century which were led by Major General Thomas Frederick Colby.

11 The Ben Alder summit plateau houses multiple stone wind shelters and a trig pillar that has seen better days. Descend gently to the south-south-west across the vast plateau, passing to the left of Lochan a' Garbh Choire. Continuing to keep right of the cliffs, drop to the bealach at 1,083m while enjoying the view south-west to the Loch Ossian Munros (pages 237, 241 and 247).

12 Pass over Sròn Bealach Beithe and resist the temptation to continue along the lip of the cliffs as this will lead you to dangerous ground. Instead, from this minor top, begin the largely pathless descent south-east. The southern end of Loch Ericht is in view as you drop gingerly over rocky, wet terrain beside a stream towards Bealach Breabag. From the bealach, a path can be picked up which leads across the lowest point of 833m before ascending Sròn Coire na h-Iolaire to the north-east.

13 When you're very close to the highest point of Sròn Coire na h-Iolaire the path swings left, just avoiding the summit. Walk north-west and then north-east along the ridge towards Beinn Bheòil, dropping slightly to a saddle. On a clear day to the distant right, as you begin the ascent of Beinn Bheòil, is the unmistakable conical profile of Schiehallion (page 277). Halfway up the short ascent, the path passes to the right slightly, offering a tremendous view down Loch Ericht, which is reputedly as deep as the Blackpool Tower is tall.

14 As you reach the summit, take in the view across the loch to Beinn Udlamain and the Drumochter Munros (page 271) which are visible on the other side of the loch to the north-east. Continue down the broad ridge to the north-north-east where the terrain is non-technical and fairly gradual. The path disappears towards the end of the ridge on soft bouncy terrain.

15 At NN521737, turn to descend north-west down the hillside; a faint path comes and goes as you navigate off the hill to meet the original approach path. From here, walk 3km back to Culra from where you can cycle or walk back to Dalwhinnie.

52 East Drumochter Munros

HUB **Pitlochry – 35-minute drive**

OVERALL 1 ② 3 4 5 TERRAIN 1 ② 3 4 5
DISTANCE 1 2 ③ 4 5 NAVIGATION 1 ② 3 4 5

DISTANCE **18.9km** ASCENT **880m**
TIMES **7:20hrs** (Walker) **6:15hrs** (Trekker)
4:45hrs (Fastpacker) **3:15hrs** (Trail runner)

START LOCATION **Large lay-by (number 87), A9, south of Dalwhinnie**
START LATITUDE/LONGITUDE **56.9088, -4.2365**
START GRID REFERENCE **NN639819**
START W3W **///voltage.cricket.suiting**

	PRONUNCIATION	NAME MEANING
A' Bhuidheanach Bheag	**'a voo-yun-uch vake'**	**little yellow place**
Carn na Caim	**'carn na came'**	**hill of the curve**

Introduction

The start point to this walk on the south-west edge of the Cairngorms National Park can be reached by car, bus or train. It's one of the least technical Munro routes with rolling grassy terrain and no scrambling required, and in a departure from many Munro routes, hikers are offered a welcome 'leg-up' as the route begins at 400 metres in elevation. Many Munro-baggers save this route for a drizzly day which is a shame as A' Bhuidheanach Bheag and Carn na Caim have a lot to offer.

Public Transport

Scottish Citylink bus services linking Glasgow and Inverness stop at Road End Dalwhinnie (NN639829, ///jousting.covers.hockey). It's a 15-minute walk to the start from here. In addition, Dalwhinnie railway station, on the Highland Main Line, is a three-kilometre walk (each way) from the start point.

Parking

Along the Drumochter Pass section of the A9 the lay-bys are conveniently numbered in ascending order from south to north. Look out for number 87, south of Dalwhinnie, where there is comfortable free parking for around seven cars. If the lay-by is full, try number 88 (400 metres to the north) or number 86 (one kilometre to the south). There are no nearby toilet facilities.

Escape Route

No escape routes; retrace your steps after the first Munro and return another day for the second.

Refreshments

Three kilometres north of the start point in Dalwhinnie you'll find The Apiary, a smashing cafe where muddy boots and paws are welcome.

The ascent track with Sgairneach Mhòr and A' Mharconaich in the background.

The Route

S The Drumochter Pass is a major road linking Inverness with the south of Scotland so be sure to cross this busy road cautiously. Walk north along the verge for 200m and turn right, passing through two gates on a wide track. After the second gate, take the right-hand fork on a track which runs parallel to the Allt Coire nan Cisteachan.

2 At 430m elevation the track veers south, traversing away from the burn. Before the track reaches the Allt Coire Bhotie, it swings left and climbs steadily up the shoulder to 902m which marks the junction between the two Munros. The navigation to this point is straightforward.

3 Take the right-hand fork to hike to A' Bhuidheanach Bheag first. The scenery here is not as dramatic as other Munro walks but it's nice to enjoy the simplicity of non-technical walking on grassy terrain every once in a while. Ben Alder (page 259) is recognisable to the south-west.

4 At a large quartz cairn, come off the main track and take the left fork. This track passes around the eastern flank of A' Bhuidheanach. From here, drop south to a peaty saddle and cross a small burn before continuing south uphill. When the ground flattens out to the west of A' Bhuidheanach Mhòr continue to walk south across the plateau to the top of A' Bhuidheanach Bheag. In low visibility, iron fence posts provide reassurance that you're heading in the right direction.

5 The summit of A' Bhuidheanach Bheag is marked by a trig pillar surrounded by quartzite stone and old iron fence posts. The summit is in the middle of a vast plateau, which means that – despite the proximity to the noisy A9 – there's a wild, remote feeling to this place. From the substantial table-like mountaintop, retrace your steps back to the split in the track at point 902m (**3**), looking out for ground-nesting birds. Ptarmigan, dotterel and plover are common sights in this upland landscape. Head north-east and then north on the gently undulating track towards Carn na Caim.

Descending A' Bhuidheanach Bheag.

6 At NN669814, leave the track and follow the iron fence posts north-east.

7 At the summit of Carn na Caim, which is marked by a cairn, there is a greater sense of being on the edge of the plateau. Fifty steps north of the summit cairn, you overlook the striking corrie of Coire Cam. Enjoy the view north-east to the Munro Meall Chuaich. As tempting as it seems, it's a reach too far for most hikers to add Meall Chuaich to the walk, so I recommend it's hiked independently (page 255). From the summit of Carn na Caim, retrace your steps back to the split in the track at 902m (**3**) and descend to the road by the route of your initial ascent.

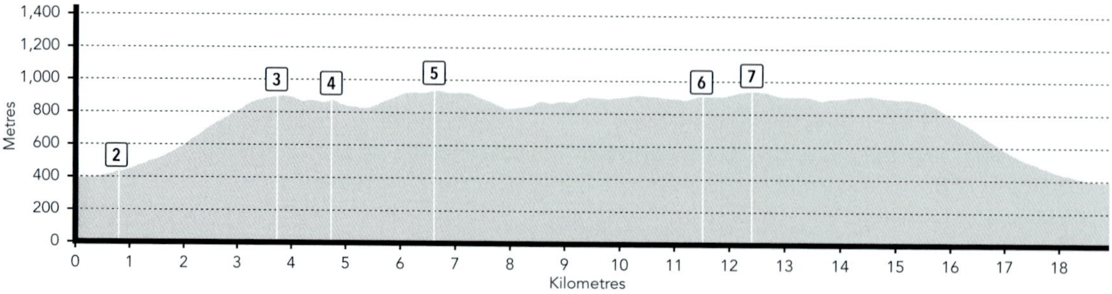

Dalwhinnie
A889
Loch Ericht
0
1 Kilometre
Allt Choire Uilleim
Allt Coire Bhathaich
Coire Cam
CARN NA CAIM
941
2
3
4
5
6
7
Allt Coire nan Cisteachan
Coire Uilleim
Coire nan Cisteachan
Truim
Glen Truim
Allt Coire Bhotie
A9
Drumochter Pass
Coire Bhotie
Coire Chuirn
Allt na Cailliche
Meall Odhar Mòr
890
Cama' Choire
A' Bhuidheanach
881
Allt Choire Chùirn
Coire Chaorainn
A' Bhuidheanach Mhòr
932
Meall a' Chaorainn
916
Creagan Doire Dhonaich
724
A' BHUIDHEANACH BHEAG
936
Glas Mheall Mòr
928
Meall Uaine
902
Fraoch Choire

SGAIRNEACH MHÒR 991m
BEINN UDLAMAIN 1,011m
A' MHARCONAICH 975m
GEAL-CHARN 917m

53 West Drumochter Munros

HUB **Pitlochry – 30-minute drive**

OVERALL 1 2 **3** 4 5 TERRAIN 1 2 **3** 4 5
DISTANCE 1 2 3 **4** 5 NAVIGATION 1 2 **3** 4 5

DISTANCE **23.9km** ASCENT **1,145m**
TIMES **9:25hrs** (Walker) **8:05hrs** (Trekker)
6:00hrs (Fastpacker) **4:05hrs** (Trail runner)

START LOCATION **Large lay-by (number 81), A9, Drumochter Pass**
START LATITUDE/LONGITUDE **56.8554, -4.2462**
START GRID REFERENCE **NN631760**
START W3W **///sobered.televise.hockey**

	PRONUNCIATION	NAME MEANING
Sgairneach Mhòr	**'skaarnyach voor'**	**big stony hillside**
Beinn Udlamain	**'beyn oodlamen'**	**gloomy hill**
A' Mharconaich	**'a varkeneek'**	**the horse place**
Geal-charn	**'ghee-yal harn'**	**white hill**

Introduction

You could split this foursome of Munros into two easier day walks, but once you're up high, it makes sense to tick them off in one stint if you're able. Located to the west of the Drumochter Pass, in a wedge of land between the A9 and Loch Ericht, these rolling hills are linked by lofty bealachs, which makes for an enjoyable, high-level tramp without any harsh ups or downs. Once you've ticked off this route, you can give the hills a nod of appreciation every time you travel past on the A9 on future Munro missions.

Public Transport

It would only be possible to complete this route using public transport with dual use of a bus or rail service coupled with a cycle. From Dalwhinnie railway station, on the Highland Main Line, cycle 6.5 kilometres to Balsporran Cottages and then walk the route from there, cycling back to the station afterwards. The total outing would be around 37 kilometres. Alternatively, Scottish Citylink bus services link Glasgow and Inverness and stop at Road End Dalwhinnie (NN639829, ///jousting.covers.hockey). This would be a total outing of 32 kilometres.

Parking

Along the Drumochter Pass section of the A9 the lay-bys are conveniently numbered in ascending order from south to north. Look out for number 81, north of the Dalnaspidal turn-off, where there is comfortable free parking for around 10 cars. If the lay-by is full, try number 80 (250 metres to the south) or number 79 (500 metres to the south). There are no nearby toilet facilities.

Allt Coire Dhomhain with Sgairneach Mhòr beyond.

Above Summit cairn on Sgairneach Mhòr with Ben Alder in the distance. **Opposite** Two walking parties descending Sgairneach Mhòr.

MTB-ability

Cyclists arriving by car can leave a bike near the track at Balsporran Cottages (NN627791, ///acquaint.official.enforced) and drive on to the start of the walk. Upon reaching Balsporran Cottages, retrieve your bike and cycle beside the A9 back to your car saving around 3.3 kilometres of walking.

Escape Route

There are a few possible escape routes.

- At the bealach between Beinn Udlamain and A' Mharconaich marked as 861 metres on OS maps, descend south-east through heather to the track in Coire Dhomhain.
- From the summit of A' Mharconaich, descend north and then follow the ridge as it curves north-east down to Balsporran Cottages.
- At the bealach between A' Mharconaich and Geal-charn marked as 739 metres on OS maps, stick to the track which descends north-east beside the Allt Coire Fhàr to Balsporran Cottages.

Refreshments

Back in Pitlochry, try Hettie's Tea Room. Their cake slices are huge, the perfect treat after a long walk.

The Route

S Leave the lay-by and walk south on the tarmac footpath which runs alongside the A9. To the right lie your Munro objectives for the day alongside the Sow of Atholl which sits nearest the road. The footpath merges right on to an access road which heads downhill slightly. After 330m, fork right on to a track which leads to a tunnel under the railway. Upon exiting the tunnel, turn right and follow the track into the valley, fording a stream as the track gently ascends.

2 At a major track junction on the 500m contour, turn left to cross a footbridge over the Allt Coire Dhomhain. After 200m turn right on to peaty ground following an all-terrain vehicle trail up the hillside. At NN617741, take the faint path to the right to traverse around to the saddle between Sgairneach Mhòr and the minor top marked 758m on OS maps.

3 Once at the saddle, walk west then south-west up to the summit of Sgairneach Mhòr following a much clearer path. Be sure to look behind you on the ascent to see how far you've come from the road and take in the view to the Munros beyond the Drumochter Pass, notably A' Bhuidheanach Bheag (page 267). On your left in the distance is the solitary pyramidal Munro of Schiehallion (page 277). The summit of Sgairneach Mhòr is comprised of a trig pillar inside a shelter cairn.

4 Descend to the south-west and begin to bear right after 500m to walk west over soft, pathless terrain. Walk across the peat hags in the saddle marked as 809m on OS maps and make your way west-north-west up and on to the southern ridge of Beinn Udlamain. Iron fence posts mark the ascent line up the ridge. Once atop the Munro, you'll find a large stone shelter in the middle of the wide plateau.

5 Continue following iron fence posts off Beinn Udlamain to the north-east. The ground is firm underfoot and good progress can be made here. Curve right to pass under the minor north-east top and follow the fence posts down to the bealach marked as 892m on OS maps. Keep an eye out for plump ptarmigan as they totter high in these Drumochter hills. Fork left on the path when it splits to bypass point 902m on OS maps and drop to the bealach at point 861m on OS maps.

6 Fence posts continue to assist in poor visibility as you ascend north-east to A' Mharconaich. At a cairned junction, continue following the iron posts to gain higher ground. It's a straightforward pull to the top of A' Mharconaich. A small cairn marks the highest point at 975m.

7 Retrace your steps for 175m then fork right. Then, at NN598758, fork right again on to a faint path which descends over straightforward terrain to meet a more prominent path which leads north-west down to point 739m on OS maps. Three stacked bodies of water are visible to the north-west as you descend – Loch Ericht, Loch Pattack and Loch a' Bhealaich Leamhain. There are some stone-pitched steps on the steepest, boggiest ground as you approach the bealach.

8 The path merges with a track from the left at the bealach. Walk northwards towards Geal-charn. Peel left off the track after 350m on to a stony path to start your ascent of the final Munro. A path on your right can be used to bypass a rocky outcrop after 500m, after which you should ascend the middle of the shoulder as the path veers off to the right.

9 In a stiff breeze, the stone shelter of Geal-charn's summit is a welcome sight. The view back down Loch Ericht is incredible, as is the handsome side profile of A' Mharconaich, but I particularly love the view to Ben Alder (page 259) and the carved-out hollow of Bealach Dubh to its right. Leave the summit, aiming for a small mound with a cairn on it before descending to the east-north-east. A collection of cairns marks the right-hand edge of the trail. From here, an indistinct path comes and goes as you descend north-east down the ridge. Motivated by the ever-present sight of the A9, descend steadily downhill over heather-spattered terrain.

Loch Ericht from Beinn Udlamain.

10 The path merges into a pronounced stony track at NN617792. Follow this for 500m until it meets a track. Turn left on to the track to walk beside the Allt Coire Fhàr and cross a bridge over the Allt Beul an Sporain. Keep right, following the track east to the railway crossing which should be crossed with caution. Pass the Balsporran Cottages and walk across a bridge over the River Truim. Pass a car park and turn right on to a path which runs beside the Drumochter Pass for 3.3km to reach the start point.

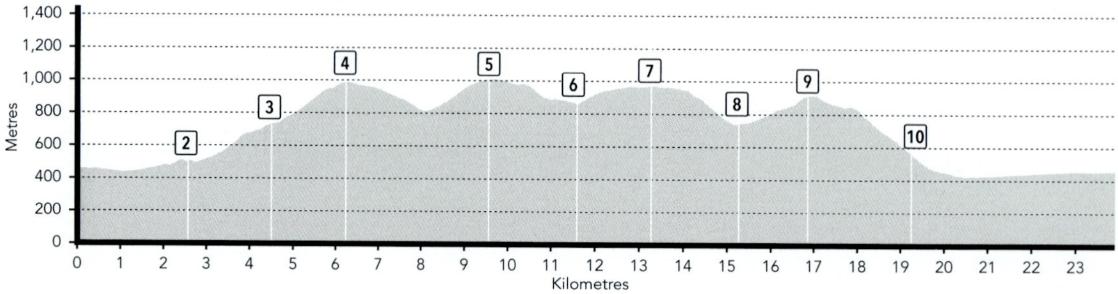

Creagan Mòr
772
Loch Ericht
Alt Beul an Sporain
Balsporran Cottages
Truim
10
GEAL-CHARN
917
9
Allt Coire Fhàr
Creagan Doire Dhonaich
724
Glen Truim
A9
8
A' MHARCONAICH
975
7
Coire an Tuirc
An Torc
739
Drumochter Pass
Allt Fraoch-choire
6
2
Coire Dhomhain
The Sow of Atholl
803
Allt Coire Dhomhain
5
BEINN UDLAMAIN
1011
758
3
SGAIRNEACH MHÒR
991
4
Coire Luidhearnaidh
0
1 Kilometre
Màm Bàn
919

54 Schiehallion

HUB **Pitlochry – 40-minute drive**

OVERALL 1 2 **3** 4 5 TERRAIN 1 2 3 **4** 5
DISTANCE 1 **2** 3 4 5 NAVIGATION 1 2 **3** 4 5

DISTANCE **10.4km** ASCENT **760m**
TIMES **5:10hrs** (Walker) **4:25hrs** (Trekker)
3:20hrs (Fastpacker) **2:10hrs** (Trail runner)

START LOCATION **Braes of Foss car park, Schiehallion Road**
START LATITUDE/LONGITUDE **56.6769, -4.0368**
START GRID REFERENCE **NN753557**
START W3W **///munch.outfitter.vote**

	PRONUNCIATION	NAME MEANING
Schiehallion	**'shee-hal-iun'**	**fairy hill of the Caledonians**

Introduction

Amble across John Muir Trust land and boulder-hop to the summit of this thrilling Munro. A short walk of just over 10 kilometres, this out-and-back is a popular route as it eases the walker in gently with a well-tended path which leads to a broad boulder-strewn ridge. There is a big reveal with a flourish at the summit as the expansive view down Loch Rannoch to the Glen Coe and Ben Nevis ranges opens up. Schiehallion has an iconic cone shape which makes it easy to spot from many other Munros, so you'll reminisce about this mountain long after you've climbed it.

Public Transport

There are no public transport services close to this walk.

Parking

Chargeable parking at Braes of Foss car park off Schiehallion Road. The car park is maintained by Forestry and Land Scotland; toilets are available. If the car park is full, use the chargeable overspill car park 600m east (NN758559, ///results.landings.radiates).

Escape Route

None.

Refreshments

Back in Pitlochry, try the Old Mill Inn. You'll find delicious food and a friendly vibe.

Walker and dog approaching the rocky section of Schiehallion.

Schiehallion from the path near to the car park.

Sunset over Loch Rannoch looking west towards the Glen Coe mountains (left) and Ben Nevis (centre).

The Route

S Exit the car park to the south and when the trail splits take the right-hand fork which passes through a gate out of the woods. Continue south alongside a drystone wall. Cross the Allt an t-Socaich using the footbridge and continue on the gravel path. Your objective, Schiehallion, is visible to your right, in contrast to many other Munro walks where the objective remains shrouded behind smaller hills until much later in the walk.

2 Pass through a gate, walk over a second bridge and take the right-hand fork at the next trail junction; a handy fingerpost confirms the right way. Pass through a third and final gate beside a sheep fank, used by farmers for shearing, dipping or marking their flock.

3 From here, the trail begins to rise; stone-pitched steps have been installed on the well-maintained path. Be sure to look behind you to enjoy the view towards Beinn a' Ghlo (page 291). As the gradient increases, so does the stone pitching. The path zigzags up the mountainside with heather on either side. Soon you can appreciate the view down Loch Tummel to the north-east, while to the south-west the tremendous Glen Lyon Munros (page 137) appear almost in touching distance on the other side of Gleann Mòr.

4 The terrain underfoot noticeably changes from around 870m elevation as small quartzite stones crunch beneath your feet. The ground steepens and becomes looser while the path winds through rocks heading west. At first it seems the path is closer to the right-hand side of the shoulder than the left and the view down Loch Tummel is clear, but the path soon weaves up the centre of the shoulder.

5 While steadily gaining elevation, the ground keeps gently flattening out, so your ascent is gradual. At this stage the summit seems so close and yet so far. Dunalastair Reservoir appears in the valley to your right, with the Corbett Beinn a' Chuallaich beyond.

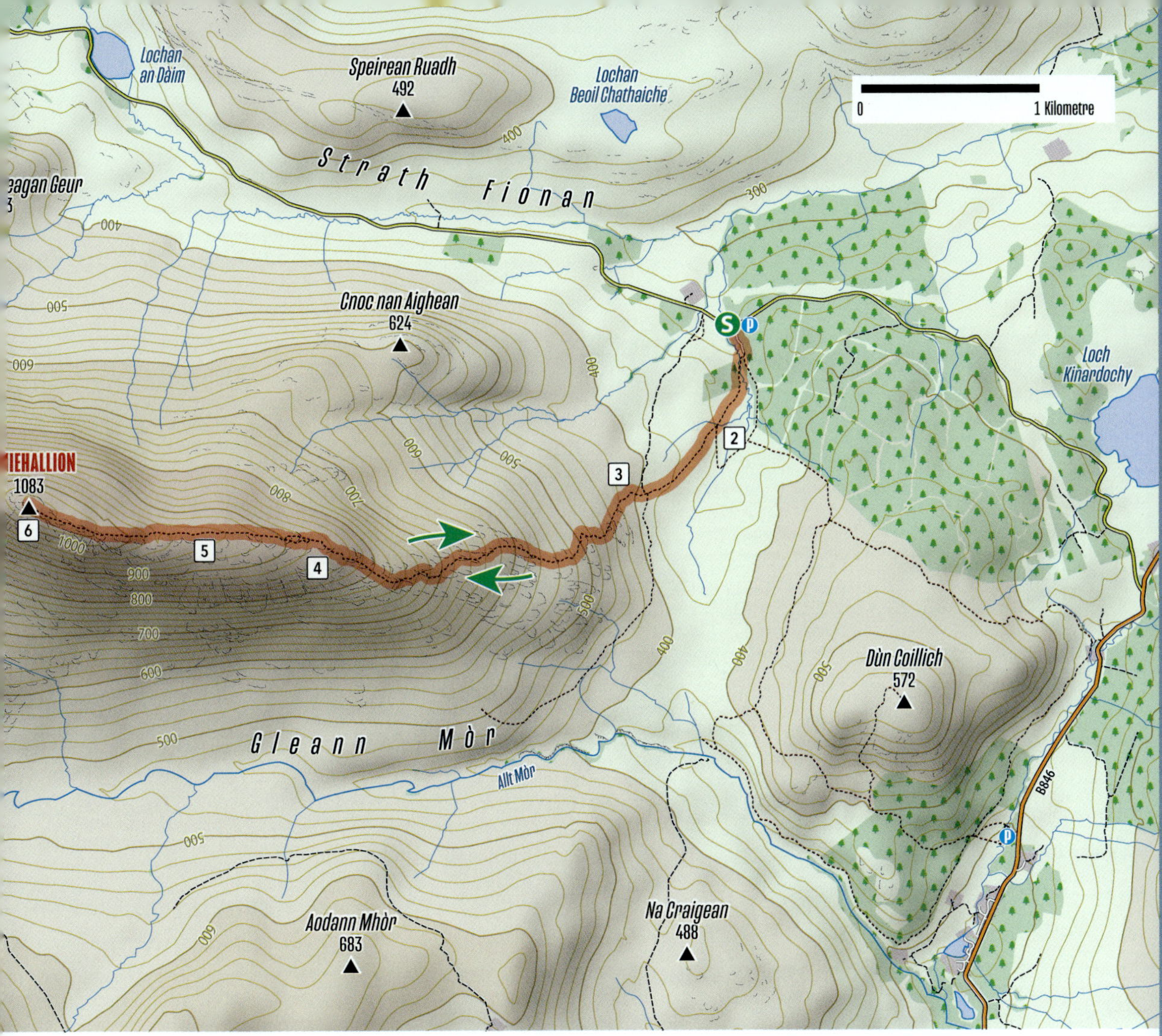

Towards the very top, the path all but disappears and blockier terrain dominates. More care is needed navigating across the boulder field as the boulders may shift beneath your feet. The gradient is mild, so while the terrain is more challenging, the steepness is behind you. The highest point of Schiehallion sits atop a rib of rock where the view west down Loch Rannoch and over to the distant Glen Coe and Ben Nevis mountain ranges is superb.

6 Retrace your steps back to the start.

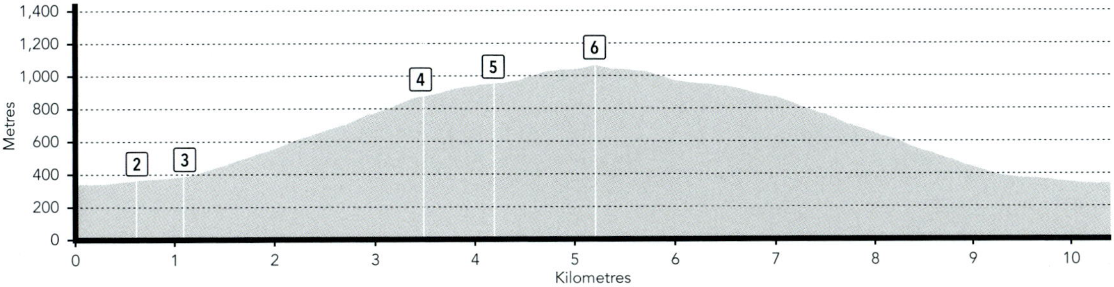

55 Beinn Dearg

HUB **Pitlochry – 15-minute drive**

OVERALL 1 2 **3** 4 5 TERRAIN 1 **2** 3 4 5
DISTANCE 1 2 3 4 **5** NAVIGATION 1 **2** 3 4 5

DISTANCE **30.2km** ASCENT **1,045m**
TIMES **10:15hrs** (Walker) **8:45hrs** (Trekker)
6:35hrs (Fastpacker) **4:35hrs** (Trail runner)

START LOCATION **Glen Tilt car park, Old Bridge of Tilt**
START LATITUDE/LONGITUDE **56.7744, -3.8433**
START GRID REFERENCE **NN874662**
START W3W **///landscape.headline.exhaled**

	PRONUNCIATION	NAME MEANING
Beinn Dearg	**'beyn jerreck'**	**red hill**

Introduction

The idea of a 30-kilometre hike for one summit will leave impatient Munro-baggers in a cold sweat – but there's no way to avoid it. Beinn Dearg is a far-flung, isolated Munro, set within a no-man's-land of windswept moorland in the southern Cairngorms. If you don't fancy the one-day hike described below, you could transform this task from a long solitary slog into a multi-sports micro-adventure. You can mountain bike the initial approach to Allt Scheicheachan bothy, before completing an out-and-back hike to the summit. Next spend a night in the charming shelter and, after coffee and a bacon sarnie the following morning, cruise back to civilisation on two wheels to complete your epic little adventure.

Public Transport

Start your walk from Blair Atholl railway station, on the Highland Main Line. Walk north for just over a kilometre past Blair Castle Caravan Park to reach the start point. It is possible to start the route by bus in the same way.

Parking

Free parking for around 30 cars in the Glen Tilt car park, Old Bridge of Tilt. Overnight parking of motorhomes and campervans is not permitted. The nearest public toilets can be found in Pitlochry.

MTB-ability

It's common to mountain bike the first and last 9.4 kilometres as far as Allt Scheicheachan bothy, leaving an 11.4-kilometre walk. Confident cyclists could continue for a further two kilometres as the track steepens.

Escape Route

None.

Allt Scheicheachan bothy from the approach to Beinn Dearg.

Refreshments

If you like 'proper chips', the Bothy Bar at the Atholl Arms Hotel in Blair Atholl is a great place for a meal after your walk.

The Route

S Leave the car park and walk west-north-west on the estate road in the direction of Old Blair for 800m until you reach a crossroads. Walk straight on here along a track from which you can enjoy the sounds of the woodland. Keep right then ignore all turn-offs for 1km until the track forks. Take the right-hand fork here to continue gently uphill.

2 Remain on the main track until NN854676 where you should fork right and exit the woodland through a gate in the deer fence. The track follows the Banvie Burn upstream for a while before curving north to follow the Allt na Moine Bàine beyond the confluence.

3 Arrive at a meticulously built cairn after which a track junction appears – take the left fork, staying on the main track, and continue north towards a bridge over the Allt na Moine Bàine. After crossing the bridge, continue uphill on the long track which leads to Allt Scheicheachan bothy, which is 5km away. As the track gently rises, Schiehallion (page 277) dominates the view to the south-west.

4 Just when you think the bothy might be a myth, you spot a glimpse of the chimney. Walk around a bend and there it is, surrounded by heather-clad hills. The bothy is a haven for many in wet weather but for others it's the destination itself. If you've cycled in you can leave your bike here. The track continues to the left of the bothy and crosses the Allt Sheicheachan. Keep the stream on your right as you follow the track uphill. Look out for birds of prey like the rare honey buzzard in this remote area.

Beinn Dearg from the approach track.

5 At the next path junction, remain on the left-hand side of the stream and take the narrow path which marks the start of a wide switchback up the heathery hillside. The path becomes wider as you gain height and is stony in places, but the gradient is gradual. A minor left-hand turn can be ignored at NN847754 as the wider more established path is followed to gain Meall Dubh nan Dearcag.

6 Turn left beside a cairn at NN851757 to approach Beinn Dearg by initially walking north-north-west for 1km over a brief boggy section. The path then turns north-north-east towards the summit of Beinn Dearg. The best view on the final part of the approach is to the right towards Càrn a' Chlamain (page 285) and Beinn a' Ghlo (page 291). The rock turns increasingly red, which is no surprise given that Beinn Dearg translates from Gaelic to 'red hill', and it is crunchy underfoot along the final part of the ascent. The very top of Beinn Dearg is bouldery and there is a trig pillar and a shelter.

7 See if you can spot Ben Macdui in the distance to the north-east before retracing your steps to the start.

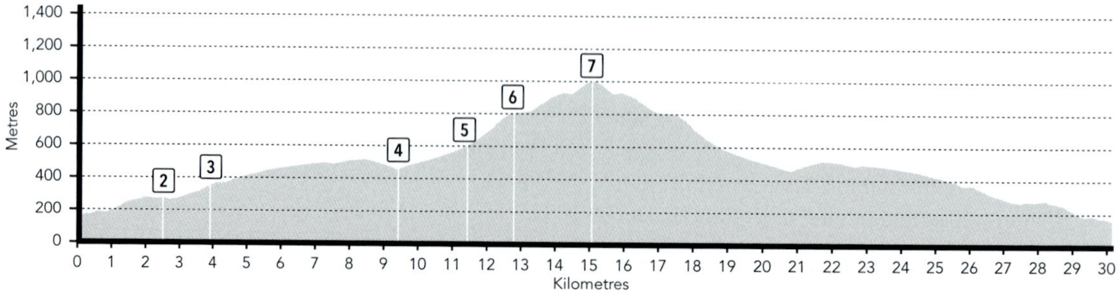

BEINN DEARG
1008
7
900
800
800
Meall Dubh
nan Dearcag
6
700
600
500
5
Allt Sheicheachan
800
700
600
4
Meall Tionail
570
Meall Dubh
569
500
500
Allt an t-Seapail
Allt Feith nam Fearna
500
400
400
Allt na Moine Baine
3
Glen Banvie Wood
Meall nan Loth
446
Glen Banvie
2
Fair Bhuidhe
462
300
Craig Urrard Wood
Baluain Wood
200
0
1 Kilometre
Old Bridge of Tilt
Tilt
Baluain
Blair Atholl
Garry
A9
B8079

56 Càrn a' Chlamain

HUB **Pitlochry – 15-minute drive**

OVERALL 1 2 **3** 4 5 TERRAIN 1 **2** 3 4 5
DISTANCE 1 2 3 4 **5** NAVIGATION 1 **2** 3 4 5

DISTANCE **25.9km** ASCENT **1,140m**
TIMES **9:45hrs** (Walker) **8:25hrs** (Trekker)
6:15hrs (Fastpacker) **4:20hrs** (Trail runner)

START LOCATION **Glen Tilt car park, Old Bridge of Tilt**
START LATITUDE/LONGITUDE **56.7744, -3.8433**
START GRID REFERENCE **NN874662**
START W3W **///landscape.headline.exhaled**

	PRONUNCIATION	NAME MEANING
Càrn a' Chlamain	**'carn a klamen'**	**hill of the kite**

Introduction
In a similar vein to neighbouring Beinn Dearg, which is located two valleys over to the west, Càrn a' Chlamain is a remote, solitary hill with a very long walk-in. But, far from being a time-wasting, energy-sapping inconvenience, the walk-in (or cycle-in, if you prefer) is arguably the best part of this route. The approach is along Glen Tilt, a hidden treasure often billed as 'one of the most beautiful glens in the Highlands'. The route passes through the Jubilee Rifle Range; you'll need to use the diversion in (**2**) on firing days: ***www.atholl-estates.co.uk***

Public Transport
Start your walk from Blair Atholl railway station, on the Highland Main Line. Walk north for just over a kilometre past Blair Castle Caravan Park to reach the start point. It is possible to start the route by bus in the same way.

Parking
Free parking for around 30 cars in Glen Tilt car park, Old Bridge of Tilt. Overnight parking of motorhomes and campervans is not permitted. The nearest public toilets can be found in Pitlochry. It used to be possible to park further up Glen Tilt, but this is now prohibited.

MTB-ability
It is possible to cycle on the private estate road through Glen Tilt to the bottom of Gleann Craoinidh and pick up the walk instructions from there (**5**). Cycle directions: leave the car park, cross the road and pass to the right of a cottage to enter woodland on to the private estate road (cyclists welcome). At NN881684 cross the bridge and continue to cycle northwards on the estate road. Cross to the other side once more at Gaw's Bridge before picking up the walk instructions from here (**5**). In total, this gives you a 16.3-kilometre cycle plus a 9.6-kilometre walk. GPX file available for download.

Càrn a' Chlamain from Beinn Dearg (see route 55).

Above Glen Tilt from the approach to Càrn a' Chlamain. **Opposite** View from the approach to Càrn a' Chlamain over the River Tilt towards Beinn a' Ghlo.

Escape Route

None.

Refreshments

If you like 'proper chips', the Bothy Bar at the Atholl Arms Hotel in Blair Atholl is a great place for a meal after your walk.

The Route

S Leave the car park to the east-north-east along a woodland footpath. Curve left once you reach the edge of the River Tilt. This part of the route is waymarked with yellow arrows and is known as the Glen Tilt Trail. The path leads to an old bridge which carries foot traffic over the road. Turn right after the bridge on to another pretty woodland trail with a gorge down on your right-hand side. When the path meets a wide track, turn right and follow it north as it traces the River Tilt from above.

2 Continue for 1.2km to reach a junction at NN877676. If Jubilee Rifle Range is active, remain on the lower track and head straight on until Gilberts Bridge (partway through **3**) where you can rejoin the route. If the rifle range is closed however, I recommend taking the left fork here to gain a bit of height through the woodland. After 1km, exit right on to a track and walk north passing through a gate beside a cattle grid. The view opens up to the right across Glen Tilt to Meall Dail Mìn. This is where you enter the rifle range. From this higher vantage point, you can see Càrn a' Chlamain further down the valley.

3 At NN877691 fork right and enjoy the views both up and down Glen Tilt in its lush green glory. At NN878694 fork right again. Cross a bridge over the Allt Slanaidh then carry on to reach two large gates next to Gilberts Bridge. Take the left-hand gate which leads uphill, remaining on the western side of the river. If you bypassed the rifle range, this is where you'll merge back on to the route. At NN881703 fork right and walk downhill on to a grassy trail, still following the yellow waymarkers. The grassy trail leads through a field where foxgloves pop up in brilliant colour in the summer. Follow the path until it merges with the track once more.

4 The track narrows as it weaves through bracken before the Ach-mhairc Bridge enables you to safely cross over the raging waters of the Allt Mhairc. At a path junction shortly after the bridge, choose the right-hand path and continue along it as it drops to run alongside the river. The path narrows further still and can resemble a tunnel of scrubland in the summer months before it crosses the Allt Sròn an Duine. Walk east-north-east until the path merges on to a track at Gaw's Bridge.

5 Remain north of the river on the track until you reach the bridge over the Allt Craoinidh – this is a great vantage point from which to appreciate Bràigh Coire Chruinn-bhalgain – Beinn a' Ghlo – which looms on the other side of Glen Tilt (page 291). After crossing the bridge, immediately turn left at NN908720 and follow the trail uphill through heather to start the main ascent. Mountain bikes can be left here.

The track up Càrn a' Chlamain.

6 At the track junction at NN908724, continue straight on along the wider gravel track which weaves up the Fàire Clach-ghlais. At NN916737 the track splits once more. Take the left-hand fork which leads up through heather and departs from the main track for a more authentic hillwalking approach to the Munro over soft peaty terrain. The trail leads north and then north-east for just over a kilometre.
7 The trail starts to curve north-north-west on the 800m contour, passing through a gap in a long-dismantled drystone wall. Continue towards the bealach between Càrn a' Chlamain and its minor south-east top, Grianan Mòr. From the broad bealach, follow the gravel track uphill until you are alongside the summit. When you cannot ascend any further on the track, peel off to the left and approach the summit heading south-south-west over a short section of loose, stony ground. The summit is marked with a cairn.
8 From the summit you can enjoy the view to Beinn Dearg (page 281). Back in 2019 I linked these Munros with a stay in Allt Scheicheachan bothy which made a fine weekend outing. Retrace your steps to Gaw's Bridge (**5**). From here you can walk back the way you came but my preference is to walk back on the other side of Glen Tilt for a bit of variety. Cross Gaw's Bridge and follow the track west-south-west. Walk through a gate beside a cattle grid and pass to the right of Marble Lodge.
9 Further down the glen, pass a small wooden hut – a river gauging station which monitors river flow. Just beyond this, take the left turn off the track to walk on a grassy path, following the yellow waymarkers once more. While you can stay on the lower track here if you wish, your feet might appreciate the break from the firm track walking as grass can be far more forgiving on a long walk like this. The grassy path merges back on to the main track.
10 After 400m on the main track, join another grassy path which appears on the left. The path passes above Auchgobhal and to the left of Croftmore before continuing south past Kincraigie to a cross-roads. Turn right here and follow the road down past the small collection of residences that sit above the Fender Burn. Cross the bridge and turn right at the next two road junctions to reach the Old Bridge of Tilt and the start point.

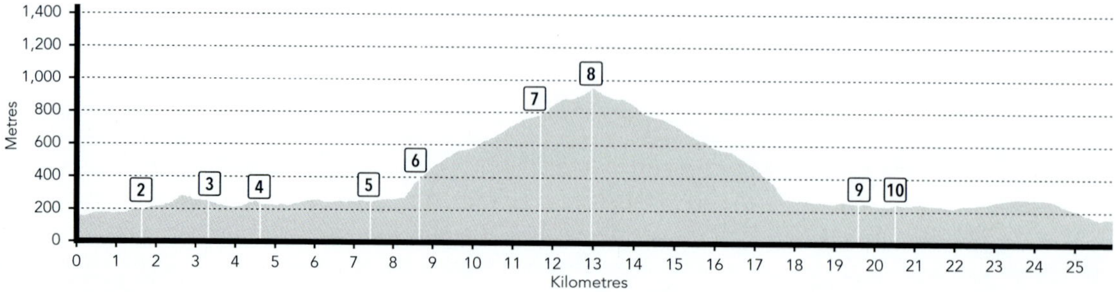

Sròn a' Chrò
855
Gleann Mhairc
Gleann Craoinidh
CÀRN A' CHLAMAIN
963
Grianan Mòr
Fàire Clach-ghlais
Tilt
Gaw's Bridge
Glen Tilt
Ach-mhairc Bridge
463
Druim a' Chait
Meall Duibhinidh
516
Auchgobhal
Gilberts Bridge
Jubilee Rifle Range
Meall Dail Mìn
536
Brae of Lude
Alternative Route
Kincraigie
Glen Fender
Loch Moraig
Creag nan Ceard
432
Meall Mòr
512
Old Bridge of Tilt
Blair Atholl
B8079
0
1 Kilometre
1
2
3
4
5
6
7
8
9
10

CARN LIATH 976m
BRÀIGH COIRE CHRUINN-BHALGAIN 1,070m
CARN NAN GABHAR 1,121m

57 Beinn a' Ghlo

HUB **Pitlochry – 20-minute drive**

OVERALL 1 2 **3** 4 5 TERRAIN 1 2 **3** 4 5
DISTANCE 1 2 3 **4** 5 NAVIGATION 1 2 **3** 4 5

DISTANCE **22.3km** ASCENT **1,290m**
TIMES **9:40hrs** (Walker) **8:20hrs** (Trekker)
6:15hrs (Fastpacker) **4:10hrs** (Trail runner)

START LOCATION **Beinn a' Ghlo car park, Loch Moraig**
START LATITUDE/LONGITUDE **56.7824, -3.7929**
START GRID REFERENCE **NN905670**
START W3W **///revamping.secret.exulted**

	PRONUNCIATION	NAME MEANING
Beinn a' Ghlo	**'beyn a glorr'**	**hill of the mist**
Carn Liath	**'carn leer'**	**grey hill**
Bràigh Coire Chruinn-bhalgain	**'bry corriya crane vallerken'**	**upland of the corrie of round blisters**
Carn nan Gabhar	**'carn nan goo-er'**	**hill of the goats**

Introduction

Every now and then, the sheer majesty of Scotland's mountains will take your breath away. It's difficult to predict when that euphoric moment will wash over you – there's probably a secret mix of factors that need to align. But, for me, during a research walk of Beinn a' Ghlo's trio of Munro summits, which rise proudly to the east of Glen Tilt near Blair Atholl, I had such an experience. Under blue skies, I was utterly mesmerised by the snaking views to Bràigh Coire Chruinn-bhalgain and Carn nan Gabhar from Carn Liath. Touch wood, you'll have a similarly uplifting experience.

Public Transport

Start your walk from Blair Atholl railway station, on the Highland Main Line. Walk north past Blair Castle Caravan Park, through Old Bridge of Tilt and north-east along minor roads to reach the start point. This would add a total of 10 kilometres to the overall walk length. It is possible to start the route by bus in the same way.

Parking

Chargeable parking for around 50 cars in Beinn a' Ghlo car park, near Loch Moraig, five kilometres north-east of Blair Atholl. The nearest toilet facilities are in Pitlochry.

MTB-ability

The first and last two kilometres can be cycled.

Approaching Bràigh Coire Chruinn-bhalgain.

Above Bràigh Coire Chruinn-bhalgain and Carn nan Gabhar from Carn Liath. **Opposite** The approach to Carn Liath.

Escape Route

It's possible to descend from the bealach between Beinn Mhaol and Bràigh Coire Chruinn-bhalgain. Pass to the west of Beinn Bheag and follow the Allt Coire na Saobhaidh to reach the descent path at NN948692.

Refreshments

The Bothy Bar at the Atholl Arms Hotel in Blair Atholl is a great place for a meal after your walk.

The Route

S Leave the car park and pass through a gate to walk north-east along the road. After 200m, fork right and pass through a gate on to a wide track. Loch Moraig, with its tree-filled islands, looks like a haven for nature if you glance over your right shoulder as you ascend gently on the track. On a clear day, the path up the south-west face of Carn Liath is visible right from the start.

2 Stay on the track until NN923678 and take the left fork on to a grassy track. Moments later pass through a gate beside a wooden estate building and follow the track north-west initially before taking a path, which is stone-pitched in places and curves north-east, avoiding the boggy ground in the depression at the foot of Carn Liath. The stone-pitched path up Carn Liath is well laid and navigation up the heather-clad slopes is of no concern as a result. To your right is Ben Vrackie, a popular Corbett at 841m.

3 Pass the trig pillar and walk on to the large cairn which marks the summit of Carn Liath. Leave the summit following a stony eroded line along a broad sweeping snake-like ridge which winds north-west, north-east, then north-north-west. This has to be one of my favourite Highland views as Bràigh Coire Chruinn-bhalgain and Carn nan Gabhar are majestic when viewed from Carn Liath and the ridge is gentle enough in gradient to really appreciate it. Bypass the summit of Beinn Mhaol by sticking to the path on the left which leads to the steeper and looser north-north-west descent on stony terrain to the bealach beneath Bràigh Coire Chruinn-bhalgain.

4 From the bealach, a clear trail of white stones leads east-north-east up Bràigh Coire Chruinn-bhalgain, swinging northwards from halfway up. Be sure to glance back at Carn Liath which looks more attractive from this side than the other thanks to its delightful curves. The terrain becomes bouldery for a short time before an obvious path on the right-hand side of the ridge guides you to the summit. Carn nan Gabhar can be seen to the right and you also get a sneak peek of the route of descent through the valley. There is a beautiful view north-east towards the scree-strewn slopes of Beinn Iutharn Mhòr and Carn an Righ (page 305) to the left of the shoulder of Carn nan Gabhar. To the north-west is the solitary Munro of Càrn a' Chlamain (page 285) on the other side of Glen Tilt.

View from Carn nan Gabhar trig pillar to the highest point of the Munro in the distance.

5 Leave the summit to the north-east initially and then descend east to point 1,016m on OS maps. From here, ascend the small knoll, from which you can see across to Carn nan Gabhar and the ant-like hillwalkers on the path ahead. Pick up the eroded trail which comprises multiple small switchbacks down to Bealach an Fhiodha.

6 Begin your final ascent of the day up Carn nan Gabhar from Bealach an Fhiodha. Continue heading east then, 300m from the bealach, note the cairn at the path fork (NN959725) as this is your route of descent into the valley. At the bealach between Carn nan Gabhar and Airgiod Bheinn walk north-east on an eroded trail of white rock. As the terrain becomes rockier, pass to the right of a pile of boulders and continue towards a second pile of boulders on the broad summit ridge. The terrain is increasingly uneven so take care on your approach. On the second small knoll you'll find the trig pillar of Carn nan Gabhar, but the highest point is further still, on the third small knoll. When you reach this feature, you're as high as you can get on Beinn a' Ghlo.

7 From the summit of Beinn a' Ghlo, retrace your steps to NN959725 and take the left fork to start your descent into the valley. Keep left of the Allt Bealach an Fhiodha on a clear path. It becomes boggy in places from around 700m in elevation.

8 At NN952711 fork right to descend over soft, wet, grassy ground and pick up another path that swiftly leads across the stream. Continue south on the right-hand side of the stream on a soft, peaty trail until the start of a manicured gravel path which continues southwards. The escape route described on page 292 merges in from the right at NN948692. Carry on for 700m to reach a path junction.

9 Turn right at the junction on to a wider track and walk west-south-west on this, ignoring all other turn-offs until you reach the start point.

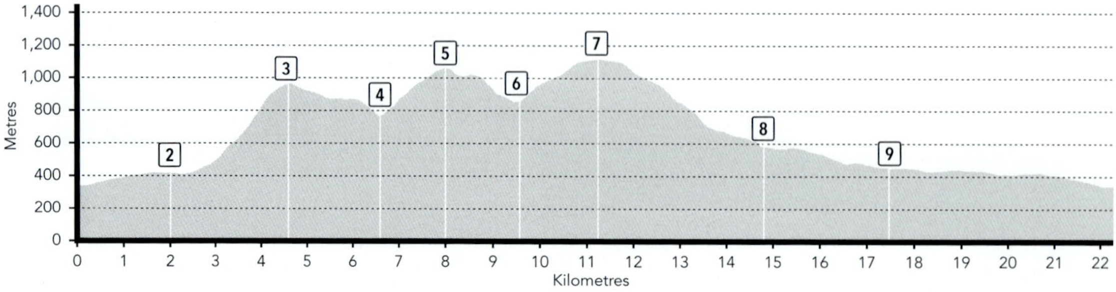

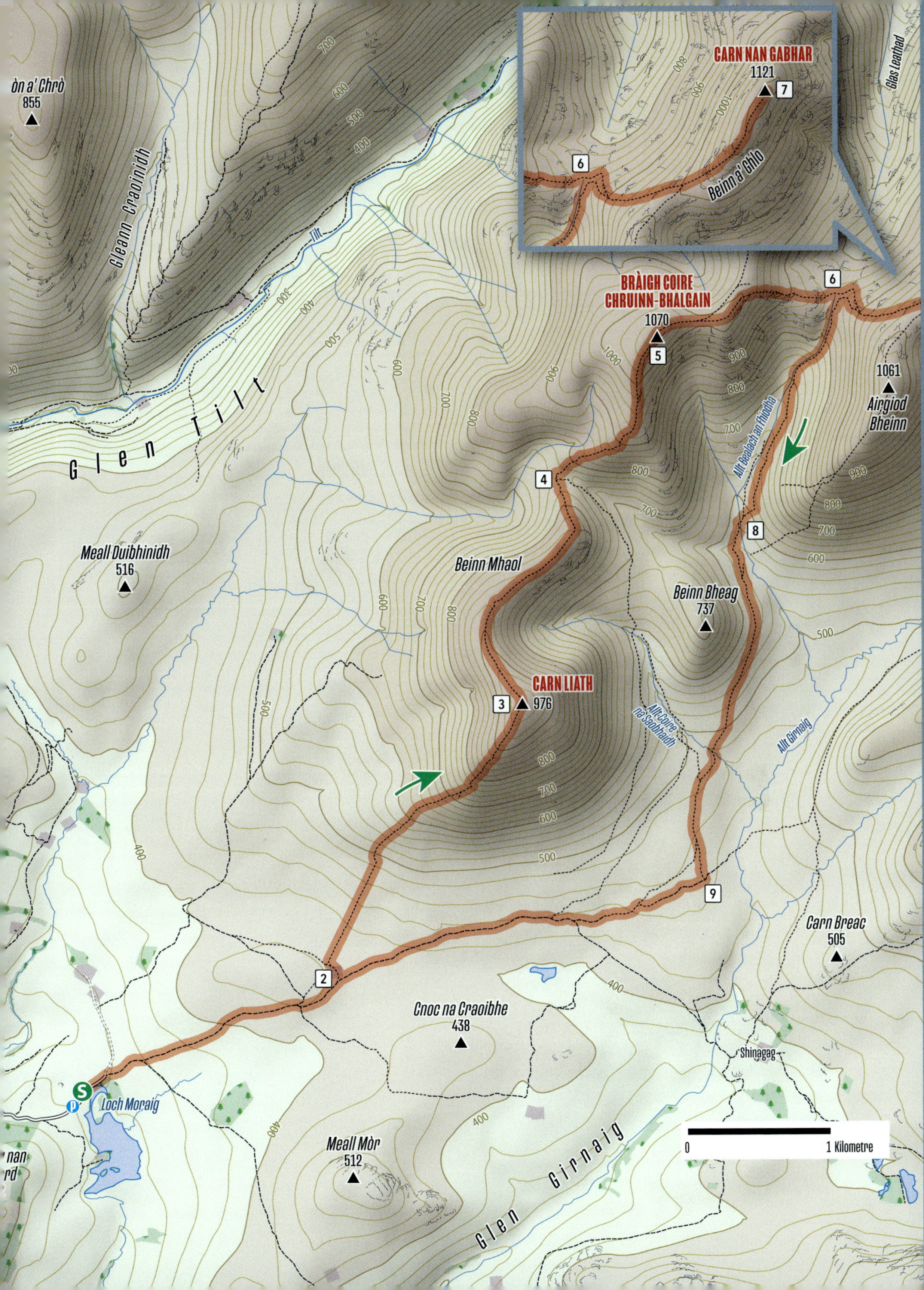

CARN NAN GABHAR
1121
Beinn a' Ghlo
Glas Leathad
BRÀIGH COIRE
CHRUINN-BHALGAIN
1070
1061
Airgiod Bheinn
Allt Bealach an Fhiodha
Glen Tilt
Tilt
Gleann Craoinidh
òn a' Chrò
855
Meall Duibhinidh
516
Beinn Mhaol
Beinn Bheag
737
CARN LIATH
976
Allt Coire na Seabhaich
Allt Girnaig
Carn Breac
505
Cnoc na Craoibhe
438
Shinagag
Loch Moraig
Meall Mòr
512
Glen Girnaig
0
1 Kilometre

HIGHLAND HUB 10

BRAEMAR

The tenth and final Highland hub is Braemar, an Aberdeenshire village at the end of the Dee valley in the magnificent Cairngorms National Park. It's the perfect size for a Highland base with a village feel but plenty of eateries plus a small supermarket.

Getting to Braemar using public transport is most easily achieved via Aberdeen, where a long, two-hour journey on the bus is the final piece of the jigsaw puzzle.

Mountain bikes can be hired from Braemar Mountain Sports which can come in really useful for Munro walks like Carn Bhac and the Lochnagar Horseshoe. The outdoor shop is also a blessing if you realise you've forgotten an item of hillwalking clothing or equipment for your trip.

I've stayed multiple times in my tent at Braemar Caravan Park. It's a lovely quiet site with excellent facilities, including a large, hot drying room – we all know how useful those can be. It's also the only place that I've used a woodpecker as a reliable alarm clock substitute.

The busiest time of the year in Braemar is the first week in September. The annual Braemar Gathering and Highland Games has attracted visitors from far and wide since the 19th century. The event is a celebration of the history, traditions, sports, language, culture and heritage of Scotland.

Mount Keen and the Glen Doll Munros of Mayar and Driesh are two outlying Munro walks which have been tagged on to the Braemar hub. To reduce the driving time, you may wish to stay in one of the towns or villages north of Dundee for these walks, such as Kirriemuir or Brechin, perhaps on your way to or from Braemar.

WHERE TO STAY

£	££	£££
Braemar Caravan Park	Braemar Youth Hostel	Braemar Lodge (log cabins)

TRAVEL TIMES

58 Carn Bhac	**10 minutes**
59 Spittal of Glenshee Munros	**25 minutes**
60 West Glen Shee Munros	**15 minutes**
61 East Glen Shee Munros	**15 minutes**
62 Glen Doll Munros	**1 hour and 35 minutes**
63 Lochnagar Horseshoe	**45 minutes**
64 Mount Keen	**1 hour and 50 minutes**

Rest day ideas

- Visit Braemar Castle which sits just to the north-east of the village. The grounds are open all year round and during the summer months you can go inside for a small fee.
- The Braemar Local History Group runs guided walks through the summer which provide insight into the heritage of Braemar.
- For a real treat, book afternoon tea at the Fife Arms, an opulent hotel built in the wake of the tourism boom resulting from Prince Albert's purchase of Balmoral for Queen Victoria.

Opposite Ben Macdui from the summit of Carn Bhac with shelter cairn in the foreground (route 58).
Overleaf View from Carn an Righ into the heart of the Cairngorms National Park with Ben Macdui on the right (route 59).

Ballater
5 LOCHNAGAR - CAC CARN BEAG
6 CARN A' CHOIRE BHOIDHEACH
7 CÀRN AN T-SAGAIRT MÒR
8 CAIRN BANNOCH
9 BROAD CAIRN
Braemar
58
CARN BHAC
5
6
63
7
8
9
MOUNT KEEN
64
CARN AN TUIRC
60
1
BEINN IUTHARN MHÒR
4
TOLMOUNT
CARN AN RIGH
TOM BUIDHE
CAIRN OF CLAISE
2
3
GLAS TULAICHEAN
GLAS MAOL
62
CREAG LEACACH
61
MAYAR
DRIESH
59
1 CARN AOSDA
2 THE CAIRNWELL
3 CARN A' GHEÒIDH
4 AN SOCACH

INOV

58 Carn Bhac

HUB **Braemar – 10-minute drive**

OVERALL 1 2 (3) 4 5 TERRAIN 1 2 (3) 4 5
DISTANCE 1 2 3 (4) 5 NAVIGATION 1 2 (3) 4 5

DISTANCE **23km** ASCENT **695m**
TIMES **7:25hrs** (Walker) **6:20hrs** (Trekker)
4:45hrs (Fastpacker) **3:20hrs** (Trail runner)

START LOCATION **Inverey car park, Inverey**
START LATITUDE/LONGITUDE **56.9852, -3.5002**
START GRID REFERENCE **NO089892**
START W3W **///mandolin.caked.tiny**

	PRONUNCIATION	NAME MEANING
Carn Bhac	**'carn vac'**	**stony hill of peat banks**

Introduction

Like most Munro names, 'stony hill of peat banks' is a very literal description of Carn Bhac. A summit plateau littered with stones, the remainder of its upland is a peat-lover's dream. In the spring you may hear the distinctive call of the oystercatcher as you walk back through Glen Ey. Alasdair Alpin MacGregor wrote in his 1937 book of folk tales, *The Peat-Fire Flame*: 'In Gaelic the cry of the oystercatcher is "Bi glic, bi glic; bi glic, bi glic!", meaning "be wise," "be prudent," "take care,"' as they warn of an approaching storm.

Walker on Carn nan Seileach – Carn Bhac is to the left in the distance.

Public Transport

There are no public transport services close to this walk.

Parking

Free parking for around 20 cars in Inverey car park. Campervans are not permitted to park overnight. The nearest public toilets are in Braemar.

MTB-ability

Those wishing to bike 'n' hike should cycle to Altanour Lodge and walk the Munro as an out-and-back from there following the descent instructions in reverse. This creates a 17.6km cycle and a 9.6km walk.

Escape Route

None.

Refreshments

Gordon's Tearoom prides itself on being one of the oldest establishments in Braemar. Try one of Mrs Sherry's fresh scones.

The Route

S Exit the car park and walk south-south-west along a well-maintained track. A fingerpost reads *To Glen Ey*. Pass multiple residences on your right and continue along the track as it enters the valley alongside the Ey Burn. The track crosses over the burn and begins to climb uphill. The hills on either side of the track are clad in heather and at certain times of year they will present like a patchwork quilt due to moorland management activities where heather is routinely burnt or cut back.

2 Just over 3km into the walk, take the turn-off to the right on to a track which leads uphill through the middle of a collection of ruined buildings at Auchelie. The ruins are visible from the main track. (Instead of taking this turning, cyclists can continue on the path next to the Ey Burn for another 5.6km to the ruins of Altanour Lodge to make an out-and-back ascent using the descent instructions, returning the same way.) Walkers ascend steeply on the track through the Auchelie ruins over grassy terrain. A glance to the right here will reveal Munro Beinn a' Bhuird in the distance. The track turns to gravel as it twists and turns up the hillside before it commits to a south-westerly approach up the shoulder of Carn nan Seileach.

3 From the summit of Carn nan Seileach you can see the summits of Ben Macdui and the other Cairngorm 4000ers to the north-west. After being obscured on the approach, Carn Bhac now looms up ahead. Look out for mountain hare here – one raced across the track in front of me during one visit. Continue until the track ends abruptly at NO067841.

4 Walk south-south-west and then south-west over soft and squelchy ground towards the saddle between Carn Bhac and Carn Creagach. There are faint all-terrain vehicle impressions to follow at first, but these may be difficult to spot in poor visibility. Make your way around peat hag obstacles, aiming for the bealach at the head of the Allt Carn Bhathaich. This largely pathless, 1.25km section is the most challenging navigational element of the walk.

5 Upon reaching the 777m saddle, which is marked on OS maps, make your ascent of Carn Bhac to the north-west. Once the initial peat hags have been negotiated a faint path appears. Carn Bhac's larger Munro neighbour, Beinn Iutharn Mhòr, can be seen to the left. Many Munro-baggers link these two mountains, but I prefer walking the latter with Glas Tulaichean and Carn an Righ (page 305). The path navigates around the now boulder-strewn slope and once atop the plateau the path veers left. A small cairn marks the true summit of Carn Bhac, 30m south of a larger shelter cairn. The most impressive aspect has to be towards the Cairngorms, with Ben Macdui commanding the skyline, but I also love the view to the complex ridgelines on Beinn a' Ghlo (page 291).

6 Leave the summit to the south, in search of the shoulder of Carn Bhac which curves from south to east above the Allt an Odhar. A small cairn indicates the start of the descent path. Navigate a short section of stony ground before reaching the peaty trail that can be followed all the way back to the main track in the valley bottom. The path is clear until it reaches a series of peat hags and becomes less distinct in places. If you lose the trail, stay in the centre of the shoulder and you'll soon pick it back up. At around 725m elevation you'll begin to pass a series of wooden grouse butts.

7 The trail loops east at around 700m in elevation and then drops to cross a stream. Follow the track uphill slightly before it continues to gently descend. There are a number of tracks through the valley, but this is my favourite as it maintains the higher ground for longer and commands better views. Pass what remains of an old settlement as the trail continues to gently descend towards the ruins of Altanour Lodge, where the track becomes more substantial. Both walkers and cyclists can now follow this track for 8.8km back to Inverey.

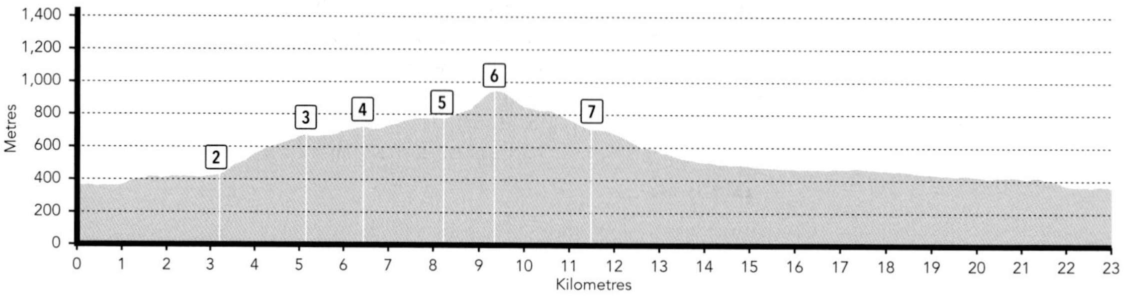

Linn of Dee
Muir
Dee
Inverey
Tom Anthon
635
Carn na Moine
755
Ey Burn
Allt Connie
Càrn Mòr
706
Carn Bhithir
573
The Colonel's Bed
Glen Ey
Auchelie
2
797
Top of the Battery
784
Carn nan Seileach
675
3
Allt Carn Bhiathaich
4
Creag an Lochain
883
CARN BHAC
946
6
Carn Creagach
894
Coire na Caillich
Creag an Fhuathais
853
5
Altanour Lodge
Allt an Odhar
Allt Beinn Iutharn
7
0
1 Kilometre

GLAS TULAICHEAN 1,051m
CARN AN RIGH 1,029m
BEINN IUTHARN MHÒR 1,045m

59 Spittal of Glenshee Munros

HUB **Braemar – 25-minute drive**

OVERALL 1 2 3 **(4)** 5 TERRAIN 1 2 **(3)** 4 5
DISTANCE 1 2 3 4 **(5)** NAVIGATION 1 2 **(3)** 4 5

DISTANCE **27km** ASCENT **1,350m**
TIMES **10:50hrs** (Walker) **9:15hrs** (Trekker)
7:00hrs (Fastpacker) **4:45hrs** (Trail runner)

START LOCATION **Dalmunzie Castle Hotel, Spittal of Glenshee**
START LATITUDE/LONGITUDE **56.8241, -3.4910**
START GRID REFERENCE **NO090712**
START W3W **///shirts.promotion.denim**

	PRONUNCIATION	NAME MEANING
Glas Tulaichean	**'glass toolican'**	**grey-green hillocks**
Carn an Righ	**'carn an ree'**	**hill of the king**
Beinn Iutharn Mhòr	**'beyn yoorhen voor'**	**big sharp-ridged hill**

Introduction

Three for the price of two – that's what you get with this walk which sits to the south of the immense central Cairngorms massif. Many guidebooks suggest walking two, not three, of these peaks together, with Beinn Iutharn Mhòr often left as an outlier to climb from the north via Glen Ey. But this route from the delightful Spittal of Glenshee, in the south, ticks off the full trio by completing a near-360-degree loop of Màm nan Càrn. It's a long, tough day in the hills, but for efficient peak-bagging it hits the mark.

Public Transport

There are no public transport services close to this walk.

Parking

There are a discretionary number of chargeable parking spaces for non-residents at the Dalmunzie Castle Hotel in the Spittal of Glenshee on a first come, first served basis. You must obtain permission and pay for parking at reception. The hotel has toilet facilities that you can use. Alternatively, there is free parking two kilometres south-east down the track from the hotel (NO105701, ///piled.pixel.pictured) – parking here adds four kilometres to the route and can be easily cycled if preferred.

Escape Route

From Carn an Righ, retrace your steps to NO050774 on the southern side of Màm nan Càrn. Walk north-east on the path to Loch nan Eun where you can pick up the descent path. Return for Beinn Iutharn Mhòr another day.

View from bridge over Ey Burn to Beinn Iutharn Mhòr (see route 58).

Above Beinn a' Ghlo from Carn an Righ with the summit cairn in the foreground. **Opposite** View to Loch nan Eun from the descent of Glas Tulaichean.

Refreshments

You can buy a drink or lounge snack at the Dalmunzie Castle Hotel after your walk. Parts of the former Laird's residence date back to the early 1500s.

The Route

S Leave the hotel using the tarmac road to the north-west. After 200m there are gates on either side of the road. Pass through the gate on the left, noting the gated grassy track on your right which will be your descent route back on to the road later. Walk west-south-west across a field and through another gate. The grassy path soon joins a rough track. Walk on the track for 700m until you reach a turn-off at NO080717.

2 Turn right off the main track here to gain the grassy higher track via a wide zigzag. This path is more scenic and avoids fording the Glen Lochsie Burn. While on the terrace track, pass through a gate and walk over a bridge across the pretty Allt a' Choire Bhric, all while enjoying the view down Glen Lochsie to your left. Cross the Allt Clais Bheag and pass through a gap in an old wall. A path leads to the right of the ruin of Glenlochsie Lodge to regain the track.

3 Turn right and walk uphill on the track, ignoring all turn-offs, into the wide heather-lined passage to gain Breac-reidh. The rounded heather-clad hills all around you are a contour-lover's dream. The most characterful set of hills at this lower altitude are from Meall a' Choire Bhuidhe to Meall Ruigh Mòr Thearlaich which are complemented by a deep gully between them with a waterfall at its base. The track passes a series of shooting butts as it steadily rises to 900m elevation.

4 The scree slopes of Carn an Righ come into view to the north-west as the track reaches a cairn at NO048759. Peel off to the right here to attain the summit of Glas Tulaichean on a grassy path. A trig pillar marks the summit. Take in the view to Carn a' Gheòidh (page 311) – this Munro is commonly hiked from the Glenshee Ski Centre.

5 Leave the summit to the north-north-east down the grassy ridge with Loch nan Eun visible in the distance. At the point marked 1,008m on OS maps the ridge splits. Follow the eroded path down the north ridge above Glas Choire Bheag. The path gradient rises slightly and then begins to descend again.

6 At NO054770, on the 900m contour, leave the path to walk north-west down pathless bilberry-filled slopes to the boggy bealach before rising up through heather to reach a peaty path on the other side at NO050774. Turn left on the path and walk west-south-west until the path merges with another from the right. Continue in the same direction along a much stonier path which you can see in the distance as it cuts a clear line through the heather up Carn an Righ. When the path splits at NO040771, fork left and drop to a boggy bealach. You'll return to this bealach after visiting Carn an Righ's summit.

7 Ascend on the stony path from the bealach. The path melds intermittently with an eroded grassy line as the route veers north-west towards the summit. With 100m vertical ascent to go, ascend over a short section of boulder scree with views to Beinn Iutharn Mhòr to your right. The trail becomes indistinct underfoot towards the top but once the summit plateau is reached, a cairn and stone shelter can be located in the centre of the plateau, from which there is a stunning view south-west over to Beinn a' Ghlo (page 291).

8 Retrace your steps to the bealach (**7**) and fork left to ascend north-east for 150m up the side of Màm nan Càrn, after which the path veers north to traverse the western slopes of this Munro Top. When the path becomes indistinct at NO046782, walk north-east to gain the bealach and pick up the ascent path for Beinn Iutharn Mhòr. This begins north-west initially before it curves and leads northwards over stonier ground on to the summit plateau. The summit is marked by a cairn and offers a great vantage point to the nearby Munro of Carn Bhac (page 301).

9 Retrace your steps back to the bealach at NO047784 then walk east on a grassy trail which passes along the northern side of Màm nan Càrn along the 910m contour. When the path fades, at NO056784, walk east-south-east towards a large boulder on the ridge-line. From here, walk south-east on to the shoulder of Màm nan Càrn and then south-south-east on pathless heathery ground, with a wonderful view over Loch nan Eun, to gain the path in the boggy bealach.

10 Walk south-east on the path which leads around to the edge of the loch and to the stream outlet on its southern border. Cross the stream and descend into the valley on the eastern side of the Allt Easgaidh.

Carn an Righ (left) and Carn nan Gabhar (right) from the ascent of Beinn Iutharn Mhòr.

The terrain from here is a mixture of stony and grassy with a handful of very wet patches. On the basis that you can't see the end of Gleann Taitneach there is nothing to incentivise this long descent over rough, uneven ground so make sure you save lots of snacks to keep you going. Remain on the left-hand side of the stream.

11 After 2.3km the path merges with a track which can be followed through a flatter section of the route into the glen. This is easier to walk on but not as nice on the eye, as is always the case with bulldozed tracks. Shortly after meeting the track, it fords a stream, the first of many until you reach a bridge over the Allt Ghlinn Thaitneich. Leave the track at NO089724 and cross the bridge. Follow the path on the western side of the stream which leads back to the access road, passing through two gates along the way. Retrace your steps to the hotel from there.

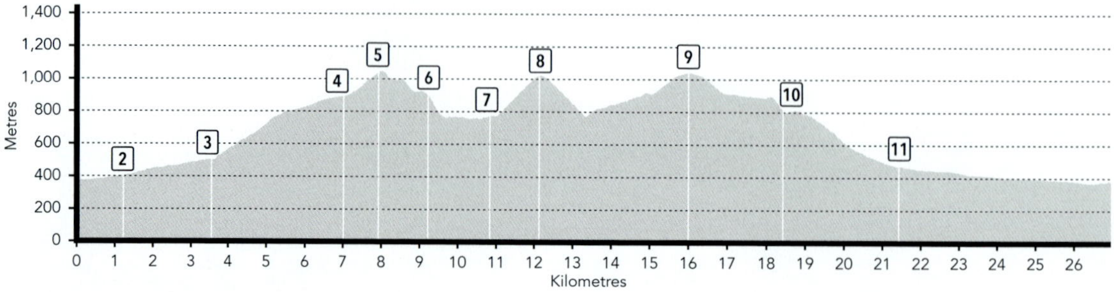

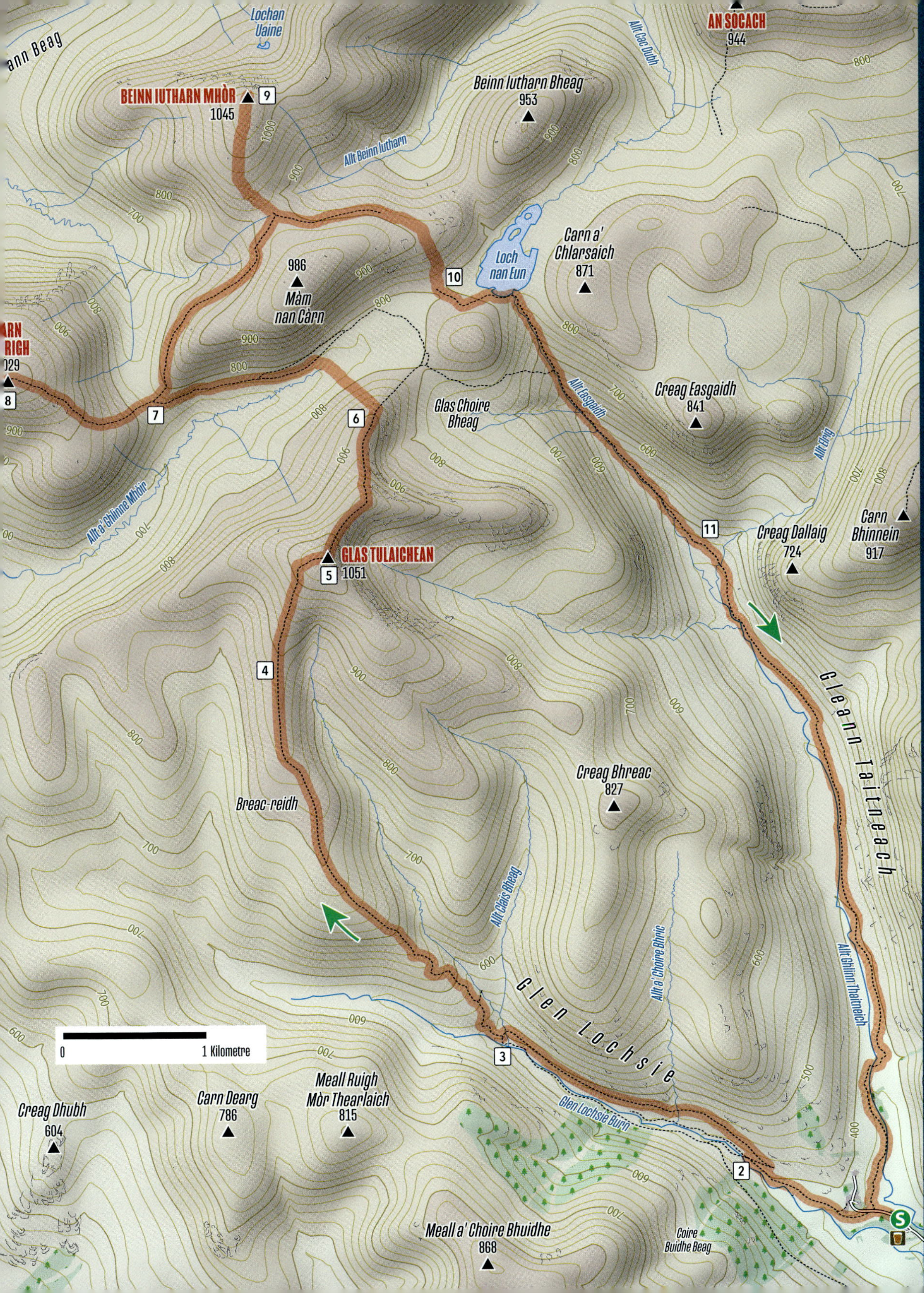

Lochan Uaine
AN SOCACH
944
BEINN IUTHARN MHÒR
1045
9
Beinn Iutharn Bheag
953
Allt Beinn Iutharn
Allt Coc Dubh
Carn a' Chlarsaich
871
Loch nan Eun
10
986
Màm nan Càrn
8
7
6
Glas Choire Bheag
Allt Easgaidh
Creag Easgaidh
841
Allt Elrig
Allt a' Ghlinne Mhòir
11
Creag Dallaig
724
Carn Bhinnein
917
GLAS TULAICHEAN
1051
5
4
Glean Taitneach
Creag Bhreac
827
Breac-reidh
Allt Clais Bheag
Allt a' Choire Bhric
Allt Ghlinn Thaitneich
Glen Lochsie
3
0
1 Kilometre
Creag Dhubh
604
Carn Dearg
786
Meall Ruigh Mòr Thearlaich
815
Glen Lochsie Burn
2
S
Meall a' Choire Bhuidhe
868
Coire Buidhe Beag

CARN AOSDA 915m
AN SOCACH 944m
CARN A' GHEÒIDH 975m
THE CAIRNWELL 933m

60 West Glen Shee Munros

HUB **Braemar – 15-minute drive**

OVERALL 1 2 **(3)** 4 5 TERRAIN 1 2 **(3)** 4 5
DISTANCE 1 2 3 **(4)** 5 NAVIGATION 1 2 **(3)** 4 5

DISTANCE **21.2km** ASCENT **1,270m**
TIMES **9:20hrs** (Walker) **8:00hrs** (Trekker)
6:00hrs (Fastpacker) **4:00hrs** (Trail runner)

START LOCATION **Glenshee Ski Centre car park, A93, Cairnwell Pass**
START LATITUDE/LONGITUDE **56.8861, -3.4142**
START GRID REFERENCE **NO139780**
START W3W **///inflation.schools.ended**

	PRONUNCIATION	NAME MEANING
Carn Aosda	**'carn oosder'**	**hill of age**
An Socach	**'an sochach'**	**the snout**
Carn a' Gheòidh	**'carn a yoye'**	**hill of the goose**
The Cairnwell	**'the cairnwell'**	**hill of bags**

Introduction
Located to the west of the Cairnwell Pass, next to Glenshee Ski Centre, a trio of easy-to-bag mountains are known collectively as the Cairnwell Munros. The two nearest to the road – Carn Aosda and The Cairnwell – are, arguably, spoilt by the skiing infrastructure. Carn a' Gheòidh feels more remote, but for a more authentic experience, I like to add in far-flung An Socach. This transforms an underwhelming route into something wilder, not least because majestic mountain hares are common here.

Public Transport
There are no public transport services close to this walk.

Parking
Chargeable parking for hundreds of cars in the Glenshee Ski Centre car park on the A93, Cairnwell Pass (Old Military Road) between the Spittal of Glenshee and Braemar. The ski centre has toilet facilities. The Cairnwell Pass can be subject to closure following heavy snow in winter.

Escape Route
Cut An Socach from the route and continue on the track to the east of Loch Vrotachan. Fork right at NO127785 and walk south, approaching Carn a' Gheòidh as an out-and-back. This reduces the walk length by eight kilometres.

Refreshments
Grab a drink and a bite at Tea@TheShee, the main cafe at the Glenshee Ski Centre. Situated along the North East 250 route, the cafe attracts both hikers and bikers in the summer.

Walkers making their way to The Cairnwell from Carn a' Gheòidh.

The Route

S Leave the car park, cross the road and walk up to the ski centre. To the right of the ski centre is a steep track which leads uphill beside a chairlift – follow this to the north-west. Pass behind the Cairnwell Cafe, an octagonal building, and continue north-west on the track between two sets of ski tows up Carn Aosda. When the track forks nearly 500m beyond the cafe, fork right and walk northwards up the gravel track. This weaves up the hillside to reach the fence line just to the west of the summit of Carn Aosda. Fork right here to ascend the final 20m in height to the top of Carn Aosda.

2 From the summit your eye is naturally drawn south to The Cairnwell, but I love the view away from the ski equipment down Glen Clunie. Retrace your steps briefly before continuing straight ahead, south-west towards Loch Vrotachan. Remain on the track until NO127786 – peel off the track here to walk west over pathless ground above the loch.

3 Pass Loch Vrotachan on its northern side over rough moorland. Once you reach the end of the loch continue west, initially keeping right of the Allt Loch Vrotachan. The ground here becomes grassier. Make your way over a short series of peat hags, keeping left of a small knoll. Cross the Allt Loch Vrotachan around NO110788, after which you should rise up on to the higher ground above the stream then follow it north-west towards the footbridge in Glen Baddoch.

4 Cross the bridge over the Baddoch Burn and follow a grassy channel uphill to reach a major track. Turn left on to this track and walk south-west along it for nearly 1km, fording a couple of minor streams along the way. At NO100789, fork right off the track to follow an all-terrain-vehicle track north-west up the peaty hillside, crossing the Allt Boruiche after 1km.

5 As the route then approaches 800m elevation the trail becomes stonier and more defined. Ascend northwards on the left-hand side of the prominent gully before the path crosses to the right-hand side of the gully further up to gain the ridge. Once on the ridge, follow the path to the west over stony ground to the summit of An Socach. There is a shelter cairn on the plateau and beautiful views down Glen Ey, panning north-west to neighbouring Munros Carn Bhac (page 301) and west across to Beinn Iutharn Mhòr (page 305).

6 Retrace your steps to the track at the foot of An Socach at NO100789. Turn left on to the track initially before forking off it to the right after 80m to follow an all-terrain-vehicle track into the valley bottom.

7 Towards the bottom, fork right off the track to make a beeline for the footbridge over the Baddoch Burn at NO101788 (not shown on the OS map). Cross the bridge and regain the all-terrain-vehicle track on the other side, which leads south-west up the hillside.

8 The track curves to the left at around 670m elevation and passes a series of shooting butts. To your right, Glas Tulaichean (page 305) is the prominent Munro visible in the distance as you ascend to the south-south-east. The path becomes indistinct in the higher reaches. When it does so, just head south-east aiming directly for the summit of Carn a' Gheòidh over cropped moorland. From the summit, my favourite view is over to Glas Tulaichean with Beinn a' Ghlo (page 291) behind.

9 Leave Carn a' Gheòidh to walk north-east, first over stony ground which becomes intermittently grassy. Stay on the main path which skirts under Carn nan Sac and heads towards a pair of small lochans. Pass to the right of the lochans and walk along Creag a' Choire Dhirich.

10 Drop to the bealach at point 808m on OS maps and turn right to walk south-east for 250m. Here the path forks; take the higher of the two options, forking left. Fork left again soon after to reach the ridge via a couple of rocky steps which don't amount to scrambling. The path then merges with a wide track which first leads south-east to a chairlift station and then south to the summit of what has to be the most cluttered summit of any Munro as it hosts multiple buildings, pylons and masts.

11 You're not likely to want to spend long on the summit here, so take in the view across to the Glen Shee Munros including Glas Maol (page 315) before retracing your steps back to the chairlift station.

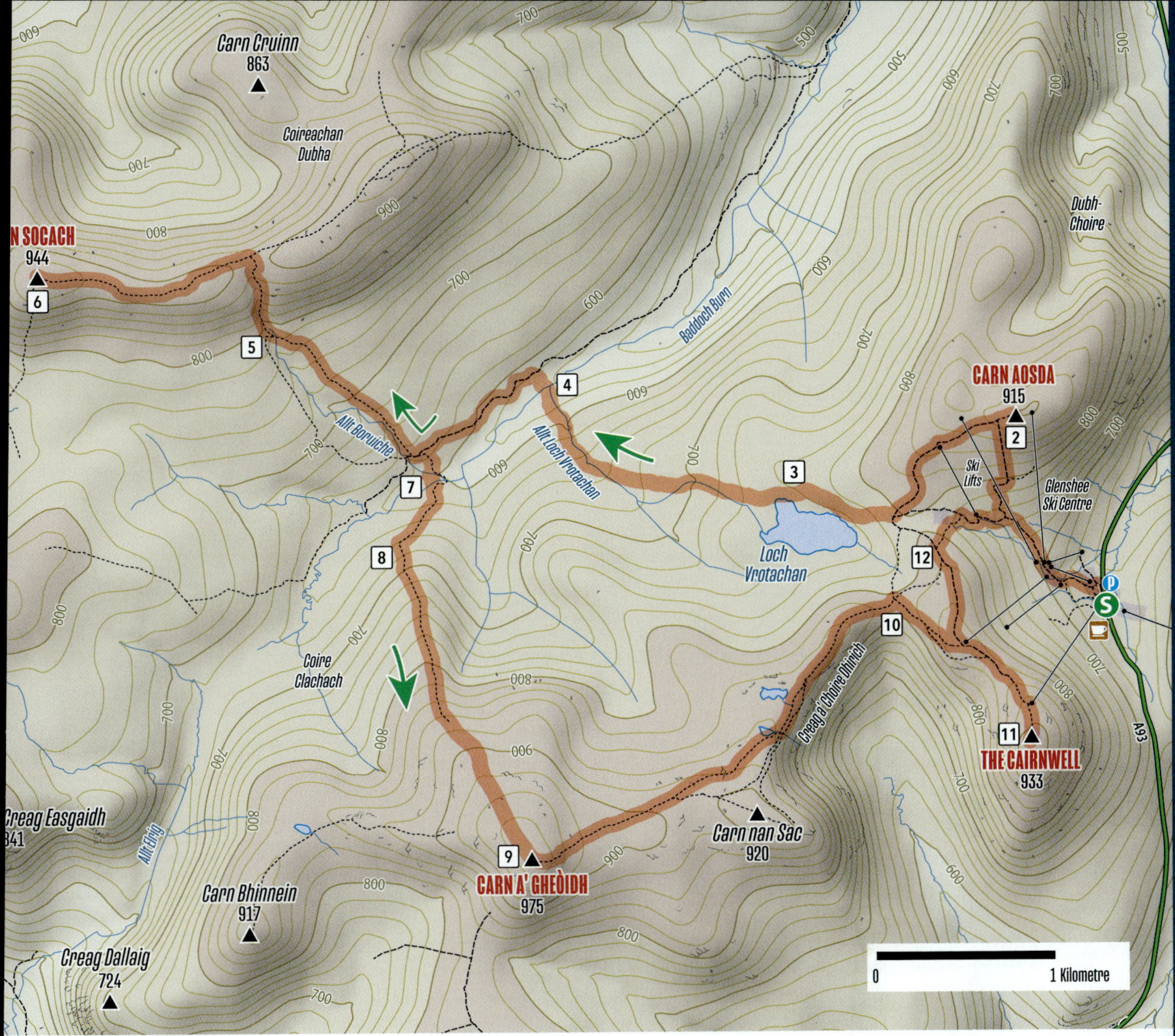

While it is possible to follow the line of the chairlift down to the start point, I don't recommend it. Instead, continue along the track as it passes over the 873m knoll marked on OS maps before descending northwards.

12 At NO129784, take the path off the track down to the right through the heather. This path meets the track lower down. Turn right here and continue your descent as the track merges with your initial route of ascent. From here, retrace your steps back to the start.

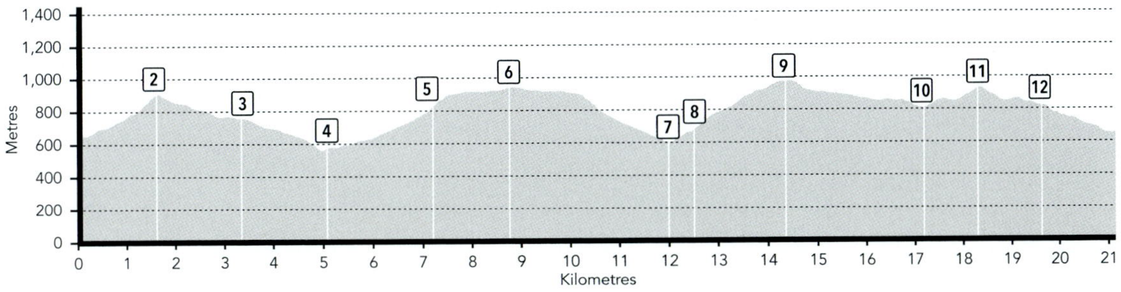

CREAG LEACACH 987m
GLAS MAOL 1,068m
CAIRN OF CLAISE 1,064m
TOM BUIDHE 957m
TOLMOUNT 958m
CARN AN TUIRC 1,019m

61 East Glen Shee Munros

HUB **Braemar – 15-minute drive**

OVERALL 1 2 **3** 4 5 TERRAIN 1 **2** 3 4 5
DISTANCE 1 2 3 4 **5** NAVIGATION 1 2 **3** 4 5

DISTANCE **25.6km** ASCENT **1,370m**
TIMES **10:35hrs** (Walker) **9:10hrs** (Trekker)
6:50hrs (Fastpacker) **4:35hrs** (Trail runner)

START LOCATION **Devil's Elbow Viewpoint car park, A93, Cairnwell Pass**
START LATITUDE/LONGITUDE **56.8646, -3.4130**
START GRID REFERENCE **NO139756**
START W3W **///solar.insurers.ironclad**

	PRONUNCIATION	NAME MEANING
Creag Leacach	**'craig leeyachach'**	**slabby crag**
Glas Maol	**'glass moel'**	**grey green hill**
Cairn of Claise	**'carn of clash'**	**hill of the hollow**
Tom Buidhe	**'tom booyer'**	**yellow hill**
Tolmount	**'tolmount'**	**hill of the valley**
Carn an Tuirc	**'carn an toork'**	**hill of the boar**

Introduction

Six Munros, one day – the peak-bagging ROI (return on investment) of this walk is phenomenal. How is this possible? First, you get a huge initial leg-up – the starting car park is on the 500-metre contour line – which kicks things off with a boost. Second, the Munros are grouped close together – the ground between them is joyously undulating. Third, while the instructions close the loop with a walk back along the verge of the Cairnwell Pass, if you're in a group with two cars there is an opportunity to walk a linear route with one car parked at each end to make the day shorter and even more gratifying. All of this makes for a superb day out in the hills to the north of Spittal of Glenshee.

Public Transport

There are no public transport services close to this walk.

Parking

Free parking for around eight cars in the Devil's Elbow Viewpoint car park on the A93, Cairnwell Pass (Old Military Road) between the Spittal of Glenshee and Braemar. The nearest toilet facilities are a short drive north at Glenshee Ski Centre. The Cairnwell Pass can be subject to closure following heavy snow in winter.

MTB-ability

Consider dropping a bike at the base of Carn an Tuirc (NO147799, ///euphoric.spending.tradition) – you can then cycle the final 4.8 kilometres of the route on the road.

Walker linking Tolmount and Carn an Tuirc with Càrn an t-Sagairt Mòr in the distance.

Escape Route

Cut Tom Buidhe and Tolmount from the route – from Cairn of Claise follow the wall north-east and then a track north-north-west before picking up the path to Carn an Tuirc. This reduces the walk length by 6.4 kilometres.

Refreshments

Grab a drink and a bite at Tea@TheShee, the main cafe at the Glenshee Ski Centre. Situated along the North East 250 route the cafe attracts both hikers and bikers in the summer.

The Route

S The route starts at a car park on the Cairnwell Pass, which used to be a droving route for getting cattle and sheep to market in the 18th century before becoming a military road. The Devil's Elbow refers to the double hairpin in the old road which was straightened out in the 1960s. The pass is now a smooth A-road connecting Braemar and the Spittal of Glenshee, although parts of the old road can still be seen. Leave the car park on the footpath which heads south and leads down to the valley bottom. Cross the stream and fork right to follow the main track as it rises up through heather. At a second fork, keep left to remain on the main track. At a third split, at NO141753, fork right off the track and on to a path which leads down to a double stream crossing at NO144751.

2 From here, walk south-south-west on a grassy slope following a clear line up the hillside. The path curves south, bypassing to the left of the bealach between Meall Gorm and Creag Leacach, before leading south-east over increasingly stony terrain to point 943m on OS maps. From here, the path levels out and curves to the left to ascend increasingly loose rocky ground to reach the highest point on Creag Leacach. A drystone wall on your right acts as a great handrail.

3 Enjoy the view across to the West Glen Shee Munros (page 311) to the north-west before continuing along the trail beside the wall as it undulates north-east. After 1km, cross the wall and handrail it on the right-hand side. After 400m, the path leaves the wall branching off to the right and then at NO161761

Above View to Sròn na Gaoithe where two streams meet at the bottom of Carn an Tuirc. **Opposite** Ascending Glas Maol with Creag Leacach behind.

it splits again. Take the right-hand fork to walk north-east over grassy terrain to the summit of Glas Maol where you'll find a somewhat battered trig pillar and a large stone shelter on a large flat summit plateau.

4 Leave the summit to the north, passing ski infra-structure on your left as you drop gently to join a track which curves east to cross a wide saddle. When the track forks take the left fork towards Cairn of Claise and remain on the main track, ignoring multiple turn-offs, as you ascend away from the saddle to the north-east. At the base of Cairn of Claise, at NO182784, fork left on to a path which swings slightly closer to the lip of the corrie of Garbh-choire. A cairn on a rocky knoll marks the summit of Cairn of Claise, which is the key decision point for shortening the walk and leaving Tom Buidhe and Tolmount for another day (see escape route).

5 Continue north-east from the summit alongside the wall for 50m, before branching off right to walk east-north-east for around 200m. Reach the path and join it, walking north-east before curving right to follow a line of intermittent old fence posts on a path to Tom Buidhe. The path forks again at NO191791 – choose the right-hand path here which continues east-south-east following the fence posts into a bowl of peat hags. Ascend grassy slopes to reach the summit of Tom Buidhe whose rounded top is rather anticlimactic.

Descending Creag Leacach towards Glas Maol.

6 Descend via the route of ascent for 150m until the path forks. Take the right-hand fork which leads north-west to a peaty depression at the foot of Tolmount. Follow the path uphill and north-north-east to the summit from which the larger Munros of the Lochnagar Horseshoe (page 325) can be admired to the north-east.

7 Retrace your steps for 100m then fork right, descending gradually to the south-west on a faint path which is wet in places. Once at the bealach, which is marked 886m on OS maps, head north-west and begin gently ascending. This section of the route involves some pathless route-finding over tufty and boggy terrain as you make your way around to the northern slopes of Cairn of Claise, being careful not to reascend the Munro too much or you'll lose the advantage gained by cutting the corner.

8 Pass 1km to the right (north) of the top of Cairn of Claise across flatter boggy ground to meet a track. Walk across the track, continuing north-west towards Carn an Tuirc on a grassy trail. Keep forking right when the trail splits to approach the summit from east to west. The final part of the ascent is very stony.

9 Leave the top to the west, passing a stone shelter, taking care over rocky terrain. This soon gives way to peaty, soft ground as you lose some height and reach the heather-clad lower slopes. The path reaches, and keeps to the right of, the Allt a' Gharbh-choire as it bends south-west to arrive at the bridge over Clunie Water, from which there is a lovely view up to Sròn na Gaoithe over the confluence of two streams.

10 Ascend to the car park, from where you can pick up your bike or second car, or walk south along the road for 4.6km back to the start.

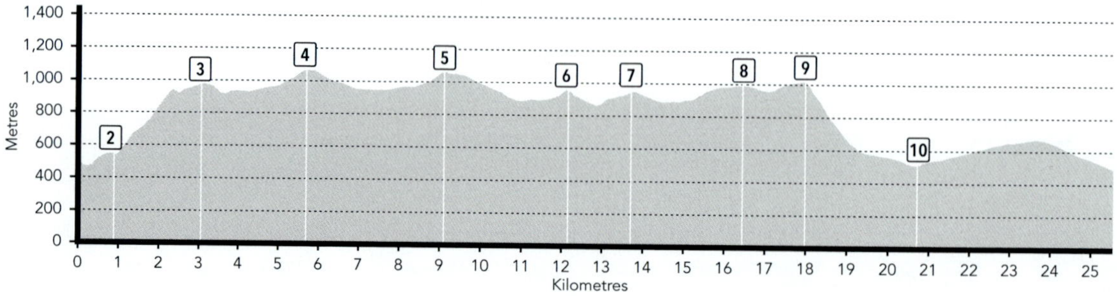

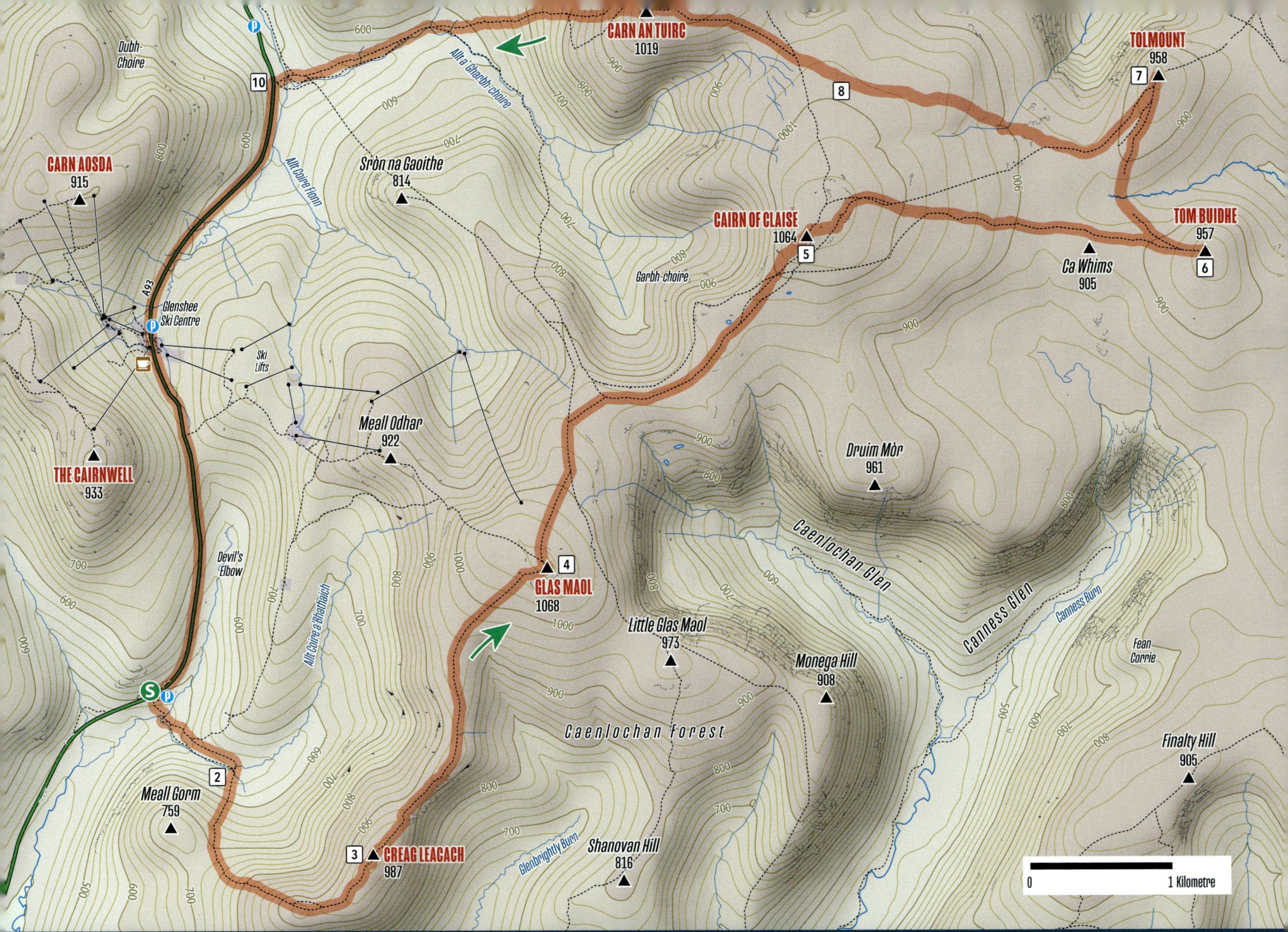
CARN AN TUIRC
1019
TOLMOUNT
958
Dubh-Choire
CARN AOSDA
915
Allt a' Gharbh-choire
Sròn na Gaoithe
814
Allt Coire Fionn
CAIRN OF CLAISE
1064
TOM BUIDHE
957
Ca Whims
905
Garbh-choire
A93
Glenshee Ski Centre
Ski Lifts
Meall Odhar
922
THE CAIRNWELL
933
Druim Mòr
961
Caenlochan Glen
Canness Glen
Canness Burn
Devil's Elbow
GLAS MAOL
1068
Little Glas Maol
973
Monega Hill
908
Fean Corrie
Allt Coire a' Bhathaich
Caenlochan Forest
Finalty Hill
905
Meall Gorm
759
CREAG LEACACH
987
Glenbrightly Burn
Shanovan Hill
816
0
1 Kilometre

62 Glen Doll Munros

HUB **Braemar – 1-hour-and-35-minute drive**

OVERALL 1 **2** 3 4 5 TERRAIN 1 **2** 3 4 5
DISTANCE 1 **2** 3 4 5 NAVIGATION 1 **2** 3 4 5

DISTANCE **14.5km** ASCENT **870m**
TIMES **6:25hrs** (Walker) **5:30hrs** (Trekker)
4:10hrs (Fastpacker) **2:45hrs** (Trail runner)

START LOCATION **Glen Doll car park, Glen Doll**
START LATITUDE/LONGITUDE **56.8711, -3.1773**
START GRID REFERENCE **NO283761**
START W3W **///doghouse.visitor.aviators**

	PRONUNCIATION	NAME MEANING
Mayar	**'may-yer'**	**high plain**
Driesh	**'dreesh'**	**a thorn bush**

Introduction

Experience the unforgettable moment of that first visit to Corrie Fee on the approach to Mayar. Imagine a giant dipped an ice cream scoop into alpine terrain and the crater left behind is the bowl-like, cliff-lined amphitheatre that is Corrie Fee. Mayar and Driesh sit to the south of The Mounth in the south-east of the Cairngorms National Park. Hiking one without the other would be like eating macaroni without cheese or fish without chips, so be sure to savour this delicious duo.

On the ascent of Mayar above Corrie Fee.

Public Transport

There are no public transport services close to this walk.

Parking

Glen Doll car park is maintained by Forestry and Land Scotland – drive through Glen Clova on the B955 to reach the car park which is just beyond Braedownie. The car park is chargeable and has over 70 spaces; it is situated next to a ranger base where there are toilets.

Escape Route

Descend Mayar as per the route description but pick up the descent early via the Kilbo path along the Shank of Drumfollow. Return another day for Driesh.

Refreshments

If you're looking for some tasty food or a pint of Clova Ale after your day in the hills, stop at the Glen Clova Hotel on your way out of the valley.

The Route

S Take the wide track to the left-hand side of the Glen Doll ranger base and head north-west. After 150m take the left fork, keeping Blair House on your left. The track, known as Jock's Road or the Tolmounth Pass, leads to multiple waymarked trails such as the Dounalt Trail and the Corrie Fee Trail. At the next major crossroads continue straight ahead, taking note that the path to the left is where you'll descend from later. The wide track continues along the right-hand side of White Water as it starts to gradually gain height. Ignore two further right-hand turns and continue straight ahead which leads to a bridge over White Water.

2 Cross the bridge; after the bridge the track veers to the right. Continue along the wide gravel track which inclines steadily through a gap in the spruce trees, ignoring another right-hand turn. The path exits into Corrie Fee National Nature Reserve. The corrie itself is a sight for sore eyes, a magnificent amphitheatre carved out by a glacier thousands of years ago at the end of the most recent ice age. A path leads from a natural viewing platform down to the left and into the bottom of the corrie. The path runs along the left-hand side of the Fee Burn and starts to ascend towards an impressive waterfall. The terrain is neither technical nor exposed.

3 Once you're just below the waterfall, follow the stone-pitched path around to the left. The path veers away from the waterfall here and you might get the sensation that you're suddenly going the wrong way, but the path finds the safest passage through the crags. Turn to climb steeply over more stone-pitched steps and rough stones before crossing a small stream and veering sharply right. The path becomes slightly indistinct here over some slabby larger rocks but only for a short time. The stone pitching then picks up again, leading uphill over grassier terrain.

4 A large cairn marks a great vantage point down into the corrie before the path reaches another impressive waterfall at 700m in elevation. From here, the path swings south, aiming directly for Mayar's summit. With 100m of ascent remaining, follow an eroded line south until the ground becomes stony and you reach the summit which is marked by a cairn. Be sure to look behind you at the looming Lochnagar Munros (page 325). The view south is as impressive as the hills diminish in size towards the major cities of Perth and Dundee, offering an expansive view.

Descending Little Driesh.

5 Take the path heading north-east away from the summit cairn which soon undulates east towards Driesh. The solitary Munro of Mount Keen (page 331) can be seen in the distance to the north-east. The ground is soft and peaty in places with grasses on either side, making navigation straightforward. The odd iron fence post provides reassurance, and these are soon replaced by wooden fence posts.

6 The path splits above an area marked Black Skellies on the OS map at NO255738. Choose the right-hand fork which leads directly to the saddle at the foot of Little Driesh, marked as 809m on OS maps. From the col, ascend via a path which climbs steeply but keeps well clear of the crags, over which the view down Corrie Kilbo is magnificent.

7 The path does not summit Little Driesh, but instead traverses the northern aspect of the top marked 915m on OS maps, aiming directly for the small saddle between Little Driesh and Driesh. From the saddle, ascend to the summit of Driesh via its broad western slopes. The summit is marked by a trig pillar inside a stone shelter.

8 Retrace your steps to the main col at point 809m and take the right-hand fork which rises briefly before it merges on to the Kilbo path which is scored into the hillside and descends in a north-easterly direction on the side of the Shank of Drumfollow. The descent is steady as the path leads into the bottom of Corrie Kilbo and to the right of the woodland, before entering it at a tall deer fence. The forest trail continues downhill and can be slippery with mud when wet.

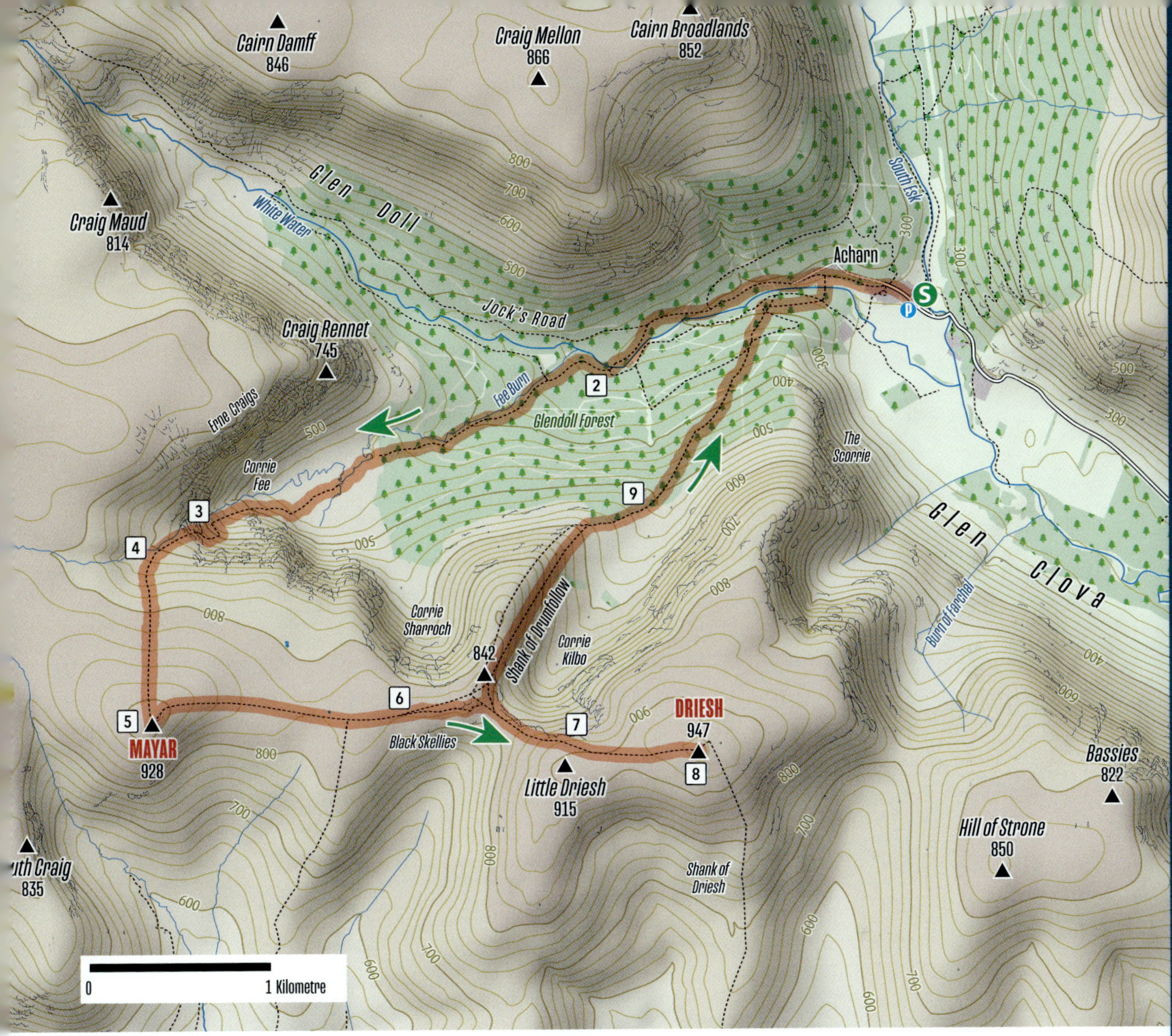

9 Cross the large stepping stones over the Burn of Kilbo as the trail opens out into a deforested area following a clear path. At a major track crossroads, continue straight ahead and, when the path meets the main track for a second time, turn left and walk around the hairpin bend. Continue east on the track until you reach a footbridge over White Water. Cross the bridge and turn right. Retrace your steps to the start from here.

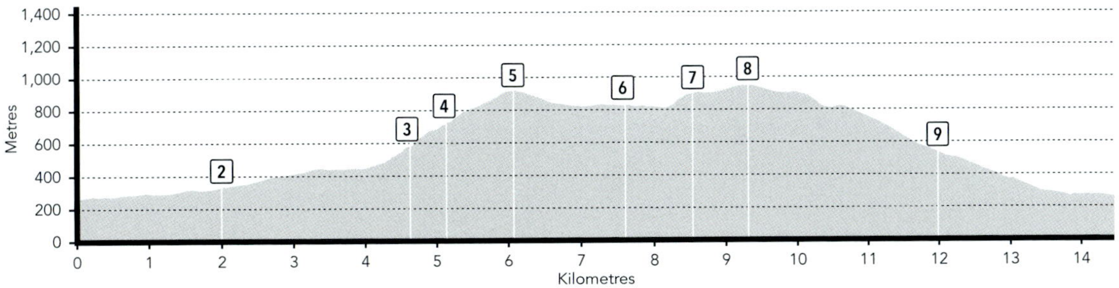

LOCHNAGAR - CAC CARN BEAG 1,156m
CARN A' CHOIRE BHOIDHEACH 1,110m
CÀRN AN T-SAGAIRT MÒR 1,047m
CAIRN BANNOCH 1,012m
BROAD CAIRN 998m

63 Lochnagar Horseshoe

HUB **Braemar – 45-minute drive**

OVERALL 1 2 **3** 4 5 TERRAIN 1 2 **3** 4 5
DISTANCE 1 2 3 4 **5** NAVIGATION 1 **2** 3 4 5

DISTANCE **32.1km** ASCENT **1,250m**
TIMES **11:25hrs** (Walker) **9:45hrs** (Trekker)
7:20hrs (Fastpacker) **5:05hrs** (Trail runner)

START LOCATION **Spittal of Glenmuick car park**
START LATITUDE/LONGITUDE **56.9524, -3.1362**
START GRID REFERENCE **NO309851**
START W3W **///captures.hoping.masks**

	PRONUNCIATION	NAME MEANING
Lochnagar	**'loch na gaar'**	**little loch of the noisy sound**
Cac Carn Beag	**'cack carn bake'**	**pass of the small cairn**
Carn a' Choire Bhoidheach	**'carn a corriya voyick'**	**hill of the beautiful corrie**
Càrn an t-Sagairt Mòr	**'carn an takarsht more'**	**hill of the priest**
Cairn Bannoch	**'cairn bannoch'**	**peaked hill**
Broad Cairn	**'broad cairn'**	**broad cairn**

Introduction

The Lochnagar Horseshoe is a rip-roaring high-level walk, bagging the five Munros above Loch Muick on the Balmoral Estate. It's a classic Highlands hill day, with tons of charm and character to keep even the grumpiest of hillwalkers happy. For many, the pièce de résistance is the dark, brooding, unfathomably huge coire of Lochnagar, but perhaps better still is Glas-allt-Shiel bothy. This quirky shelter is attached to a large house, built for Queen Victoria as a country escape from which to mourn her late husband, Albert. The location is spectacular, and the bothy is a perfect base for tackling the Lochnagar Horseshoe.

Public Transport

There are no public transport services close to this walk.

Parking

Chargeable parking in the large Spittal of Glenmuick car park which also has overflow spaces. If your intention is to stay overnight at the bothy, the pay-and-display machines allow you to pay for multiple days. There are toilets beside the visitor centre.

MTB-ability

The first and last five kilometres are cyclable.

Escape Route

From Carn a' Choire Bhoidheach or Càrn an t-Sagairt Mòr, drop to NO214842 between Càrn an t-Sagairt Mòr and Càrn an t-Sagairt Beag. Descend alongside the Allt an Dubh-loch to Dubh Loch. Pass the loch on its northern side and return to Glas-allt-Shiel from there.

View from the summit of Lochnagar - Cac Carn Beag to Loch nan Eun, Carn a' Choire Bhoidheach and Càrn an t-Sagairt Mòr.

Above Approaching the bridge over the Glas Allt on the ascent of Lochnagar - Cac Carn Beag. **Opposite** Ascending Lochnagar - Cac Carn Beag with Loch Muick behind.

Refreshments

If you've got a sweet tooth, stop at Shorty's ice cream parlour in Ballater on your way back to Braemar. Home-made ice cream is always a good idea.

The Route

S Leave the car park and walk south-west along the track past the toilet facilities and visitor centre. Continue on the wide track towards Loch Muick, ignoring a turn-off to the left. Fork right when the path splits at the head of the loch.

2 Cross the bridge over the River Muick and walk west and then north-west along the loch shore. Cross in front of a boathouse and ascend steps to gain the main track. Walk south-west along the track for almost 3km until you enter the woodland and reach Glas-allt-Shiel, the royal lodge built in Queen Victoria's reign. An outhouse to the rear was donated to Dundee University and has been used as a bothy for decades. Many will choose to stay here in order to split this walk into smaller chunks, either in the bothy or in a tent nearby. If you've cycled the route so far, leave your bike at Glas-allt-Shiel.

3 Walk in front of the lodge heading uphill and keeping the Glas Allt on your right. Ascend atop the pine-needle-carpeted trail before veering left away from the stream to walk west-north-west to the boundary wall. Exit the woodland on to the open hillside through a gap in the wall and follow the path back to the edge of the Glas Allt.

4 The trail leads uphill past the stunning Falls of the Glasallt and continues to be a well-maintained stone-pitched path as you gain elevation. Cross a bridge over the Glas Allt at NO267834. The gradient eases significantly for over 1km on this section of the ascent which enables swift progress up towards the first Munro. Follow the path past the cairned summit of Cac Càrn Mòr and walk north on to the main rock tor summit of Lochnagar - Cac Carn Beag. The highest ground is next to the trig pillar and from this vantage point it is possible to view much of the route ahead. Look over Loch nan Eun to the wide eroded path that runs along the lip of the bowl.

5 Retrace your steps to Cac Càrn Mòr and fork right to walk for 2km along the edge of the cliffs which embrace the coire. Loch nan Eun and Sandy Loch are both visible on the way to Carn a' Choire Bhoidheach. At NO226849, at a cairn marking a path junction, turn left and walk south on the very short out-and-back to the top of Carn a' Choire Bhoidheach. From the summit of the second Munro, the rock tors of Lochnagar - Cac Carn Beag and Cac Càrn Mòr are pointy protrusions on an otherwise vast, rounded plateau.

View from Càrn an t-Sagairt Mòr towards Lochnagar - Cac Carn Beag.

6 Return to the path junction and turn left to continue towards Càrn an t-Sagairt Mòr. Ignore a minor turn-off to the right after 150m and continue on the main path heading downhill. The downhills never last long though on this high-level circuit and the path soon ascends again. Glance left down to Dubh Loch which acts as an escape route if required. Turn right at the path crossroads at NO210837 for another short out-and-back up Càrn an t-Sagairt Mòr, whose summit is marked by a cairn beside a large boulder.

7 Return to the crossroads as the final two Munros of the day – Cairn Bannoch and Broad Cairn – fill your field of vision up ahead. Continue straight ahead towards Cairn Bannoch, the ascent of which is straightforward over grassy terrain. Take a moment to sit atop the rocky summit mound with views to the East Glen Shee Munros (page 315) to the south-west.

8 Leave Cairn Bannoch's summit and drop on to an eroded path which continues along the broad ridge to a grassy saddle. From here, pass to the right of a minor top (marked as 991m on OS maps) and follow the path as it descends south-east to another saddle. The path then leads to the rocky-tipped Broad Cairn, your final Munro of the day, which offers an incredible view down the length of Loch Muick.

9 Leave the summit to the east and take care over the rockier terrain which persists for the first part of the descent before reaching a flat section of shoulder – Little Craig. The good path veers south-east to a corrugated hut and eastwards for 170m where you should turn left on a path which leads north-east. Descend gradually towards the picturesque loch, with land falling away from right to left high above the Corrie Chash Burn.

10 Reach the bottom path and walk clockwise around Loch Muick using a series of footbridges over the streams until you reach the bothy once more (**3**). Reunite with your bike or sleeping equipment if necessary and either stay another night or retrace your steps back to the start.

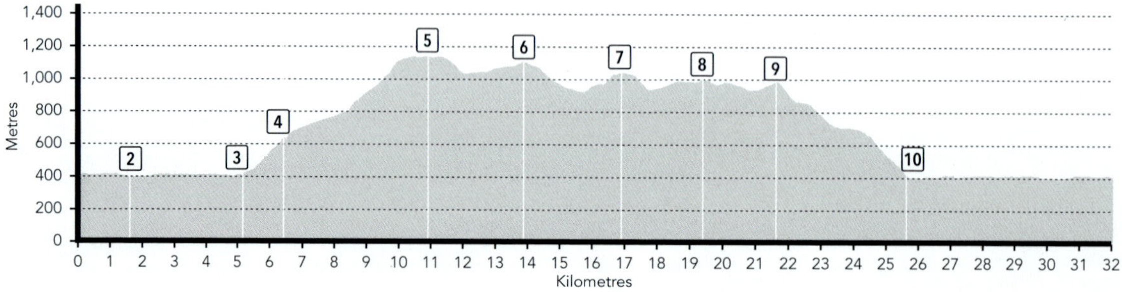

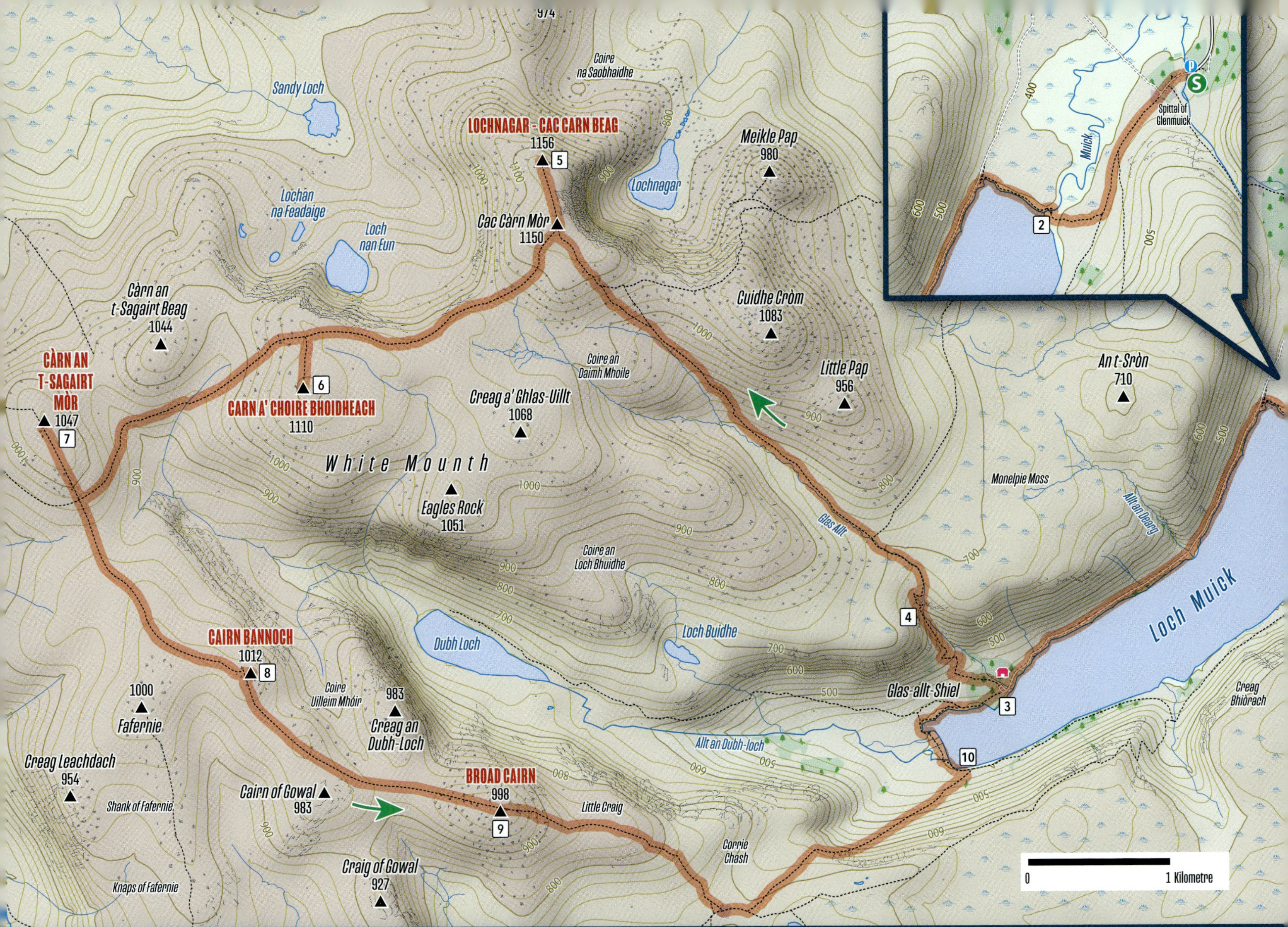

LOCHNAGAR - CAC CARN BEAG
1156
5
Coire na Saobhaidhe
Sandy Loch
Lochan na Feadaige
Loch nan Eun
Lochnagar
Meikle Pap
980
Cac Càrn Mòr
1150
Càrn an t-Sagairt Beag
1044
CÀRN AN T-SAGAIRT MÒR
1047
7
CARN A' CHOIRE BHOIDHEACH
1110
6
Coire an Daimh Mhoile
Creag a' Ghlas-Uillt
1068
Cuidhe Cròm
1083
Little Pap
956
An t-Sròn
710
White Mounth
Eagles Rock
1051
Monelpie Moss
Allt an Dearg
Glas Allt
Coire an Loch Bhuidhe
4
Loch Muick
Dubh Loch
Loch Buidhe
Glas-allt-Shiel
3
CAIRN BANNOCH
1012
8
Coire Uilleim Mhóir
983
Creag an Dubh-Loch
Allt an Dubh-loch
10
Creag Bhiorach
1000
Faferniè
Creag Leachdach
954
Shank of Fafernie
Cairn of Gowal
983
BROAD CAIRN
998
9
Little Craig
Corrie Chash
Craig of Gowal
927
Knaps of Fafernie
0
1 Kilometre
Spittal of Glenmuick
Muick
2

64 Mount Keen

HUB **Braemar – 1-hour-and-50-minute drive**

OVERALL 1 **2** 3 4 5 TERRAIN **1** 2 3 4 5
DISTANCE 1 2 **3** 4 5 NAVIGATION **1** 2 3 4 5

DISTANCE **17.8km** ASCENT **690m**
TIMES **6:20hrs** (Walker) **5:25hrs** (Trekker)
4:05hrs (Fastpacker) **2:50hrs** (Trail runner)

START LOCATION **Invermark car park, Glen Esk**
START LATITUDE/LONGITUDE **56.9112, -2.9100**
START GRID REFERENCE **NO446803**
START W3W **///developed.bravo.clan**

	PRONUNCIATION	NAME MEANING
Mount Keen	**'mount keen'**	**smooth or gentle hill**

Introduction

Mount Keen is an outlier in the east of the Cairngorms National Park; in fact, it is the most easterly of all Munros. Mount Keen is a fine beginner Munro as a wide, well-maintained track leads right from the car park to the summit plateau. As a result, navigation is easier than most other Munro walks. The Queen's Well, a landmark built in honour of Queen Victoria, adds historical interest to the walk.

Public Transport

There are no public transport services close to this walk.

Parking

From the B966 follow the minor road into Glen Esk. There is ample free parking for over 30 cars at Invermark car park in Glen Esk. The car park has rubbish bins and picnic benches. The nearest public toilet facilities are in Tarfside, which you pass in the last ten minutes of your journey to the start point.

MTB-ability

A mountain bike would be handy for the first and last 4.4 kilometres of the route. Experienced cyclists may be able to cycle almost all the way to the top.

Escape Route

None.

Refreshments

Stop for a hearty breakfast at Sinclair's Larder in Edzell on your way into the glen.

The track through Glen Mark which follows the Water of Mark.

The Route

S Leave the car park to the west. Walk over the bridge and pass Lochlee Parish Church. After 100m turn right where a fingerpost reads *Mount Keen*. Follow the rough track for 150m until the track splits in front of House of Mark. Fork left and pass through two gates as the track opens out on to moorland.

2 The pleasant track through Glen Mark is comprised of rough stones and gravel and features occasional muddy patches after heavy rain. The track crosses a few minor streams with no difficulty. Down to your left is the Water of Mark which the track loosely follows through the glen, while to your right is a sea of heather.

3 The track merges with two others; one appears from the right at the Burn of Glascorrie and another appears from the left from the other side of the river. Ignore these and continue north-west through Glen Mark. A landmark in the shape of a crown will appear 4km into the walk on the right-hand side of the track. Take the grassy right fork to visit the Queen's Well, built in 1861 to honour Queen Victoria who loved the area. She visited and drank the spring water – although I don't recommend you do that today.

4 Return to the track and continue along it, passing Glenmark Cottage. On a clear day you can see the track ascending the hillside in front of you. The track crosses two wide streams, the Easter Burn and the Ladder Burn; the latter can be fast-flowing in wet weather and features a makeshift footbridge. As the ascent steepens, you'll notice Carlochy to your left, a small lochan below the craggy northern slopes of Hard Hill.

5 There are a couple of switchbacks on the steepest part of the climb, which continues to be a bulldozed, wide track. Many will curse the installation of a bulldozed track on a beautiful wild hillside, but it certainly helps provide easy navigation to those at the beginning of their Munro-bagging journey. A large cairn marks a clear crossroads in the track. Take the right-hand path. On a clear day you'll see the Mount Keen summit mound in front of you.

Walker descending with Mount Keen summit mound in the distance.

6 At the next junction, continue straight on, ignoring the path on the left. Enjoy the view to the higher peaks in the heart of the Cairngorms National Park over in the distance on your left. As the track becomes more eroded, there are a few stone-pitched steps on the higher reaches of the Munro. At the very top, Mount Keen has a trig pillar which sits atop boulders. It's a simple scramble to the summit.

7 There is a handy sunken shelter to the east of the trig pillar which provides excellent shelter for a lunch or snack break before the descent. Retrace your steps back to the start.

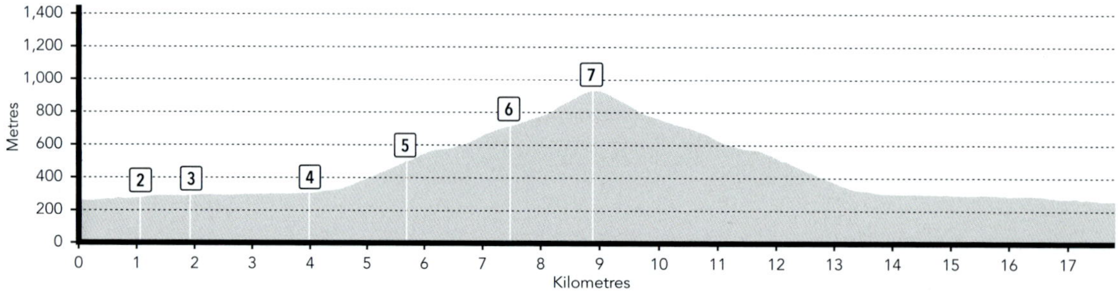

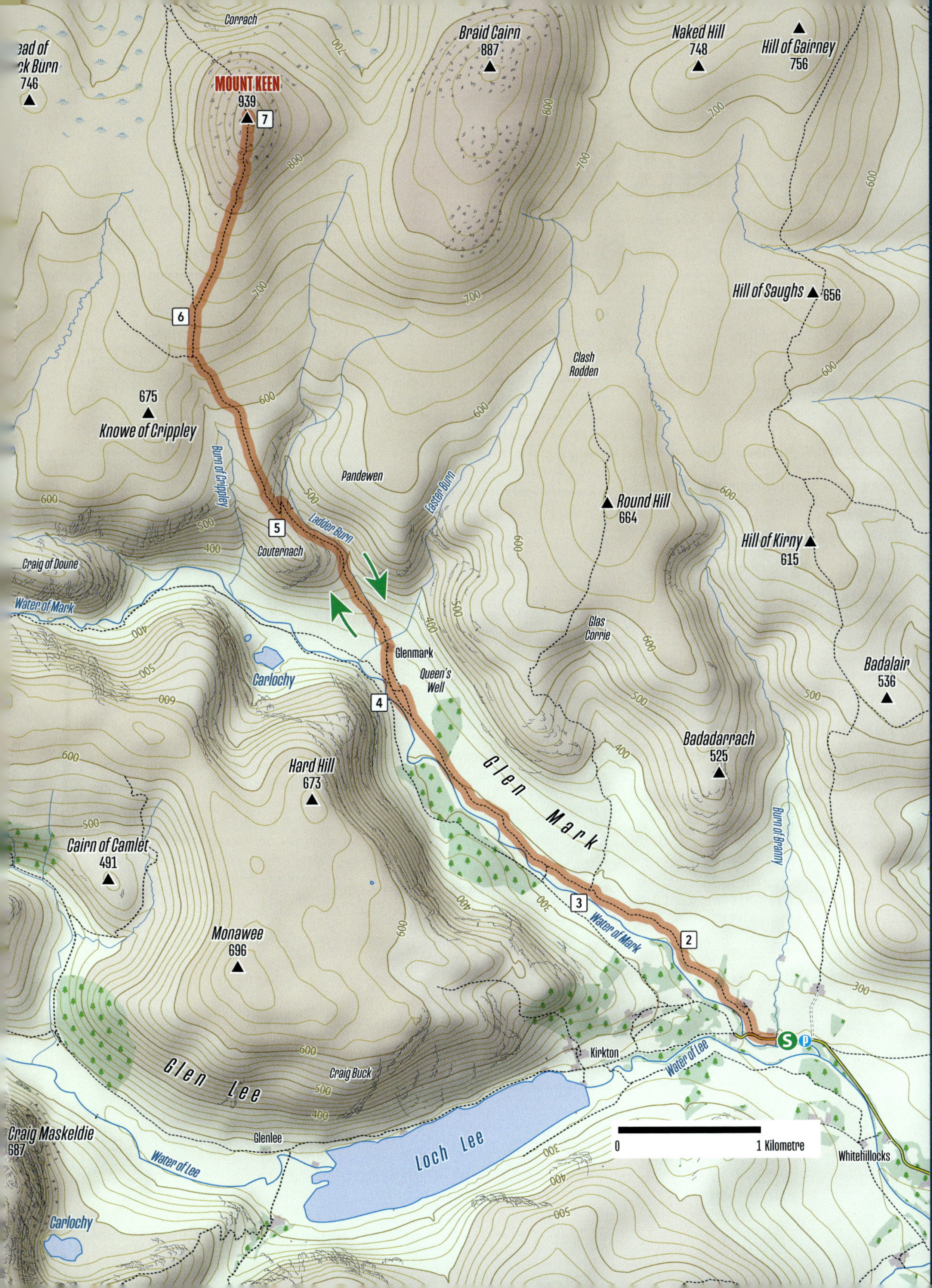
Corrach
Braid Cairn
887
Naked Hill
748
Hill of Gairney
756
MOUNT KEEN
939
7
Hill of Saughs
656
6
Clash
Rodden
675
Knowe of Crippley
Burn of Crippley
Pandewen
Easter Burn
Round Hill
664
5
Ladder Burn
Couternach
Hill of Kirny
615
Craig of Doune
Water of Mark
Glenmark
Glas
Corrie
Carlochy
Queen's
Well
4
Badalair
536
Badadarrach
525
Glen Mark
Hard Hill
673
Burn of Brawny
Cairn of Camlet
491
3
Water of Mark
2
Monawee
696
Kirkton
Water of Lee
Craig Buck
Glen Lee
Craig Maskeldie
687
Glenlee
Water of Lee
Loch Lee
0
1 Kilometre
Whitehillocks
Carlochy

Appendix

The following is a list of Tourist Information Centres, shops, websites and other contacts that might come in handy.

Tourist Information

The existing network of VisitScotland information centres (iCentres) will close before the end of 2025. The following websites should be helpful for planning your trip:

www.visitscotland.com – official website of Scotland's national tourist board.

www.lochlomond-trossachs.org – official website of the Loch Lomond & The Trossachs National Park.

www.cairngorms.co.uk – official website of the Cairngorms National Park.

Accommodation

There are lots of accommodation options in the Scottish Highlands, from camping to B&Bs and hotels. For specific information, we recommend you visit the following website:
www.visitscotland.com/places-to-go/highlands

Hostelling Scotland can also be found in the following locations. For more information, please visit **www.hostellingscotland.org.uk**

Braemar Youth Hostel	**T** 01339 741 659
Crianlarich Youth Hostel	**T** 01838 300 260
Glencoe Youth Hostel	**T** 01855 811 219
Glen Nevis Youth Hostel	**T** 01397 702 336
Loch Ossian Youth Hostel	**T** 01397 732 207
Oban Youth Hostel	**T** 01631 562 025
Pitlochry Youth Hostel	**T** 01796 472 308
Rowardennan Lodge Youth Hostel	**T** 01360 870 259
Tobermory Youth Hostel	**T** 01688 302 481

Weather & Ground Conditions

www.metoffice.gov.uk
www.mwis.org.uk
www.sais.gov.uk

Outdoor & Running Shops

Braemar Mountain Sports, Braemar
www.braemarmountainsports.com
T 01339 741 242

Cotswold Outdoor, Fort William
www.cotswoldoutdoor.com
T 01397 719 118

Ellis Brigham, Fort William
www.ellis-brigham.com
T 01397 706 220

INOV8 (online)
www.inov8.com

Killin Outdoor Centre, Killin
www.killinoutdoor.co.uk
T 01567 820 652

Nevisport, Fort William and Pitlochry
www.nevisport.com
T 01397 780 011 (Fort William)
T 01796 473 118 (Pitlochry)

The Outdoor Store at the Green Welly Stop, Tyndrum
www.thegreenwellystop.co.uk
T 01301 702 089

Tiso, Perth
www.tiso.com
T 01738 634 464

Other useful websites

www.themunrosociety.com – join the Munro Society here.

www.smc.org.uk/hills/compleators – register your Munro round here.

Vertebrate Publishing

At Vertebrate Publishing we publish books to inspire adventure.

It's our rule that the only books we publish are those that we'd want to read or use ourselves. We endeavour to bring you beautiful books that stand the test of time and that you'll be proud to have on your bookshelf for years to come.

The Peak District was the inspiration behind our first books. Our offices are situated on its doorstep, minutes away from world-class climbing, biking and hillwalking. We're driven by our own passion for the outdoors, for exploration, and for the natural world; it's this passion that we want to share with our readers.

We aim to inspire everyone to get out there. We want to connect readers – young and old – with the outdoors and the positive impact it can have on well-being. We think it's particularly important that young people get outside and explore the natural world, something we support through our publishing programme.

As well as publishing award-winning new books, we're working to make available many out-of-print classics in both print and digital formats. These are stories that we believe are unique and significant; we want to make sure that they continue to be shared and enjoyed.
www.adventurebooks.com